Ecophysiology of Fungi

Ecophysiology of Fungi

R.C. Cooke
Department of Animal and Plant Sciences
University of Sheffield

J.M. Whipps
Horticulture Research International
Littlehampton

OXFORD
Blackwell Scientific Publications
LONDON EDINBURGH BOSTON
MELBOURNE PARIS BERLIN VIENNA

Editorial Offices:
Osney Mead, Oxford OX2 0EL
25 John Street, London WC1N 2BL
23 Ainslie Place, Edinburgh EH3 6AJ
238 Main Street, Cambridge
Massachusetts 02142, USA
54 University Street, Carlton
Victoria 3053, Australia

Other Editorial Offices:
Librairie Arnette SA
2, rue Casimir-Delavigne
75006 Paris
France

Blackwell Wissenschafts-Verlag GmbH
Meinekestrasse 4
D-1000 Berlin 15
Germany

Blackwell MZV
Feldgasse 13
A-1238 Wien
Austria

First published 1993

Set by Excel Typesetters Company, Hong Kong
Printed and bound in Great Britain at the University Press, Cambridge

DISTRIBUTORS

Marston Book Services Ltd
PO Box 87
Oxford OX2 0DT
(*Orders*: Tel: 0865 791155
Fax: 0865 791927
Telex: 837515)

USA
Blackwell Scientific Publications, Inc.
238 Main Street
Cambridge, MA 02142
(*Orders*: Tel: 800 759-6102
617 876-7000)

Canada
Oxford University Press
70 Wynford Drive
Don Mills
Ontario M3C 1J9
(*Orders*: Tel: 416 441-2941)

Australia
Blackwell Scientific Publications
Pty Ltd
54 University Street
Carlton, Victoria 3053
(*Orders*: Tel: 03 347-5552)

A catalogue record for this title is available from the British Library

ISBN 0-632-02168-3

Library of Congress
Cataloging-in-Publication Data

Cooke, R.C. (Roderic C.), 1936–
Ecophysiology of fungi/
R.C. Cooke, J.M. Whipps.
p. cm.
Includes bibliographical references and index.
ISBN 0-632-02168-3
1. Fungi – Ecophysiology.
I. Whipps, J.M. II. Title.
QK604.2.E28C66 1993
589.2'045222 – dc20

Contents

Preface

In his foreword to Professor Lilian Hawker's *Physiology of Fungi*, published in 1950 by the University of London Press, Professor William Brown FRS warned that: 'The emphasis in mycology, or to use the wider and rather ill-defined word microbiology, is coming more and more to be physiological and biochemical ... but it would be regrettable if a tendency to make microbiology an appanage of organic chemistry should go too far.'

An examination of current microbiology research publications leads to the ineluctable conclusion that now, in many areas, this tendency has gone far enough. Fungi increasingly are being employed merely as convenient tools for investigating biochemical processes, justification for using particular species often being little more than their immediate availability in culture. This dismays some of us who believe that mycology is a distinct subject, and that mycologists should be concerned primarily with those special characteristics of these eukaryotic, predominantly filamentous, heterotrophic entities which facilitate their incredible variety of life styles. Admittedly this is a personal view, but it is based on what we trust is a more widely-held conviction that the ultimate goal of fundamental biological research is an understanding of how organisms function in their natural surroundings.

Within mycology much physiological, biochemical and genetic research is, quite properly, directed towards industrial exploitation of fungi: that is the optimizing or maximizing of one or more of their natural attributes under defined conditions. However, a growing amount of activity in these same fields is not related to exploitation, nor does it expand our appreciation of natural behaviour. Whilst we accept that what may seem irrelevant now may assume relevance in the future, we deplore this apparently accelerating movement and plead for attention to be refocused on the whole fungus. To do so requires the establishment of some common ground within which ecologists at one extreme and physiologists and biochemists at the other can work to mutual advantage. We have tried to do this by examining the existing links between fungal ecology and fungal physiology and also, perhaps, by forging a few others.

The difficulties inherent in this task became apparent at the planning stage. Anonymous comments on the projected synopsis revealed the fear amongst ecologists that there would be far too much physiology. Physiologists urged the inclusion of vastly more physiology in greater detail to offset what they saw as a strong ecological bias. We have called down a plague on both their houses by writing a book largely about those areas of physiology that we think have a direct bearing on ecological behaviour, given general ignorance of the latter. Its themes are the fundamental interaction of the fungus with its abiotic and biotic environments and the reconciliation of laboratory studies with observations made under more natural conditions.

We have been highly selective. For instance, we have given no special treatment to lichens and other mutualistic associations or to plant pathogens, but have instead drawn them into a more general ecological context, at the same time paying somewhat greater attention to fungus–heterotroph relationships. Although we have tried to remain objective, our views are often partisan. This is because we feel strongly that there are some areas where, if synthesis is impossible, it is appropriate for opinions to override facts, such as they are.

Our critics will ensure that these opinions will not be the last word on the subject. We wish to draw to their attention the following passage from Ralph Waldo Emerson's essay *Nature* (1836): 'Nature hates calculators; her methods are saltatory and impulsive. Man lives by pulses; our organic movements are such; and the chemical and ethereal agents are undulatory and alternate; and the mind goes antagonizing on, and never prospers but by fits. We thrive by casualties . . .'

Rod Cooke
Peak Forest

John Whipps
Littlehampton

April 1992

Nomenclature and terminology

The classification, nomenclature and mycological terminology used are as set out in the *Dictionary of the Fungi* (7th edn, 1983), edited by D.L. Hawksworth, B.C. Sutton and G.C. Ainsworth, and published by the Commonwealth Mycological Institute. To avoid clumsiness, species of Ascomycotina, Basidiomycotina and Deuteromycotina (Fungi Imperfecti) are often referred to as being 'ascomycetes–ascomycetous', 'basidiomycetes–basidiomycetous' and 'imperfect' respectively. Where appropriate the affinities of the latter are indicated. In cases where fungi have both a perfect stage (teleomorph) and an imperfect stage (anamorph) the most familiar binomial is usually used, the alternative usually being given in brackets at first mention but not subsequently. There is strict usage of those basic ecological terms that are frequently incorrectly employed. Here, 'habitat' is the place where a fungus lives; 'substratum' is the medium within the habitat that physically supports the fungus; 'substrate' is a specific biochemical component of the substratum; 'ecological niche' is the functional role of a fungus within a particular habitat.

Acknowledgements

We would like to thank the many authors, editors and publishers who granted us permission to use copyright material. We are in particular grateful to numerous colleagues who generously supplied us with illustrations and information. We also wish to thank David Wood for critical appraisal of parts of the manuscript.

We are also indebted to Glyn Woods and Andy Smith for photography, and to Jane Bird and Angela Doncaster who, over a long period, have patiently typed and retyped draft chapters and who made a valiant final effort to produce a finished version of the book. We would also like to thank Bob Grange for his computer wizardry during the compilation of the bibliography.

Finally, we are especially grateful to Judy Cooke and Sue, Tom and Jenny Whipps for their considerable support and patience throughout the hours of neglect during this undertaking. We dedicate this book to them.

PART ONE
Concepts

1 Evolution, ecology and physiology

The morphological and behavioural diversity of fungi is reflected in their spectacular success in terms of biomass, species numbers, variety of habitats occupied, and capacity for symbiosis with other micro-organisms, plants and animals (Hawksworth, 1991). Although it is impossible to define exactly the boundaries of physiological ecology, essentially it is concerned with the elucidation of life style determinants, that is the mechanisms controlling responses to the abiotic and biotic environments in which fungi develop in nature. This yields information not only on their present capacity for survival but also may permit an occasional glimpse into the history of this ability. Therefore, despite the fragmentary evidence for it, and the resultant controversies surrounding it, the evolution of fungi is an appropriate starting point for a consideration of their physiological ecology.

Fungi have uncertain and varying origins and genealogies and there has been constant revision of schemes for the circumspection of the fungal Kingdom and the arrangement of groups within it (see Hawksworth *et al.*, 1983; Cavalier-Smith, 1987). Such shifts in taxonomic thinking are stimulated by new insights into evolutionary pathways and interrelationships, these traditionally being gained through structural studies, but currently with an increasingly important physiological and biochemical contribution.

Origins and phylogeny

Although there has been a long-standing preoccupation with phylogeny, there is still no totally acceptable scheme for the emergence of the major fungal groups. Problems in establishing such a sequence stem from the obscure origins of extant fungi and from the difficulty in deciding which of their properties are primitive or advanced. Furthermore, because of their relatively simple structure, there are a limited number of ways by which fungi can attain important ends, so that there has been a high degree of parallel or convergent evolution (Savile, 1968). This also applies to their physiology which, in many respects, has a remarkable degree of uniformity; in comparison with organizational and behavioural development, biochemical innovation amongst fungi has been limited (Carlile, 1980). This is not the place to rehearse phylogenetic arguments but rather to review the evolution of life styles, many aspects of which, as will be seen shortly, can be divorced from the difficulties that beset phylogeny. However, some comment might be made on the kind of biochemical evidence that has been used to support ideas on phylogeny.

Leaving aside the highly problematical Myxomycota, evolution of fungi and allied organisms possibly has taken place along two distinct lines, each arising from perhaps separate flagellate ancestors (Table

Table 1.1 Distribution of biochemical characteristics amongst Subdivisions and Classes of fungi and their allies (after Shaw, 1965, 1966; Bartnicki-Garcia, 1970; Pfyffer *et al.*, 1986, 1990).

	Main cell wall polysaccharides	Tryptophane biosynthesis enzymes	Fatty acid synthesis pathway	Acyclic polyol pattern*
KINGDOM FUNGI				
Basidiomycotina				
Hymeno-Gasteromycetes	Chitin, glucan	T I, T VI	ω3	P_2
Ustilaginomycetes	Chitin, mannan	T III	ω3	P_2
Ascomycotina : Deuteromycotina	Chitin, glucan, mannan	T I, T II	ω3	P_1, P_2
Zygomycotina	Chitin, chitosan	T III	ω6	P_1
Mastigomycotina				
Chytridiomycetes	Chitin, glucan	T I	ω3 + ω6	P_1
KINGDOM CHROMISTA				
Oomycetes	Cellulose, glucan	T IV	ω6	P_0
Hyphochytriomycetes	Cellulose, chitin	—	—	—

* P_0, polyols absent; P_1, polyols except mannitol; P_2, mannitol and other polyols.

1.1). The vast majority of fungi (Eumycota) seem to belong to a single phyletic series, the main stream of evolution being broadly in the direction Mastigomycotina (Chytridiomycetes) → Ascomycotina: Deuteromycotina → Basidiomycotina (Hymenomycetes: Gasteromycetes), with tributaries giving rise to Zygomycotina and other Basidiomycotina (Urediniomycetes: Ustilaginomycetes). Other 'fungi' (Hyphochytriomycetes: Oomycetes), which comprise relatively few species, probably evolved along a separate line and are now placed in the Kingdom Chromista (Förster *et al.*, 1990). They nevertheless fall within the remit of mycologists and their study is inseparable from that of 'true fungi', although such a term may have no strict validity. The intricacies of the pathways between groups are a matter of debate but, whatever the route of their arrival, these Subdivisions and Classes constitute a definite hierarchy within which biochemical distinctions can be made with respect to lysine synthesis, cell wall composition, tryptophane and long-chain fatty acid synthesis, and distribution of acyclic polyols (Table 1.1).

All organisms capable of lysine synthesis do so via either the *meso*-α,ε-diaminopimelic acid pathway (DAP pathway) or the L-α-aminoadipic acid pathway (AAA pathway), none being known that utilize both (Table 1.2). The DAP pathway occurs in Hyphochytriomycetes and Oomycetes but the AAA pathway is used by all other fungi (Vogel, 1964; Léjohn, 1974; Ragan & Chapman, 1978). Since there are no steps in common between the two routes, this difference in distribution would seem to reflect a dichotomy of some magnitude. The point

Table 1.2 Steps in the *meso*-α,ε-diaminopimelic acid (DAP) and L-α-aminoadipic acid (AAA) pathways of lysine biosynthesis.

DAP pathway	AAA pathway
L-aspartic-β-semialdehyde + pyruvate	α-ketoglutarate + acetyl co-enzyme A
↓	↓
2,3-dihydrodipicolinic acid	Homocitric acid
↓	↓
Δ^1-piperideine-2,6-dicarboxylic acid	Homoisocitric acid
↓	↓
N-succinyl-L-α-amino-ε-ketopimelic acid	α-ketoadipic acid
↓	↓
N-succinyl-L-α,ε-diaminopimelic acid	L-α-aminoadipic acid
↓	↓
L,L-α,ε-diaminopimelic acid	L-α-aminoadipic semialdehyde
↓	↓
meso-α,ε-diaminopimelic acid	Saccharopine
↓	↓
L-lysine	L-lysine

might also be made here that, whilst the DAP pathway is additionally found in prokaryotes and algae, the AAA pathway is almost exclusive to fungi; this supports suggestions that Oomycetes have algal ancestry. In addition, Oomycetes contain either cholesterol or alkylidine cholesterols as their dominant sterols and lack ergosterol, the latter typically occurring in other fungi (Coffell *et al.*, 1990). Divergence also seems to be indicated by the distribution of ATP-dependent hexokinases. Hyphochytriomycetes and Oomycetes have either glucokinase or mannofructokinase or both, with other major groups having low-specificity hexokinase, sometimes together with gluco-mannokinase (Delvalle & Asenio, 1978).

Although data are very limited, there is reason to suppose that, in the main, cell wall composition is constant amongst related fungi and that this is a reliable indicator of kinship (Bartnicki-Garcia, 1968, 1970, 1987). The distribution of major structural polysaccharides again distances Hyphochytriomycetes and Oomycetes from other fungi in that the latter do not possess cellulose (Table 1.1).

In all organisms so far studied, the same enzyme series is involved in L-tryptophane synthesis. In many instances, following purification, some of these enzymes exhibit associations in sedimentation patterns, perhaps reflecting clustering and co-transcription of the genes for them. Eight patterns have been identified (Types 0–VIII) but only five have been found in fungi (Table 1.3). Although relatively few

Table 1.3 Enzymes and intermediates in the biosynthesis of L-tryptophane, and enzyme sedimentation patterns found amongst fungi (from Hütter & DeMoss, 1967; Crawford, 1975).

	Chorismic acid
1 Anthranilate synthase	↓
	Anthranilic acid
2 Phosphoribosyl transferase	↓
	N-(5′-phosphoribosyl) anthranilic acid
3 Phosphoribosyl anthranilate isomerase	↓
	1-(ortho-carboxyphenylamine)-1-deoxyribulose-5-phosphate
4 Indole glycerophosphate synthase	↓
	Indole-3-glycerol phosphate
5 Tryptophane synthase	↓
	L-tryptophane

Sedimentation patterns

Type I : 1,3,4 co-precipitated, 2,4 separate
Type II : 1,4 co-precipitated, 2,3,5 separate
Type III: as for Type I in presence of L-glutamine and EDTA, in absence of these, 1 unaffected, 3,4 co-precipitated, 2,5 separate
Type IV: 3,4 co-precipitated, 1,2,5 separate
Type V : 1,2,3,4 co-precipitated, 5 separate

fungi have been examined, major differences between some groups are indicated. For instance, the Type IV pattern is restricted to Oomycetes, Type II to Ascomycotina and Type VI to the higher Basidiomycotina (Hütter & DeMoss, 1967; Crawford, 1975).

Most eukaryotes synthesize unsaturated long-chain fatty acids from linoleic acid by one of two pathways (Table 1.4). One is via α-linolenic acid and gives rise to the ω3 group of fatty acids; the other, through γ-linolenic, results in ω6 fatty acids, the two groups being distinguished by the position of the double bonds along their carbon chains. The ω3 acids appear to be typical of Ascomycotina and Basidiomycotina, but it is difficult to see what might be firmly deduced from either this or the possession of ω6 acids by both Oomycetes and Zygomycotina. Chytridiomycetes, in common with some zooflagellates, have both groups of fatty acids, which suggests that these fungi may have originated from primitive ciliate or flagellate organisms (Shaw, 1965, 1966; Ellenbogen *et al.*, 1969).

Acyclic polyols are widely found within the fungi, and the polyol character seems to be conservative, being thus a potentially reliable marker of evolutionary relationships (Pfyffer *et al.*, 1986). Three states have been distinguished: P_0, polyols absent; P_1, polyols except mannitol present; and P_2, mannitol and other polyols present (Table 1.1). Lack of polyols again distances Oomycetes from other groups, whilst the possession of mannitol appears to be a property of septate fungi.

Table 1.4 Pathways for synthesis of unsaturated fatty acids. The structure of each acid is indicated by three sets of figures. The first gives the number of carbon atoms making up the molecule; the second indicates the number of double (unsaturated) bonds between carbon atoms; the third, in parentheses, gives the position of the double bonds, counting from the carboxyl end of the molecule (from Korn *et al.*, 1965).

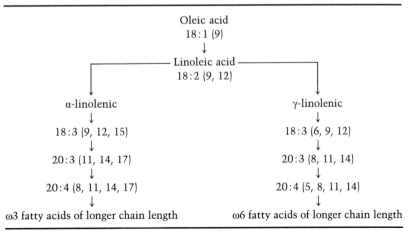

Oleic acid
18:1 (9)
↓
Linoleic acid
18:2 (9, 12)

↓	↓
α-linolenic	γ-linolenic
↓	↓
18:3 (9, 12, 15)	18:3 (6, 9, 12)
↓	↓
20:3 (11, 14, 17)	20:3 (8, 11, 14)
↓	↓
20:4 (8, 11, 14, 17)	20:4 (5, 8, 11, 14)
↓	↓
ω3 fatty acids of longer chain length	ω6 fatty acids of longer chain length

Although distribution of such physiological or biochemical features can be used in support of phylogenetic proposals, their adaptive significance – if indeed they always have one – is not invariably clear. It is impossible to discern with certainty which of them might be more evolutionarily advanced than others, except in terms of their occurrence within taxa that are considered, on other grounds, to be either relatively primitive or relatively advanced. This raises the important question of whether it is at all feasible to define, as primitive or advanced, a sufficient number of characteristics from which to identify possible trends in the physiological evolution of fungi. The aforementioned general lack of biochemical innovation amongst fungi, combined with the evening-out effects of parallel, convergent and retrograde evolution, would seem to make this unlikely. However, the dilemma can be resolved through a slight but significant shift of emphasis towards their ecophysiological traits. That is by attempting to identify physiological and morphological characters, the acquisition of which, either singly or in combination, has promoted ecological success. Before discussing this in more detail it is necessary first to outline the environmental background to fungal evolution.

Territorial expansion

According to the most recent line of thought, after a possibly Precambrian origin, fungi probably became widespread in the sea and subsequently acquired the ability to live first in brackish and then in fresh water (Hallbauer & van Warmelo, 1974; Pirozynski, 1976). From here, over a period stretching from the late Cambrian to beyond the Devonian, they were in a position to move to the land. In order to do so they had to obviate the serious problem of desiccation, probably by emerging as endophytes of early plants. This would have allowed them to survive the critical transition period until such a time as terrestrial conditions permitted them to escape from the protection of their hosts. Then, as vegetation spread and diversified, a range of habitats was created within which fungi could occupy new ecological niches. The history of the fungi thus parallels, and is inseparable from, that of plants (Table 1.5).

The highly fragmented fossil record of fungi provides only a fragile basis for the chronology of their terrestrialization (Stubblefield & Taylor, 1988). However, by taking account of evolution of plants, a convincing picture has been drawn of some major ecophysiological events that probably occurred during invasion of the land by fungi (Lewis, 1987). Evidence from microfossils suggests the existence of an Ordovician bryophyte-like land flora which by the Silurian had given rise to primitive vascular plants (Edwards & Fanning, 1985; Gray, 1985). These possibly emerged from a ground cover of cyanobacteria and algae, some of which may have been lichenized; fungal remains

Table 1.5 Presumed or established occurrence of fungi during evolutionary time in relation to the development of terrestrial plant life (after Lewis, 1987).

Period	Age (yr × 10^6)	Major features of plant life in relation to fungi	Possible features of fungi
Tertiary	10–75	Modern flora established Emergence of Ericales	Development of endomycorrhizas (ericoid) with Ascomycotina and possibly some Deuteromycotina and Basidiomycotina
Cretaceous	75–100	Conifers dominant, appearance of *Pinus* spp. and potentially mycorrhizal Angiosperms	Ascomycotina increasingly evident
Jurassic	130–140	Luxuriant conifer and fern forests	Gradual emergence and finally wide establishment of ectomycorrhizas formed with Basidiomycotina; Polyporales (Basidiomycotina) also well established
Triassic	160–180	Climatic fluctuations leading to a sparse desert flora; evidence for appearance of Coniferales (conifers)	
Permian	190–200		
Carboniferous	200–250	Development of swamp forests of tree ferns and early Gymnosperms	Basidiomycotina on woody plant residues
Devonian	260–300	Vascular plants, *Rhynia* and *Asteroxylon*, but lacking root systems	Endophytes of subterranean and aerial plant organs; some evidence for lignolytic fungi; Zygomycotina and Basidiomycotina
Silurian	300–350	Evidence for lichenized algae, a bryophyte-like land flora and emergence of vascular plants	Occurrence of septate fungi in association with plants and cyanobacteria
Ordovician	425		
Cambrian	500		Marine Oomycetes and Chytridiomycetes associated with algae and free living

dating from the Silurian having been found which have affinities with Ascomycotina (Sherwood-Pike & Gray, 1985). There would then have been opportunities for endosymbioses with early land plants by means of either mycophylla, that is non-pathogenic infections of aerial structures, or mycorrhizas of absorbing organs. Present-day leafy liverworts have mycophyllous fungi, so that this type of endophytism may even predate the appearance of vascular plants (Pocock & Duckett, 1985). Certainly, endophytic hyphae occur in aerial plant parts from the Devonian onwards, and mycophylla are abundant throughout the modern vascular flora, where they have a probably mutualistic role (Pirozynski, 1976; Carroll, 1986; Clay, 1986). However, it is from mycorrhizas that the possibilities for further radiation can be traced.

Early vascular plants possessed vesicular–arbuscular type (Zygomycotina) mycorrhizas, which has led to the suggestion that the first land plants were not only obligately mycorrhizal, but that their very origins were dependent on this symbiosis (Pirozynski & Malloch, 1975; Pirozynski, 1976, 1981; Malloch *et al.*, 1980). Strong arguments have been put forward giving physiological grounds for belief in this view (Lewis, 1987). First, early land plants would have been subjected to exactly the same selective forces that extant plants respond to by fungal symbiosis, that is appropriation of minerals and water, exclusion of toxic elements, and maintenance of a compatible acid–base balance. Second, even by the early Devonian, plants still lacked roots. Third, there is no reason to believe that elements were more abundant in Ordovician, Silurian and Devonian soils than they are at present, so that frequently phosphorus would have been growth limiting, a situation which was probably alleviated then, as now, by acquisition of the mycorrhizal state. Oomycetes, with their limited capacity to store phosphorus, would then (as they still are) be excluded from this symbiosis (Chilvers *et al.*, 1985).

A later, major influence on fungal behaviour was the growing availability of lignin as plants developed an increasingly woody habit. Basidiomycetous wood-inhabitants existed as early as the late Devonian, and mull and moder soils had developed by the early Carboniferous (Stubblefield *et al.*, 1985; Wright, 1985). The scene was thus set for the emergence of ectomycorrhizal Basidiomycotina from lignolytic relatives, and for the subsequent co-evolution of these symbionts with potentially compatible hosts. Principally the latter were Pinaceae, which arose during the late Triassic, and *Pinus* itself which, together with some notable angiospermous taxa, dates from the Cretaceous and early Tertiary (Malloch *et al.*, 1980). A major spur to the establishment of such associations was possibly the onset of climatic fluctuations during those times. Fungus–root symbioses confer on the plant an ability to obtain nitrogen from otherwise inaccessible organic compounds and also facilitate water uptake in arid conditions (Read, 1984; Abuzinadah & Read, 1986a,b). Finally, whilst life styles continued to diversify in relation to the multiplying complexity of available microbial habitats and niches, the geologically recent Ericaceae became hosts for those Ascomycotina and Basidiomycotina with a propensity for the endomycorrhizal way of life. This again permitted their hosts to tap organic sources of nitrogen, and conferred on them an ability to colonize organic soils of high acidity (Read, D.J., 1983).

Determinants of success

It is now pertinent to consider a number of features that fungi developed which expanded their capabilities and so multiplied the evolutionary options available to them. Events leading to acquisition of these features can only be guessed at, but their importance is manifest in the success of extant Eumycota. In later chapters attention will be focused on details of these characteristics, so that comment here will be confined to their relevance to some aspects of general fungal behaviour. Fresh insights into this subject have been obtained by examining those fungal attributes that represent clear gains over the prokaryotic condition, this approach being particularly enlightening when comparisons have been made between the performance of fungi and heterotrophic bacteria (Carlile, 1979, 1980).

Irrespective of somatic form (that is whether they have thalli of limited extent, are unicellular or pseudomycelial, or possess extensive mycelia), all fungi are massive relative to bacterial cells and have a much lower surface area to volume ratio. Consequently there is decreased efficiency of diffusion into and out of fungal cells-as a result of which, in general, fungi have much lower metabolic rates than bacteria, this being reflected in reduced vegetative growth rates. For example, under optimum conditions bacterial generation times of around 20 minutes are common, and can sometimes be half this, whilst the shortest generation times recorded for mycelial fungi and yeasts are in the region of 1 hour.

However, a significant advantage of larger cell size is improved performance in directed swimming or growth either towards desirable sites or away from harmful conditions (Carlile, 1975). Unlike motile bacteria, zoospores are too big to exhibit Brownian motion and are thus able to exercise greater control over their swimming movements. Also unlike bacteria, they are massive enough to detect differences in the intensity of attractants or repellants at different points on their surface. Increased control of movement, coupled with sensory systems, allows spatial information concerning a stimulus to be directly translated into advance or retreat with respect to that stimulus. In this way asexual planospores are enabled to select favourable settling sites and motile gametes are drawn to their appropriate partners. Similarly, in hyphae, a gradient in the intensity of any extrinsic stimulus is created across the cell, so that the direction of the former may be determined and growth oriented relative to it. This facilitates ordered exploration of the environment by vegetative and reproductive hyphae, and in particular can bring about a range of interhyphal meetings so providing the means for somatic and sexual fusions (Gooday, 1975; Carlile & Gooday, 1978).

The vast majority of fungi have mycelia of tubular, branched, apically growing hyphae, a habit that has a number of distinct advantages over those of thalloid and yeast forms, and the emergence of which must have been a key event in fungal evolution. The hypha allows rapid extension over and penetration of substrata, whilst branching increases the volume of material exposed to extracellular enzymes, permits efficient nutrient uptake, and maintains a favourable surface area to volume ratio as the mycelium gains in size. Over and above these advantages, the filamentous habit facilitates internal communication and creates potential for co-ordination of metabolic processes in separate regions of the mycelium, division of labour between its various parts, and a consequent increase in behavioural complexity.

These possibilities are further enhanced, in septate fungi, by anastomoses between hyphae of either the same mycelium or of different but genetically compatible mycelia. Fusions enable hyphal aggregation and so allow the construction of massive vegetative and reproductive organs, morphogenesis of which involves complex cognitive and co-ordinated growth processes. The development of fusion mechanisms would also have given added impetus to evolutionary advance by opening up new ways of acquiring and storing genetic information. In particular, heterokaryosis would have aided the conservation and utilization of mutations, and permitted the survival of recessive genes that were not immediately advantageous (Carlile, 1980). However, the co-existence of different nuclear types is in most, if not all, cases strictly controlled by somatic and sexual incompatibility systems, so that the developmental versatility which is of great survival value in fluctuating conditions is due to differential gene expression rather than selection of nuclei appropriate for one mode of life or another.

Most filamentous fungi are able to maintain their form during periods of either intermittent or continuous stress. Furthermore, a relatively few species are known to be able to continue vegetative growth in conditions severe enough to preclude development by adopting a reversible yeast phase. Depending on circumstance, the temporary or permanent yeast habit has its own selective advantages. It can be maintained over a wider range of physical conditions than can filamentous growth, and the more favourable surface area to volume ratio of yeast cells permits their multiplication where nutrient supplies would be too poor to support hyphal development. For similar reasons some thalloid fungi are well adapted to take advantage of ephemeral or limited nutrient availability, their rapid achievement of full size being quickly followed by total commitment to reproduction.

Another salient feature of the hypha is the possibility for the occurrence within it of cytoplasmic streaming. As well as alleviating size-imposed diffusion problems, this has several other important functions. Despite their relatively long generation times, fungi exhibit high localized growth rates; extension of hyphal apices and development of multicellular organs being generally too rapid to be accounted for by autonomous synthetic processes. Streaming provides a mechanism for bulk flow of materials to growth sites from other areas of the mycelium. The ability of hyphal contents to migrate also serves to conserve cytoplasm. For example, when mycelia are growing in nutrient-poor conditions, cytoplasm may be continually evacuated from older hyphae which, when empty, then become walled off through either septum formation or the plugging of existing septa. Similarly, damaged hyphae may be evacuated and sealed off, thus conferring an ability for repair and defence. This kind of behaviour, coupled with the added rigidity given to hyphae by cross walls, indicates some of the advantages gained by septate fungi over aseptate forms. By their capacity for closure, septa probably also influence the ways in which adjacent hyphal compartments develop, and so may exert control over, and greatly enlarge possibilities for, differentiation and organogenesis. It should also be noted that interhyphal fusions, and most interhyphal antagonisms, are characteristic of compartmented hyphae. Such interactions between vegetative hyphae are rare in Oomycetes and Zygomycotina, and whilst fusions are commonplace between their specialized gametangia, the latter are invariably delimited by septa.

Finally, some points might be made briefly with respect to dissemination. Successful emergence onto, and radiation upon, the land required increasingly efficient dispersal systems. The wide adoption of conidia indicates their adaptive value not only as units of propagation but also as agents for the fusion of compatible haploids. The origins of air-borne propagules are obscure, but a possibility is that some arose from the increasing need of some mycophyllous fungi to infect aerial plant parts directly rather than via the substratum. Echoes of this beginning may perhaps be found amongst extant endophytic parasites in which melanization of either the spore or the infective appressorium confers protection during their critical epiphytic phase (Parbery & Emmett, 1977).

Nutritional traits When contemplating fungal evolution the nature of nutritional patterns should also be taken into account. Any particular nutritional trait is an overt summation of a great number of cryptic physiological mechanisms arrived at via natural selection. It might be argued, therefore, that modes of nutrition can be evaluated in an evolutionary context in terms of which of them are relatively primitive and which are relatively advanced. Although distinct nutritional trends undoubtedly occur within major phyletic assemblages, such arguments are not without controversy.

Three nutritional modes are recognized amongst fungi according to the manner in which they are able to exploit external resources (Thrower, 1966). These are: saprotrophy, where non-living organic substrata, other than those that may have been killed by the fungus itself, are utilized; necrotrophy, in which host tissues are first killed then utilized saprotrophically; and biotrophy, in which nutrients can be obtained only from living cells, death of the latter terminating biotrophic activity. On the basis of apparent relative specialism, saprotrophy has been envisaged as being the most primitive state from which necrotrophy was derived, giving rise in turn to biotrophy. This is in harmony with the long-standing view that parasites evolved from free-living ancestors, at the same time becoming increasingly specialized physiologically. There is no doubt that this has occurred frequently, but whether it has been an unvarying trend is questionable. For instance, the albeit imperfect geological evidence for the emergence of fungi suggests that this has not been the case. Furthermore, nutritional patterns within some extant assemblages indicate possible alternative routes. For example, within Chytridiomycetes and Basidiomycotina, evolutionary advance, as judged by morphological criteria, is sometimes accompanied by a trend from biotrophy towards saprotrophy (Savile, 1968; Barr, 1978).

These conflicts cannot be resolved in detail here but some important, and largely overlooked, points might be made. Many fungi are not fixed within a single mode but show various degrees of flexibility during their life cycle. Examination of these shifts suggests that spatial and temporal interfaces between all three nutritional modes can occur, and have done so in the past, in a variety of combinations (Luttrell, 1974; Cooke & Whipps, 1980, 1987). This implies that the three modes have arisen on numerous occasions throughout a range of assemblages as a response to changing environmental conditions, the previously mentioned capacity to store genetic information, of no immediate utility, for more than one mode facilitating the retention of potential for change. It follows that the direction of nutritional evolution taking place at any time would have been determined by the nature of the environmentally imposed options presented to fungi, including, where appropriate, narrowing specialism within a single mode.

Less contentiously, nutritional modes have been used as a means of delimiting econutritional categories of behaviour according to whether a fungus is obligately dependent on a single mode or can adopt another mode in order to complete its life cycle in nature. This can provide a valuable summary of the results of nutritional evolution and has given great impetus to studies on many aspects of nutritional behaviour (Lewis, 1973, 1974). Nine principal categories may be recognized, but it must be emphasized that, as with the modes themselves, these do not carry hierarchical connotations and, in addition, represent a series of continua (Table 1.6 & Fig. 1.1). Where the domains of modes overlap there is the opportunity for facultative behaviour with respect to an adjacent mode or modes. On the basis of observed frequency of facultative behaviour it appears that there is a much greater overlap between saprotrophy and necrotrophy than between either of these modes and biotrophy (Fig. 1.1). In addition, many fungi would seem to have an equal ability for both saprotrophy and necrotrophy. Two

Table 1.6 Econutritional groups of fungi according to nutritional mode and ecological behaviour (after Lewis, 1973, 1974; Luttrell, 1974; Cooke, 1977; Cooke & Whipps, 1980). Abbreviations in parentheses indicate the location of the various groups within the econutritional scheme shown in Fig. 1.1. Groups are listed in a clockwise sequence from Fig. 1.1 beginning at biotrophy.

Obligate biotrophs (OB)	No capacity for saprotrophy or necrotrophy
Hemibiotrophs (HB)	Initially biotrophic but then becoming necrotrophic; saprotrophic potential as for obligate necrotrophs (ON) below
Facultatively saprotrophic hemibiotrophs (FSH)	Initially biotrophic but then becoming necrotrophic; a final saprotrophic phase then occurs
Obligate necrotrophs (ON)	Normally necrotrophic; any saprotrophic ability severely limited or restricted to survival in dead tissues
Facultatively saprotrophic necrotrophs (FSN)	Normally necrotrophic but with some ability to become saprotrophic
Facultatively necrotrophic saprotrophs (FNS)	Normally saprotrophic but with some ability to become necrotrophic
Obligate saprotrophs (OS)	No capacity for necrotrophy or biotrophy
Facultatively biotrophic saprotrophs (FBS)	Normally saprotrophic but with some ability to become biotrophic
Facultatively saprotrophic biotrophs (FSB)	Normally biotrophic but with some ability to become saprotrophic

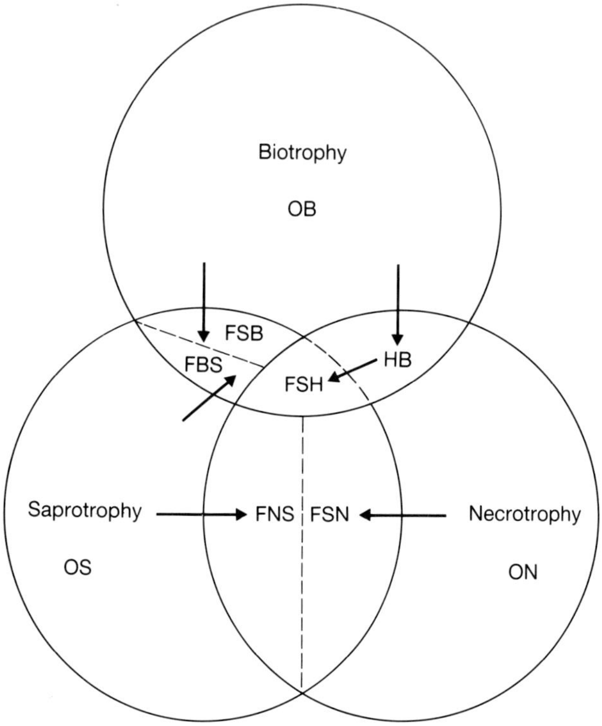

Fig. 1.1 Model of relationships between the three modes of fungal nutrition and the location of possible kinds of econutritional behaviour with regard to them. Letters refer to the econutritional groupings described in Table 1.6. Arrows indicate the direction of the nutritional shifts occurring. The number of arrows equals the number of additional nutritional domains arising from overlap of the three modes.

further points should be noted. First, it is possible to distinguish two kinds of hemibiotrophy depending on whether or not the shift from biotrophy to necrotrophy is followed by a further phase of saprotrophy. Secondly, whilst the occurrence of predominantly saprotrophic fungi that are facultative for biotrophy is a possibility, there is only slight evidence for their existence. They may be represented by yeasts which can maintain occupancy of animal cells and by the symptomless endophytes of grasses and other plants (Carroll, 1986; Clay, 1986; Cooke & Whipps, 1987).

Strategy theory

If, in the long term, studies on physiological ecology are to achieve anything more than a record of observations on isolated phenomena, then it becomes essential to erect a functional context within which to assess results and view progress. Econutritional groupings provide one kind of framework; another is presented by a consideration of life strategies. As it pertains to fungi, a strategy may be defined loosely as a grouping of similar or analogous physiological characteristics occurring between species, or within communities, which cause them to exhibit ecological similarities. The importance of strategy theory in attempts to construct a unifying classification of organisms into functional types is well established in plant and animal ecophysiology but has been applied to fungi much less widely (Pugh, 1980; Andrews & Rouse, 1982; Cooke & Rayner, 1984; Cooke & Whipps, 1987).

The fundamental premise of strategy theory is the occurrence, as a result of natural selection, of a spectrum of strategies for survival at the opposite poles of which are two types of organisms (Harper & Ogden, 1970; Pianka, 1970; Gadgil & Solbrig, 1972). The first, *R*-selected, have a short life expectancy and rapidly commit much or all of their available resources to reproduction. The second, *K*-selected, have a long life expectancy and either devote only a small proportion of available resources to reproduction at any one time or only commit themselves to reproduction at the end of their life span. Within the *R–K* continuum three distinct forms of selection have operated to give three primary strategies (Grime, 1977, 1979). These also form a continuum of overlapping domains. Selection forces and the distribution of species within strategy domains can be represented in a simple model (Fig. 1.2). Thus *C*-selection has resulted in a combative or competitive strategy which maximizes the ability to occupy and exploit resources in conditions of low stress and low disturbance. A ruderal strategy, emerging via *R*-selection, is characterized by a short life span associated with high reproductive capacity, and determines success in highly disturbed, but nutrient-rich, conditions. Finally, *S*-selection has culminated in a stress-tolerant strategy in which there has been adaptation to conditions of continuous environmental stress. Where primary strategies merge, secondary strategies may arise that combine their features (Fig. 1.2). Four main types of the latter have been proposed. First, there is a competitive ruderal strategy, *C–R*, where there is adaptation to circumstances in which stress has a low impact and the incidence of competitors is moderated via disturbance. Secondly, there is a stress-tolerant ruderal strategy, *S–R*, characterized by adaptation to lightly disturbed, unproductive habitats. Thirdly, there is a stress-tolerant competitive strategy, *C–S*, in which adaptation is to relatively undisturbed conditions where there is moderate intensity of stress. Finally, *C–S–R* strategists are adapted to habitats where the level of competition is restricted by moderate intensities of both stress and disturbance.

Another theoretical model which, whilst being compatible with that just described, stems from a rather different approach, is the habitat templet. This has been used as an aid with which to study ecological

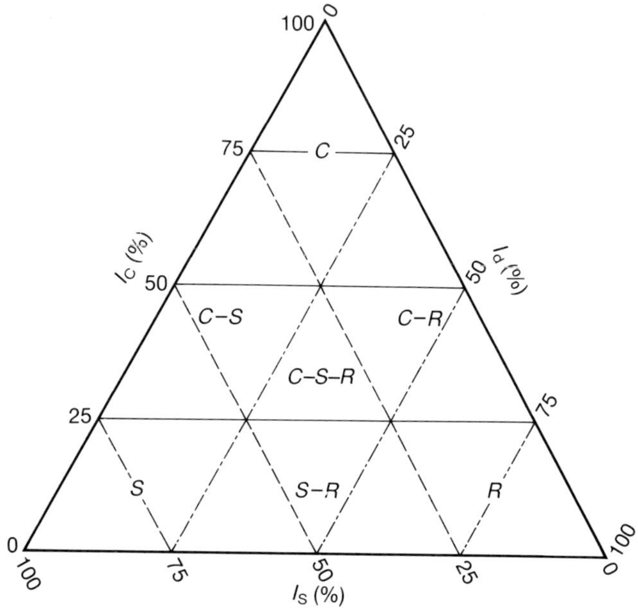

Fig. 1.2 Model of location of primary and secondary strategies in relation to selection forces: I_c, relative importance of competition; I_d, relative importance of disturbance; I_s, relative importance of stress. For a full explanation of other symbols see text (from Grime, 1977, *American Naturalist*, **111**, by permission of University of Chicago Press, © 1977, University of Chicago Press).

strategies in animals, and consists of a rectangular figure defined by quantifiable attributes of habitat (Southwood, 1977). The proportions of the templet will depend on the scaling of the axes and on the relationships between the chosen variables. A simple version of it with some relevance to fungi consists of a square (Fig. 1.3) with axes representing the favourableness or predictability of habitat conditions, and a diagonal vector indicating the complexity of community properties and processes, and hence biotic unpredictability (Greenslade, 1983). Certain areas of the templet are dominated by particular selection processes (Fig. 1.3). Habitats with low predictability, that is prone to disturbance, will favour *R*-selected species, whilst *K*-selected species will predominate in habitats where predictability and favourableness and biotic unpredictability are all high. Predictably unfavourable habitats, that is those characterized by continuous stress, encourage adversity selection, *A*-selection, which results in the conservation of adaptations to consistently adverse environments (Whittaker, 1975; Southwood, 1977). Although somewhat differently defined, there would seem to be agreement that *S*-selection and *A*-selection are virtually synonymous.

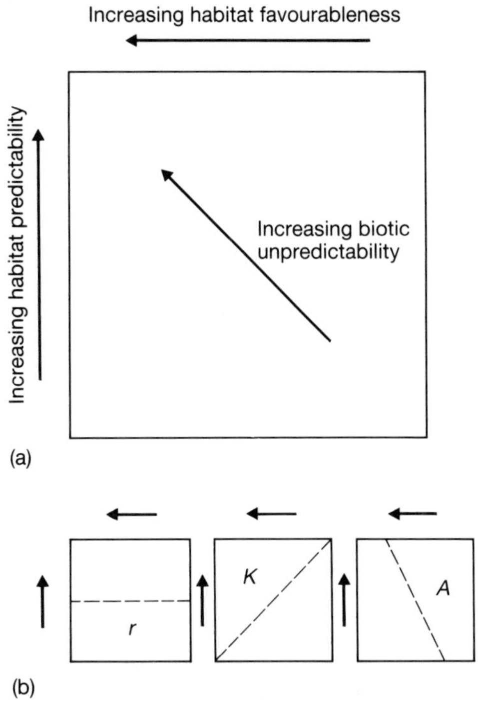

Fig. 1.3 The habitat templet. (a) Templet showing habitat conditions and biotic properties; (b) location of dominant selection processes (from Greenslade, 1983, *American Naturalist*, **122**, by permission of University of Chicago Press, © 1983, University of Chicago Press).

Niche determinants

Strategy theory provides predictive models within which ecophysiological hypotheses may be erected and tested, common goals of both strategy theory and experimental physiological ecology being the identification of niche determinants and the exposition of relationships between fundamental and realized niche. However, some general ecological concepts inherent in strategy theory, although appropriate for autotrophs and animals, require some modification with respect to fungi. These alterations centre around the importance of substrate availability and the definition of competition, stress and disturbance (Cooke & Rayner, 1984).

Fungi depend for their activities on reduced carbon compounds, usually carbohydrates, for which they commonly have a high demand. There is thus a requirement for intimate association with carbon sources, and the most important single activity during vegetative development is the acquisition and utilization of these sources. Taking into account the varying degrees of enzymic competence exhibited by fungi and the assimilable or refractory nature of substrates available to them, then the latter have a profound effect on development patterns. Substrate characteristics determine not only what kinds of fungi will grow in any particular situation but also the duration of the life span of those that do so. Further to this, the successful capture and consumption of energy resources are strongly influenced by the additional impact of competition, stress and disturbance.

Competition, as it relates to filamentous fungi, is defined as an active demand by two or more individuals of the same or different species for the same resource. Competition has diverse forms involving a variety of mechanisms, but two main aspects have been distinguished: primary resource capture and combat. The former is the process of gaining initial access to, and influence over, an available resource, that is occupation of territory. The second is either the defence of captured territory or the wresting of territory, secondary resource capture, from an established occupant. The use of the term combat in this context serves to emphasize the active, antagonistic means through which both defence and aggression are mediated.

Stress describes a state in which there is some form of continuously imposed environmental extreme that tends to restrict biomass production by the majority of fungi. It can have a nutrient or non-nutrient basis and at its most severe renders growth impossible. However, it is tolerated at lesser levels by those fungi that have acquired the physiological characteristics necessary to do so and, indeed, some have developed fitness traits which enforce dependence on specific stress factors that limit development of unadapted species.

Disturbance occurs when all or part of total fungal biomass is either destroyed or subjected to new selection pressures through a drastic, but transient, alteration in environmental conditions. As with competition, this has two distinct aspects. First, there is destructive disturbance via a violent perturbation. Events of this kind occur in nature and affect fungal habitats but are relatively rare. Secondly, there is enrichment disturbance which, by contrast, is a common, natural, non-destructive perturbation occurring when there is an input of organic material containing a low indigenous fungal biomass (Pugh, 1980).

The behavioural concepts embodied in econutritional groupings and strategy theory, together with what is known of the evolutionary history of fungi, provide a broad, coherent background against which it is possible to view many ecophysiological themes. A large part of what follows is concerned with outlining these themes and with examining details of function in relation to this theoretical background.

PART TWO
Growth and Development

2 Resource acquisition and utilization

Having proposed an evolutionary framework within which to view ecological behaviour it is now appropriate to provide information on some basic physiological and biochemical attributes of fungi. Selected major characteristics are described here, largely in outline, in relation to the functional continuum of substrate utilization, nutrient uptake, translocation, metabolism and storage. Only passing reference will be made at this stage to the possible ecological significance of particular processes, or to the ways in which these may be affected by environmental constraints; such matters are for subsequent chapters. However, immediate comment is necessary concerning the source material for much of our understanding of fungal physiology and metabolism.

Despite its great size, the physiological literature contains detailed studies of relatively few species, with the balance being firmly in favour of yeasts at the expense of filamentous fungi. Furthermore, although there has been a recent tendency to study fungi under ecologically meaningful conditions (which often means oligotrophic conditions), most observations have been and still are being made using unrealistically high nutrient levels and temperatures. Consequently, much of what is known of functions, whilst having value in itself, can be only tenuously related to natural behaviour. What follows is therefore frequently no more than an eclectic summary of functional capabilities under ideal nutritional and environmental conditions in the laboratory. Where possible, filamentous species are brought to the fore because of their predominant roles as decomposers and as symbionts with other organisms.

Substrate relationships

Fungi vary widely in their minimum requirements for sources of carbon, nitrogen, phosphorus and other minerals, vitamins and other growth factors. It might seem reasonable to suppose that laboratory-defined substrate requirements have direct relevance to econutritional behaviour. Whilst this is commonly the case, they do not provide an infallible guide to natural nutritional traits. For example, in axenic culture, the nutritional requirements of some highly specialized biotrophs may be no more exacting than those of much less fastidious saprotrophs, and do not reflect the complex and finely balanced nutritional relationship that exists between symbiont and host. However, on a more general level, it is salutary to consider common substrates that are available as carbon sources in relation to their assimilability and persistence, and to which fungal groupings can utilize them (Table 2.1). Naturally occurring non-polymeric compounds, which are readily assimilable, can be utilized by virtually all fungi, as can those polymers that are relatively easily hydrolysed.

Table 2.1 Some substrates available to fungi as major carbon sources together with some prominent utilizers (from Cooke & Rayner, 1984, © Longman).

Degree of assimilability and persistence	Substrate	Major sources	Prominent utilizers
Readily assimilable, non-persistent	Glucose, fructose, mannose (and other hexoses), xylose	Living or dead plant and animal tissues, living or dead microbial cells	All fungi except certain Oomycetes
	Sucrose, maltose (and other hexose oligomers)	Living or dead plant and animal tissues, living or dead microbial cells	Probably most fungi with some notable exceptions, e.g. sucrose not utilized by many Chytridiomycetes, Zygomycotina, and some Ascomycotina
	Organic acids	Living or dead plant and animal tissues, living or dead microbial cells	Some fungi, but ability highly specific
	Fatty acids	Living or dead plant and animal tissues, living or dead microbial cells	Some Chytridiomycetes (Blastocladiales) and Oomycetes (Leptomitales)
Readily to fairly readily assimilable, non-persistent to fairly persistent	Starch	Plant tissues	A wide range of fungi including some Oomycetes
	Inulin	Plant tissues (Compositae, Liliaceae)	A wide range of fungi including some Oomycetes
	Glycogen	Animal tissues, and microbial cells	Probably a wide range of fungi
	Hemicelluloses	Plant cell walls, especially higher algae	A wide range of fungi, possibly particularly marine Ascomycotina
	Pectins	Plant primary cell walls and middle lamellae	A wide range of fungi
	Lipids (fats and oils)	Animal and plant tissues, microbial cells, animal secretions	A wide range of fungi
	Proteins (non-keratinized)	Living or dead plant and animal tissues, microbial cells	Many fungi, but release of ammonia may be quickly inhibitory

Table 2.1 *Continued*

Degree of assimilability and persistence	Substrate	Major sources	Prominent utilizers
Slowly or very slowly assimilable, fairly persistent to very persistent	Cellulose	Plant cell walls	A wide range of fungi, but few Zygomycotina
	Cutin	Plant cuticles	Little information, but cutinases possibly widespread among Ascomycotina
	Lignin	Plant cell walls	Some Ascomycotina (Xylariaceae), many Basidiomycotina
	Suberin	Plant bark and endodermises	Undoubtedly some fungi, but information scanty
	Chitin	Arthropod exoskeletons, fungal cell walls	Many soil-borne fungi, and species of Ascomycotina that inhabit living arthropods
	Keratin	Dermal tissues of animals; hair, fur and feather, hoof and horn	Many Ascomycotina, including species that inhabit living animals
	Waxes	Plant cuticles	Many leaf-surface fungi
	Other hydrocarbons (oil, bitumen, kerosene, petrochemical products)	Mainly industrial	Many Ascomycotina

Complex, refractory, persistent substrates are utilized much more selectively, with the fungi able to do so often exhibiting specificity for substrata rich in such substrates. These polymers constitute some of the most abundant organic resources on Earth, there being a vast annual input of cellulose, hemicellulose, lignin, suberin and waxes from primary producers together with lesser, but nevertheless important, amounts of chitin and keratin from consumers. Fungi appear to be the principal organisms involved in degrading these materials, so it is apposite to begin with this aspect of resource acquisition.

Utilization of polymeric compounds

An important distinction is made here between storage and structural compounds in terms of their availability to fungi in both time and space. Starch, inulin, glycogen and lipids are abundant in plant or animal cells and can be readily utilized by a wide range of fungi. However, they probably only become accessible in quantity to either pathogens or to non-pathogenic primary colonizers of dead tissues. By contrast, plant and microbial cell walls and extracellular structural components of plants and animals, whilst also acting as carbon sources for pathogens and primarly colonizers, have greater persistence and so become available to a greater spectrum of fungi over a longer period of time.

Storage polymers

Although starch and glycogen are considered to be the distinctive polysaccharide reserves of plants and animals respectively, both have been reported as occurring in fungi (Blumenthal, 1976). Each has a basic structure of repeating units of α-1,4-linked D-glucose. Starch consists of two polymers: amylose, a linear molecule of mainly α-1,4-linked units, and amylopectin which makes up 75–85% of most starches and which is highly branched with α-1,6-links along the main α-1,4-chain (Fig. 2.1). Glycogen has a similarly branched structure but the α-1,6-links are more frequent.

A range of enzymes is involved in degradation and their co-operative action may be required for complete breakdown to be achieved (Table 2.2). Modes of attack are dependent on the nature of the substrate and environmental conditions. Furthermore, large variations in enzyme patterns can occur even amongst different isolates of the same species. Probably α-amylase is the most widely distributed of these enzymes. It hydrolyses α-1,4 linkages but leaves α-1,6-linkages unaffected, so producing α-limit dextrins. Amyloglucosidase and α-glucosidase remove glucose residues via hydrolysis of α-1,4-linkages, doing so most rapidly from longer and shorter chain oligosaccharides respectively. The former enzyme also has some activity towards α-1,6-linkages.

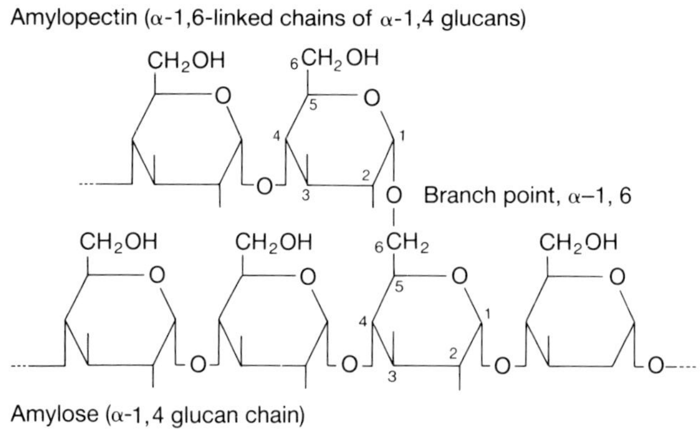

Fig. 2.1 Basic structure of amylopectin and amylose.

Table 2.2 Fungal enzymes degrading starch and glycogen (after Fogarty, 1983, © Elsevier Applied Science).

Enzyme	EC number	Action and bonds attacked	Products	pH optimum	Temperature optimum (°C)
α-amylase (α-1,4-D-glucan glucanohydrolase)	3.2.1.1	Endo α-1,4-links	α-D-glucose, maltose, maltotriose, α-limit dextrins	3–6	40–60
Amyloglucosidase (α-1,4-D-glucan glucohydrolase)	3.2.1.3	Exo α-1,4-links, with some α-1,6-links	β-D-glucose	4.5–5.0	40–60
α-glucosidase (α-D-glucoside glucohydrolase)	3.2.1.20	Exo α-1,4-links with some α-1,3- and α-1,6-links	α-D-glucose	3.0–7.5	35–60
Isoamylase (amylopectin 6-glucanohydrolase)	3.2.1.68	Endo α-1,6-links	α-1,4-glucans of varying chain length	5.0–6.5	25–60
Phosphorylase (α-1,4-D-glucan : orthophosphate α-D-glucosyltransferase)	2.4.1.1	Exo α-1,4-links	Glucose-1-phosphate	—	—

Amyloglucosidases are glycoproteins containing 5–20% carbohydrate and appear to be almost exclusive to fungi (Fogarty, 1983). The limit dextrins produced by α-amylase, α-glucosidase and, to a lesser extent, amyloglucosidase are degraded further by debranching enzymes. Amongst fungi the most common of these is isoamylase which hydrolyses α-1,6-linkages so allowing attack by the other enzymes. Phosphorylase is important in the breakdown of endogenous glycogen and starch.

In addition to polysaccharides, plant, animal and microbial cells may contain large quantities of lipids in the form of both storage compounds and membrane components. The constituent fatty acids vary in aliphatic chain length, degree of unsaturation and position of double bonds (Table 2.3). Some may be phosphorylated at the ester link and others may form lipoprotein complexes (Macrae, 1983). A single enzyme, lipase (glycerol ester hydrolase EC 3.1.1.3), can catalyse the hydrolysis of most water-insoluble triglycerides to release fatty acids, partial diglycerides and glycerol.

In fungi, lipases are generally wall bound and repressed by the presence of monosaccharides, disaccharides or glycerol, and are induced by triglycerides, fatty acids and lecithin. Two groups of lipases can be distinguished with differing positional specificity (Fig. 2.2). Different forms of the enzyme may also have different substrate specificities, but absolute specificity towards one triglyceride type is rare as the reaction catalysed remains the same. The majority of lipases are glycoproteins which can become tightly bound to hydrophobic surfaces. In *Geotrichum candidum* the active site of lipase lies at the centre of the enzyme in a hydrophobic area large enough to accommodate the triglyceride molecule; this may explain the ability of the enzyme to attack insoluble substrata. Lipid utilization may be of particular importance for some entomogenous fungi since lipid, together with chitin and protein, is a major constituent of arthropod cuticle (Charnley, 1984).

Structural components of plants

In nature eventual colonization by fungi is the fate of all plant organs, virtually every structural component being susceptible to degradation, however slowly. Putting aside details of the ways in which fungi colonize living and dead plant parts, some major structural materials are described here, together with a necessarily compressed general account as to how they may be utilized.

Waxes and suberin. Epicuticular wax occurs to varying degrees on leaves, fruits and stems. When in sufficient quantity, it presents an infection barrier to potential pathogens; its presence in litter provides a slowly assimilable, persistent resource for those fungi able to utilize

Table 2.3 Structure and components of triglycerides and phosphoglycerides.

Generalized structure	Fatty acids			Alcohols*	
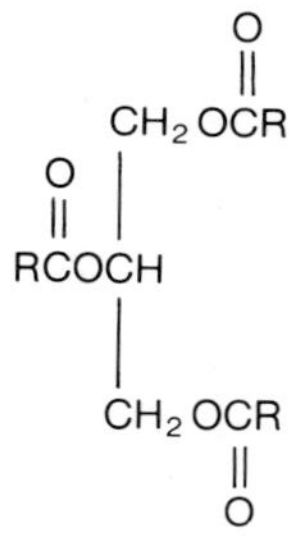	*Saturated*			Ethanolamine	$\underline{HO}CH_2CH_2NH_2$
	Lauric	12:0	$CH_3(CH_2)_{10}COOH$	Choline	$\underline{HO}CH_2CH_2N(CH_3)_3$
	Myristic	14:0	$CH_3(CH_2)_{12}COOH$	Serine	$\underline{HO}CH_2CHNH_2COOH$
	Palmitic	16:0	$CH_3(CH_2)_{14}COOH$		
	Stearic	18:0	$CH_3(CH_2)_{16}COOH$		
	Arachidic	20:0	$CH_3(CH_2)_{18}COOH$		
Triglycerides	*Unsaturated*			Inositol	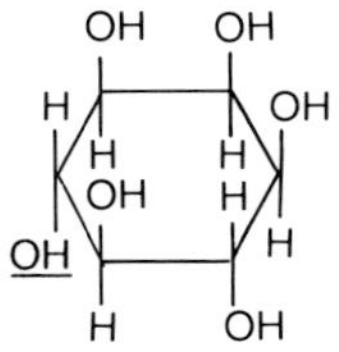
	Palmitoleic	16:1	$CH_3(CH_2)_5CH{=}CH(CH_2)_7COOH$		
	Oleic	18:1	$CH_3(CH_2)_7CH{=}CH(CH_2)_7COOH$		
	Linoleic	18:2	$CH_3(CH_2)_4CH{=}CHCH_2CH{=}CH(CH_2)_7COOH$		
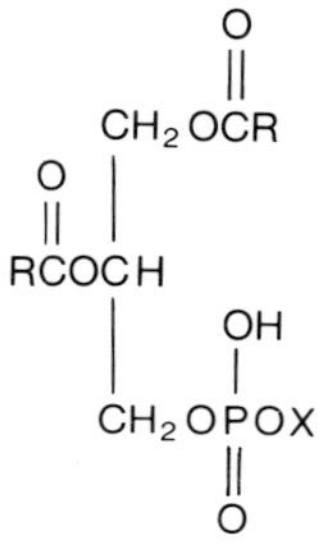	Linolenic	18:3	$CH_3CH_2CH{=}CHCH_2CH{=}CHCH_2CH{=}CH(CH_2)_7COOH$		
	Arachidonic	20:4	$CH_3(CH_2)_4CH{=}CHCH_2CH{=}CHCH_2CH{=}CHCH_2CH{=}CH(CH_2)_3COOH$		
Phosphoglycerides					

R = fatty acid; X = alcohol.

Figures after names of fatty acids give number of carbon atoms in the chain and number of double bonds present.

* The hydroxyl group esterified to phosphoric acid is underlined.

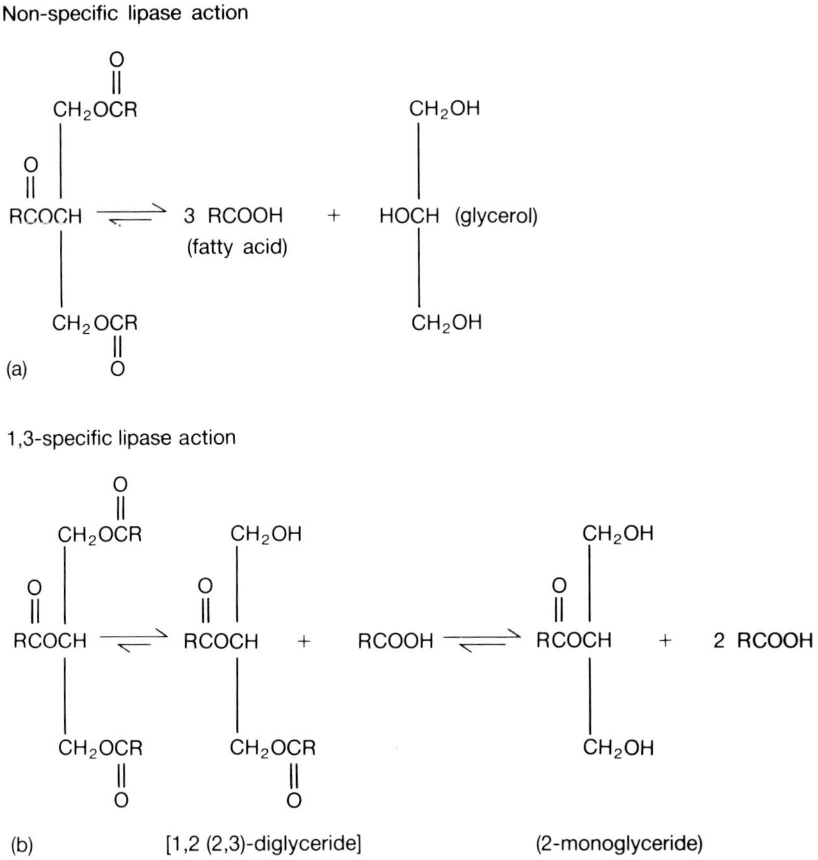

Fig. 2.2 Action of lipase on triglycerides (from Macrae, 1983, © Elsevier Applied Science Publishers).

it. Despite their ubiquity, decompositions of plant waxes have been little studied, although wax-degrading ability seems to be well developed amongst phylloplane inhabitants. Erosion of wax to expose the underlying cuticle can be effected by both saprotrophs, for example *Sporobolomyces roseus*, and biotrophs, notably *Erysiphe graminis* (McBride, 1972; Staub *et al.*, 1974).

A similar situation obtains with respect to suberin, an important component of secondary thickening, which consists of long-chain fatty acids and phenolic compounds embedded in wax (Kolattukudy, 1985). Detail of its utilization by fungi is scanty. In culture, *Fusarium solani* f. sp. *pisi* will grow on suberin-enriched wall fractions obtained from potato tuber periderm, an esterase being induced with identical properties to a cutinase (see below); however, mechanisms by which phenolic compounds might be utilized remain unexplored. Suberin breakdown in litter is presumed to be slow but may not be inordinately so in the light of the observation that *Armillaria mellea* can reduce the suberin content of root bark by 60% over a 10-month period (Swift, 1965).

Cutin. The extracellular cutin of aerial plant parts is more easily degradable than its overlying waxes. This complex polymer of hydroxy and epoxy fatty acids, bonded via ester linkages of primary alcohol groups, can be broken down relatively quickly by a wide range of fungi. The cutin esterases involved are inducible glycoproteins which have specificity for the dominant primary alcohol–ester linkages, the released monomers then being rapidly utilized (Kolattukudy, 1985).

Pectin. The importance of pectic materials in plant structure and as nutrient sources for a great many fungi with widely differing life styles needs no emphasis here. For pathogens in particular, an ability to degrade pectin would seem to be the key to gaining access to other cell wall components. When fungi are cultured on isolated cell walls, pectin-degrading enzymes are the first polysaccharidases to be induced, followed in turn by hemicellulases and cellulases. Monomers released by an initially low level of pectic enzymes are taken up, and this results in further enzyme production via autocatalytic induction. At higher concentrations these readily assimilable carbon compounds repress enzyme production (self-catabolite repression), as can the presence of glucose and other sugars. This induction–repression process controls production of many other polymer-degrading enzymes.

Pectic substances are composed mainly of polymerized α-1,4-linked galacturopyranose, which may have methylated carboxyl groups (Fig. 2.3). Interspersed with this may be α-1,2-linked rhamnose (rhamnogalacturan) with the uronide residues acetylated at the 2 and/or 3 position. Pectin may also contain polymers of α-1,3- and α-1,5-linked arabinofuranose (araban) and β-1,4-linked galactopyranose (galactan), possibly covalently linked to rhamnogalacturan. The structural complexity of pectin is matched by the array of enzymes that are necessary to bring about its complete degradation (Table 2.4, Fig. 2.3). In addition to these there are also enzymes that hydrolyse minor components: namely endo and exo β-1,4-galactanase and β-1,4-galactosidase, which act on β-1,4-galactan, and exo α-L-arabinofuranosidase, which attacks arabans.

Hemicelluloses. Hemicelluloses consist of several heterogeneous polymers and can be divided into three major types: β-1,4-linked D-xylans, β-1,4-linked D-mannans, and β-1,3- and β-1,4-linked galactans, with xylans being the most common form present in wood. These can join to other sugars to form heteropolymers such as xyloglucans,

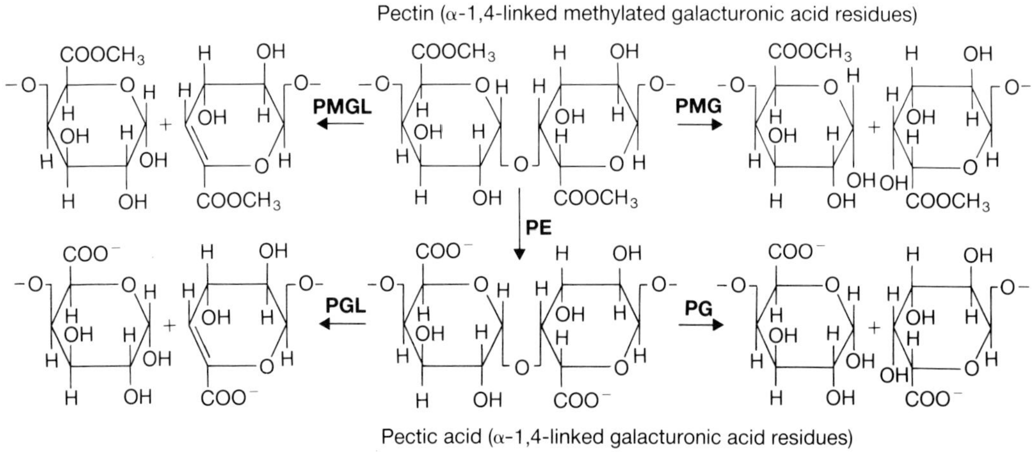

Fig. 2.3 Basic structure of pectin and pectic acid with mode of action of pectinases. Enzymes: PMGL, polymethylgalacturonate lyase; PMG, polymethylgalacturonase; PE, pectinesterase; PGL, polygalacturonate lyase; PG, polygalacturonase (from Fogarty & Kelly, 1983, © Elsevier Applied Science Publishers).

arabinogalactans, rhamnogalacturonans and glucogalactomannans, which may also carry a range of side chain constituents.

As with pectin, a range of enzymes is necessary to degrade hemicelluloses completely (Table 2.5). In strict terms, hemicellulases (glycan lyases, EC 3.2.1) are those enzymes which degrade the glycan chains that form the backbone of the polymer. This excludes exo-glucosidases which hydrolyse only low molecular weight glycosides and short-chain or monosaccharide side branches. Nevertheless, these aid the action of hemicellulases by removing steric hindrance (Dekker, 1985).

The best studied of hemicellulases are the β-1,4-D-xylanases which are produced both by yeasts and filamentous fungi. Two types of endo action are known. The first is debranching, by which L-arabinose is liberated from arabinoxylans and arabinoglucuronoxylans; the second is non-debranching, and L-arabinose is not released (Dekker, 1985). There are numerous forms of each enzyme, with different products frequently resulting from the action of each. The substrate binding site is often complex, having up to seven subsites for the binding of β-1,4-D-xylopyranosyl residues with the active catalysis site being present near the central subsites. Xylanases are commonly induced by short-chain oligomers of xylose. In the yeast *Cryptococcus albidus*, xylobiose and xylotriose are taken up by an active β-xyloside permease, and this induces xylanase production and secretion, and endogenous synthesis of xylosidase (Krátký & Biely, 1980).

Table 2.4 Pectin-degrading enzymes, with alternative terminology and mode of action (after Fogarty & Kelly, 1983, © Elsevier Applied Science Publishers).

Enzyme	Substrate and optimum pH	Reaction or bond attacked
Pectinesterase PE (EC 3.1.1.11) (polymethylgalacturonate esterase PMGE, pectinmethylesterase)	Pectin pH 4–7	De-esterifies to pectic acid by removal of methoxyl residues
Polymethylgalacturonase PMG (pectinmethylgalacturonase)	Pectin	
Endo PMG		Random hydrolysis of α-1,4-linkages
Exo PMG		Sequential hydrolysis of same
Polymethylgalacturonate lyase PMGL (pectin transeliminase, pectin lyase)	Pectin	
Endo PMGL (EC 4.2.2.10)	pH 5–8	Random cleavage of α-1,4-linkages by transelimination
Exo PMGL		Sequential cleavage of same
Polygalacturonase PG	Pectic acid	
Endo PGL (EC 3.2.1.15)	pH 4–5.5	Random hydrolysis of α-1,4-linkages
Exo PG1 (EC 3.2.1.67)	pH 4–6	Sequential hydrolysis of same to release galacturonic acid
Exo PG2 (EC 3.2.1.82)		Sequential hydrolysis of same to release digalacturonic acid
Polygalacturonate lyase PGL (pectate lyase)	Pectic acid	
Endo PGL (EC 4.2.2.2)	pH 8–10	Random cleavage of α-1,4-linkages via transelimination
Exo PGL (EC 4.2.2.9)		Sequential cleavage of same

Two basic processes are thought to be induced in hemicellulose degradation. In the first, the action of exoglycosidases precedes that of some hemicellulases, thus removing side chains to expose the polymer's backbone. In the second, endohemicellulases are the first to act, and attack parts of the main glycan chain at relatively unbranched points, so producing oligosaccharide mixtures. The latter then require further action by exoglycosidases and hemicellulases, the sugars produced being rapidly taken up.

Cellulose. Cellulose is the most abundant organic resource on Earth; there being 7×10^{11} tonnes in existence at any time, with an annual turnover rate of 4×10^{10} tonnes (Coughlan, 1985). It is a linear polymer of up to 14 000 β-1,4-D-glucose residues with a basic repeating

Table 2.5 Hemicellulose-degrading enzymes and their modes of action.

Enzyme	Major substrate	Reaction or bond attacked
Endo β-1,4-D-xylanase (EC 3.2.1.8)	β-1,4-D-xylan	Hydrolyses 1,4-links in main chain
Endo β-1,3-D-xylanase (EC 3.2.1.32)	β-1,3-D-xylan	Hydrolyses 1,3-links of side chain
Exo β-1,3-D-xylanase (EC 3.2.1.72)	β-1,3-D-xylan	As for the endo enzyme
β-D-xylosidase (EC 3.2.1.37)	Oligo-β-1,4-D-xylan	Hydrolyses terminal 1,4-links
Endo β-1,4-mannanase (EC 3.2.1.78) (β-1,4-D-mannan mannanhydrolase)	β-1,4-D-mannan	Hydrolyses 1,4-links
D-galactanase(s)	β-1,3- and β-1,4-D-galactan	Endo and exo enzymes attack both linkages
α-L-arabinase(s)	α-1,3- and α-1,5-L-arabans	Endo and exo enzymes attack both linkages
L-arabinofuranosidase (EC 3.2.1.55)	α-1,3-L-araban	Exo enzymes acting on α-1,3-links
α-D-glucuronidase(s)	4-*O*-methylglucuronic acid-substituted xylo-oligomers	Exo enzyme cleaving terminal 4-*O*-methyl-glucuronic acid side chains
Esterases	Acetyl xylo-oligomers	Cleaves *O*-acetyl groups from C-2 and C-3 positions on xylose residues

unit of cellobiose. The chains are orientated in parallel, and are so arranged that intramolecular hydrogen bonds stabilize each chain, with intermolecular bonds tightly cross-linking adjacent chains (Fig. 2.4). Approximately 40 double chains when bonded form an inert, rigid microfibril which can become aggregated with others to form a macrofibril. Macrofibrils may then become associated in bundles. Within each macrofibril, well-ordered crystalline regions lie amongst more poorly organized amorphous areas (Fig. 2.4). Macrofibrils are generally associated with pectin, hemicelluloses or lignin, cell walls usually containing 30–50% cellulose.

Much of what is known of the enzymology of cellulose utilization has been obtained from studies made *in vitro*. Although it is not always possible to do so, this information should be viewed in relation to natural situations where cellulose forms only part of intermixed macromolecular structures. Fungi differ in their ability to degrade complexed cellulose, notable examples being found amongst wood decay species. For example, some Basidiomycotina and certain

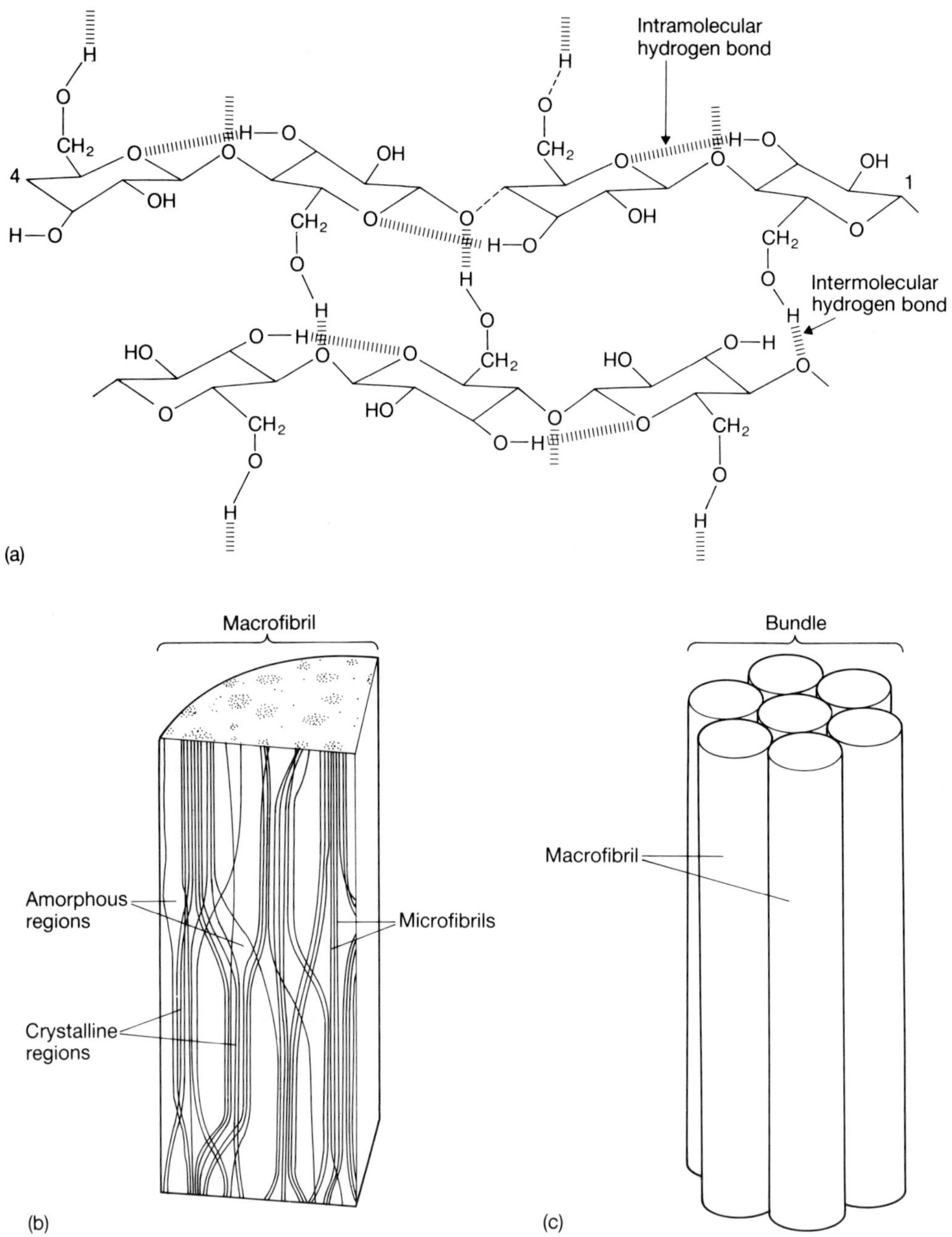

Fig. 2.4 Basic structure of cellulose. (a) Chains of β-1,4-linked glucose residues with hydrogen bonding within and between cellulose chains (from Alberts *et al.*, 1983, © Garland Publishing); (b) approximately 40 double chains bond together to form microfibrils which associate with other microfibrils within a macrofibril; (c) macrofibrils associate in bundles or fibres (from Sihtola & Neimo, 1975, © SITRA).

xylariaceous Ascomycotina (white rots) decay all components of wood at a similar rate. Other Basidiomycotina (brown rots) and Ascomycotina (soft rots) utilize hemicellulose and cellulose but only slightly modify lignin (Eriksson & Wood, 1985). Such behavioural differences are not apparent amongst inhabitants of other lignocellulosic materials, for example those decomposing cereal straw.

A number of enzymes bring about cellulose breakdown (Table 2.6). The three major enzymes (endo- and exocellulase, and β-glucosidase) are hydrolytic and are able to cause considerable degradation. However, additional processes may be of importance, and two are of particular interest. The first involves cellobiose dehydrogenase, which can use either lignin or its degradation products as electron acceptors, and which consequently may act as a link between cellulose and lignin breakdown. The second involves a non-enzymic iron-containing factor which, possibly acting together with endo- and exocellulase, produces short microfibrils of cellulose without releasing soluble sugars (Griffin *et al.*, 1984). Single strains commonly produce multiple forms of cellulase but the reason for this, apparently energetically wasteful, production of enzymes having the same function is not known.

Table 2.6 Some cellulose-degrading enzymes and their modes of action.

Enzyme	Substrate	Reaction or bond attacked
Endocellulase (EC 3.2.1.4) (β-1,4 (1,3 : 1,4)-D-glucan 4 glucanohydrolase)	Cellulose	Internal hydrolysis of β-1,4-linkages
Exocellulase (EC 3.2.1.91) (β-1,4-D-glucan cellobiohydrolase)	Cellulose	Hydrolyses terminal β-1,4-linkages to release cellobiose
β-glucosidase (EC 3.2.1.21) (β-D-glucoside glucohydrolase)	Cellulo-oligosaccharides including cellobiose	Removes glucose from non-reducing ends of chains
Glucan β-1,4 glucosidase (EC 3.2.1.74) (β-1,4-D-glucan glucohydrolase)	β-1,4-D-glucan	Removes glucose units
Cellobiose dehydrogenase (EC 1.1.5.1) (Cellobiose: quinone 1-oxidoreductase)	Cellobiose	Converts cellobiose to cellobiono-1,5-lactone
Cellobiose oxidase	Cellobiose	Converts cellobiose to cellobiono-1,5-lactone

Synthesis of cellulases *in vitro* is induced by cellulose and repressed by glucose or other readily metabolizable sugars, although which compound causes induction *in vivo* is not clear. Again, *in vitro*, endocellulases are produced before exocellulases, so that enzyme secretion seems to be well co-ordinated. Subsequent action of the two enzymes is synergistic, and at least two action sequences probably occur. The first involves the binding and release of enzymes. In *Trichoderma reesei*, endo- and exocellulases bind to different sites on the cellulose molecule, but each may be able to affect the other by hastening catalysis and thus bringing about enzyme release. A 'competitive' cycle of release and binding therefore takes place. The second sequence depends on the need for removal of cellobiose units by exocellulase following the action of endocellulase. If exocellulase were absent, the glycoside linkages broken by endocellulase would re-form rapidly. Furthermore, due to the position of hydrogen bonding in the cellulose chains, the stereoscopic position of each β-1,4-linkage would necessitate the existence of at least two endocellulases of different specificities and, consequently, at least as many stereo-specific forms of exocellulase. Repeated action of one stereospecific form of exocellulose would produce terminal groups suitable for the other. This might explain observed endocellulase–exocellulase synergism in *Trichoderma reesei* and, additionally, in *Penicillium funiculosum* (Wood, 1985, 1989).

The overall effect of enzyme action on cellulose may be summarized as follows (Fig. 2.5). After enzyme induction, crystalline cellulose is rendered more accessible to hydrolysis via the process of amorphogenesis, which may involve the iron-containing factor mentioned above. Whilst being poorly understood, amorphogenesis results in some depolymerization, and consequent loss of strength, of stranded microfibrils. Endocellulase then attacks the main cellulose chains, and either exocellulase releases cellobiose from the exposed ends or, if present, β-glucosidase similarly releases glucose and additionally converts cellobiose to glucose.

Lignin. Lignin is the most complex of plant polymers, and it occurs mainly as lignocellulose attached to the hydroxyl groups of cellulose microfibrils via ether linkages. It is a three-dimensional molecule made up of phenylpropane-based monomers linked via a variety of bonds, with some lignins additionally containing phenolic acids (Fig. 2.6). Lignin composition varies widely between plant groups.

Lignin is highly refractory and persistent, with only Basidiomycotina and some xylariaceous Ascomycotina being able to degrade it extensively. This requires an exogenous carbon source, for example that provided by free sugars from cellulose breakdown, because lignin

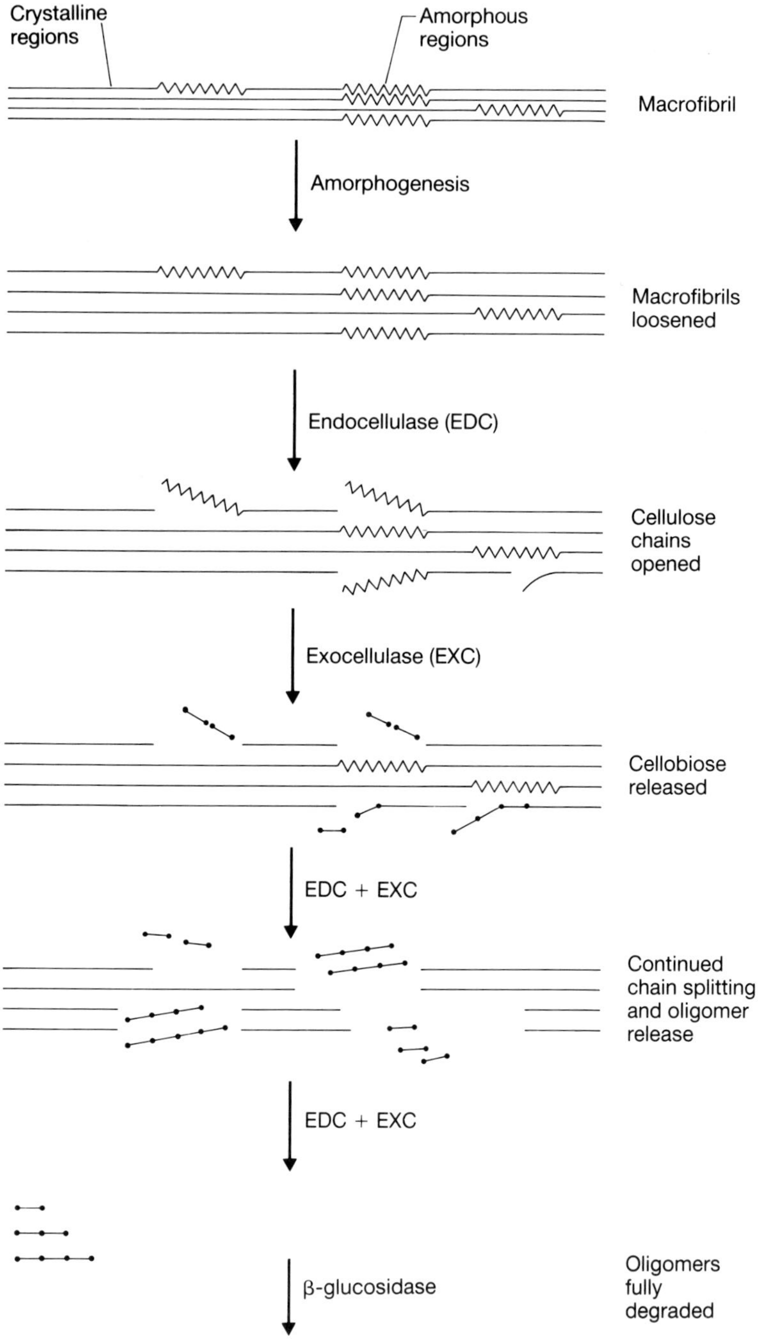

Fig. 2.5 Possible events during sequential enzyme action on a cellulose macrofibril; (●), glucose (from Montencourt & Eveleigh, 1979, © TAPPI).

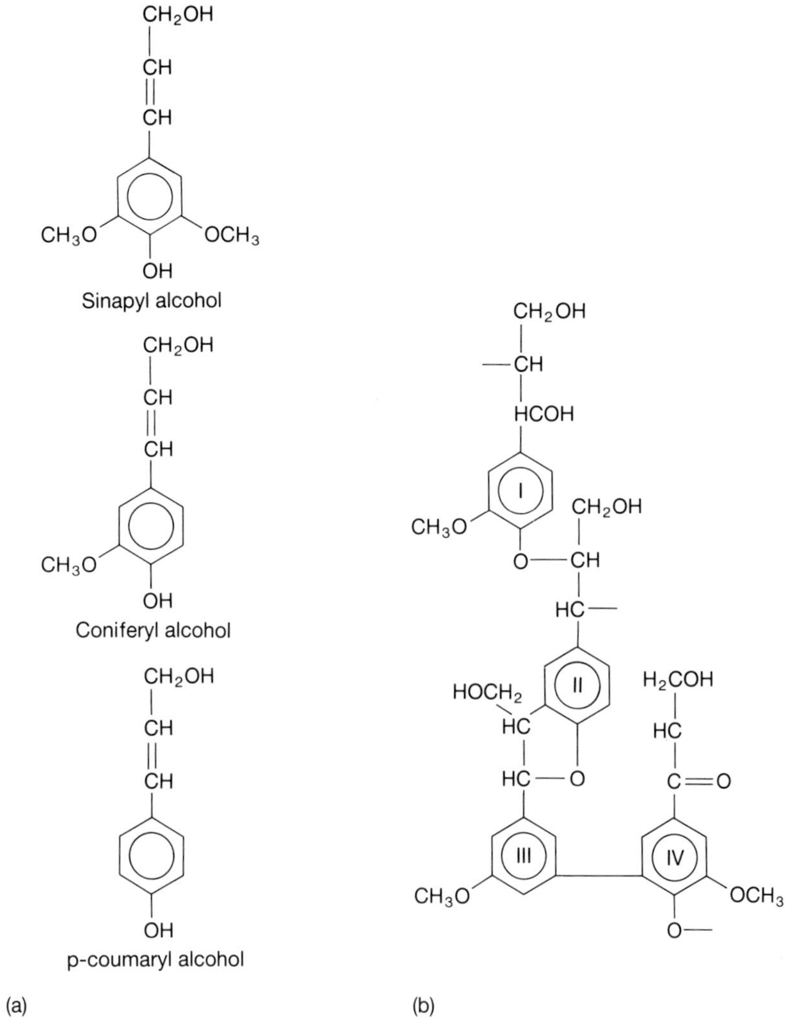

Fig. 2.6 Basic structure of lignins. (a) Common phenolic subunits; (b) part of lignin polymer showing predominant linkage types between subunits: I–II, arylglycerol-β-aryl ether bond; II–III, phenylcoumaran bond; III–IV, biphenyl bond (Adler, 1977, © Springer-Verlag; Kirk & Fenn, 1982, © Cambridge University Press).

itself is thought not to supply sufficient energy to maintain its own degradation (Kirk *et al.*, 1978). Nevertheless, ericoid mycorrhizal fungi *Hymenoscyphus ericae* and *Oidiodendron griseum*, and the facultative ectomycorrhizal basidiomycete *Paxillus involutus*, are able to degrade lignin as the sole carbon source (Haselwandter *et al.*, 1990). In culture, ligninolytic enzymes may be synthesized only under conditions of nitrogen starvation, and induced by lignin degradation products and the secondary metabolite veratryl alcohol. The suggestion has therefore been made that lignin breakdown is a secondary metabolic function that mainly serves to unmask cellulose and expose it to cellulases to which it would otherwise be inaccessible (Tonon *et al.*, 1990). Release of highly reactive, low molecular weight molecules, such as H_2O_2 or veratraldehyde, which can readily diffuse away from hyphae and penetrate into lignocellulose, may initiate degradation by making cellulose and lignin more susceptible to enzymic attack. As lignocellulolytic enzymes do not appear to diffuse far from hyphae in wood, the process would serve to facilitate degradation with a minimum of energy expenditure on enzyme production (Ruel, 1990).

Lignin degradation involves numerous chemical reactions and releases a variety of breakdown products, which suggests the existence of an array of ligninolytic enzymes. However, a single enzyme with a multiplicity of forms is produced *in vitro* by *Phanerochaete chrysosporium* which, depending on the substituents on the aromatic ring, will carry out many different reactions identified during degradation. It is an H_2O_2-dependent peroxidase that functions by accepting an oxygen atom from H_2O_2 and so possesses two oxidizing equivalents more than the native enzyme. The active oxyferryl centre so formed has a sufficiently high redox potential to oxidize phenolic and non-phenolic substrates (Harvey *et al.*, 1985). In a first, one-electron oxidation of a non-phenolic substrate, a radical cation intermediate is formed which then breaks down in different ways depending on the substituent groups present (Fig. 2.7). Radical cations themselves might then act as further one-electron oxidants, and so mediate single electron transfer between aromatic groups in areas of the lignin substratum at a distance from the mycelium, thus continuing and enlarging the degradation process. Veratryl alcohol is thought to be a co-factor for ligninase in this type of reaction, in which it acts as an electron carrier (Faison *et al.*, 1986). In the presence of suitable oxidants, aromatic radical cations can also hydrate reversibly to produce phenolics, which explains the common appearance of the latter in media where fungi are degrading lignin. Ligninase can additionally function as a phenol-oxidizing enzyme carrying out the same kinds of reaction as laccase (oxygen requiring) and peroxidase (peroxide requiring). These enzymes have themselves been implicated in lignin breakdown, but they differ from ligninase in being unable to oxidize non-phenolic moieties, and are so unable to effect complete degradation.

Chitin Chitin is second only to cellulose in abundance, its main sources being the exoskeletons of terrestrial and marine invertebrates and fungal cell walls (Berkeley, 1979). It consists of polymerized α-1,4-linked *N*-acetyl-D-glucosamine, and together with chitosan, its

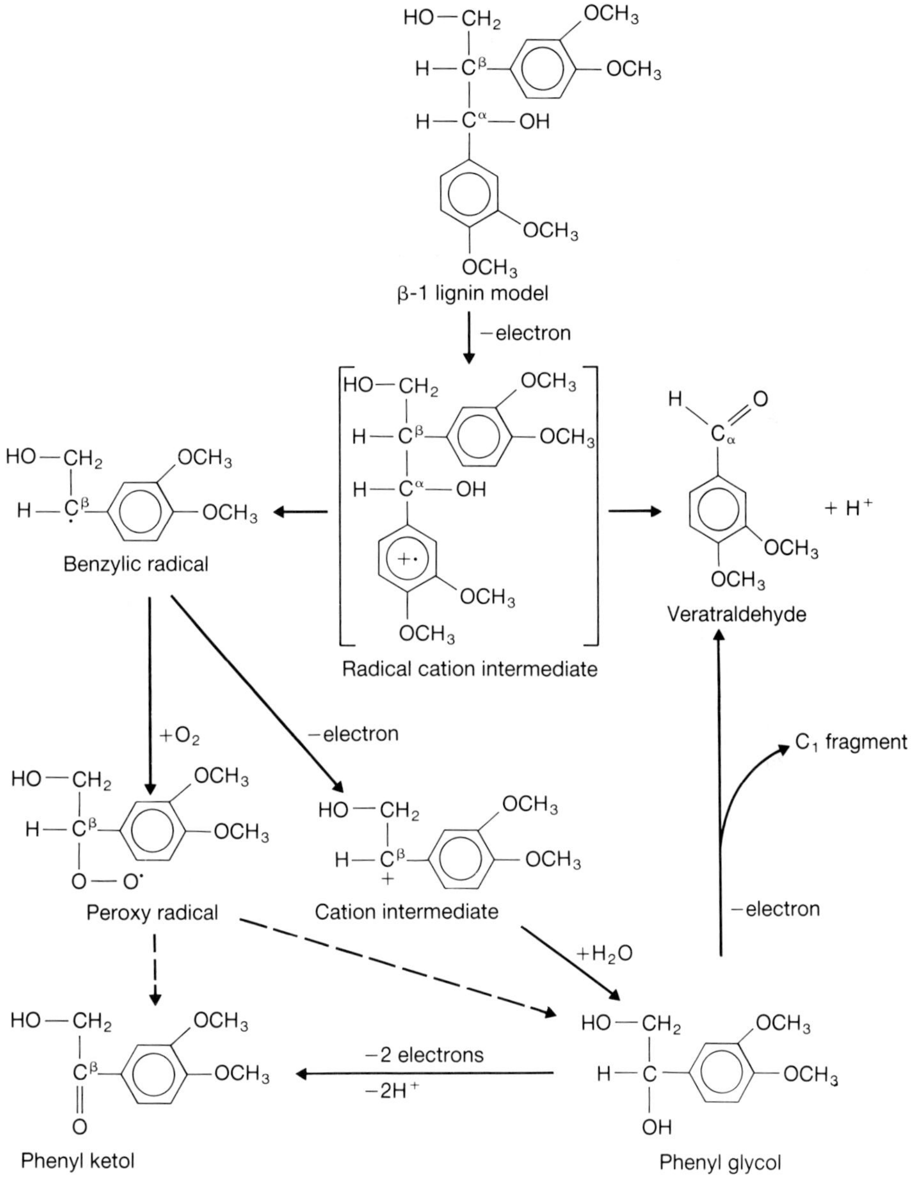

Fig. 2.7 Degradation pathway for a non-phenolic β-1 lignin model compound via a radical cation intermediate (from Harvey *et al.*, 1985, by permission of Oxford University Press).

deacetylated form may constitute 4–60% of cell wall dry weight in fungi and up to 80% of invertebrate cuticle (Fig. 2.8). In marine habitats, chitin decomposition appears to be brought about mainly by bacteria and Actinomycetes, but in fresh water and on land fungi assume major importance. Chitinolytic ability is widespread, and is

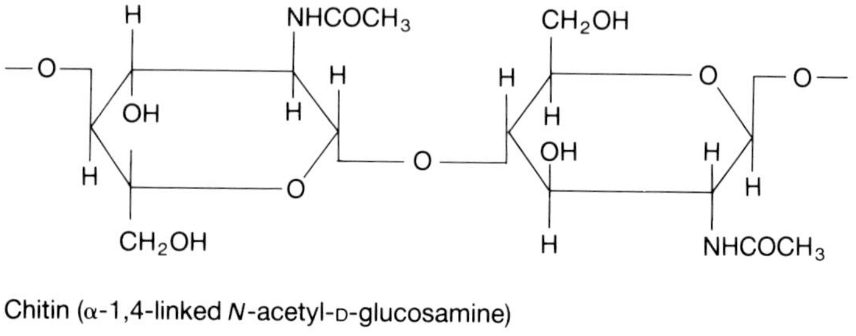

Fig. 2.8 Basic structure of chitin.

crucial to the success of particular ecological groups, for instance insect pathogens and non-pathogenic inhabitants of insect cuticle.

Chitinase (endo β-*N*-acetylglucosamidase, EC 3.2.1.14) randomly attacks the α-1,4-linkages to release oligosaccharides, principally di-*N*-*N*′-acetylchitobiose, which are then further degraded to *N*-acetylglucosamine. Invertebrate chitin is complexed with proteins and lipids, so that protease and lipase activity is often required to prepare chitin for chitinase action. Chitosan is broken down by an endo-acting chitosanase to produce oligosaccharides and some *N*-acetylglucosamine, but this enzyme shows little or no activity towards chitin. Glucosamine is taken up intact to be then converted to glucose via intracellular deamination.

Proteins So far, descriptions of polymer utilization have centred on its value for carbon provision. With regard to nitrogen, the most abundant potential source is probably protein of plant, animal and microbial origin. This includes not only cytoplasmic protein but also structural forms such as those bound to lignin or complexed into keratin. Some of the selective advantages accruing from an ability to utilize proteins as sources of carbon and nitrogen, and sometimes sulphur, have been disclosed only relatively recently. For instance, it may confer advantage to decomposer basidiomycetes inhabiting protein-rich litter, and on the ericoid mycorrhizal species *Hymenoscyphus ericae* in peat soils (Bajwa & Read, 1985; Kalisz *et al.*, 1987). Ectomycorrhizal fungi can use protein as a sole nitrogen source to the nutritional benefit of their plant symbionts (Abuzinadah & Read, 1986a,b). Similarly, some plant pathogens (*Colletotrichum lagenarium, Monilinia fructigena*), entomogenous pathogens (*Beauveria bassiana, Metarhizium anisopliae*), dermatophytes (*Trichophyton* species) and mucosal yeasts (*Candida* species) may depend on proteins during exploitation of their hosts (North, 1982).

Fungi produce a range of enzymes capable of degrading most forms

Table 2.7 Peptide hydrolases produced by fungi.

Enzymes	Location of action	pH optimum
Exopeptidases		
α-aminoacylpeptidases (EC 3.4.11)	N terminus	7.0–10.0
Serine carboxypeptidases (EC 3.4.16)	C terminus with serine at active site	4.5–6.0
Metallo-carboxypeptidases (EC 3.4.17)	C terminus, Zn^{2+} and Co^{2+} required	—
Endopeptidases		
Serine proteinases (EC 3.2.21)	Serine and histidine at active site	7.0–12.0
Cysteine (SH or thiol) proteinases (EC 3.4.22)	Cysteine at active site	4.0–8.0
Aspartic (carboxyl or acid) proteinases (EC 3.4.23)	Acid residue at active site	3.0–6.0
Metallo-proteinases (EC 3.4.24)	Metal ion, often Zn^{2+}, required	4.0–8.0

of protein (Table 2.7). Most species possess at least one kind of peptide hydrolase, which may be either wall bound or freely extracellular. Depending on the cultural methods employed, species may produce either several types of enzyme or multiple forms of the same enzyme (North, 1982). Endoproteinases are most widely found in fungi, and of these, serine proteinases and carboxyl proteinases are common, thiol proteinases apparently being rare. In *Neurospora crassa* and *Aspergillus* species extracellular proteinase production in the presence of protein is controlled via derepression in the absence of an additional source of carbon, nitrogen or sulphur (Hanson & Marzluf, 1975; Cohen, 1981). By contrast, in the basidiomycetes *Agaricus bisporus*, *Coprinus cinereus* and *Volvariella volvacea*, extracellular proteinase activity in the presence of protein is not repressed by additional sources of these three elements (Kalisz *et al.*, 1987). This may indicate an important ecophysiological difference between Ascomycotina and Basidiomycotina.

Acquisition of low molecular weight nutrients

All nutrients entering the cytoplasm must cross the plasmalemma and cell wall. The latter has a mainly structural role and is permeable to most small molecules. There are, however, enzymes within it that can affect nutrients en route, and certain wall components have a non-specific cation-binding capacity. Notwithstanding these wall properties, it is the plasmalemma which is the functional barrier to solute movement.

There is a vast corpus of data on acquisition of low molecular weight nutrients by fungi in axenic culture. However, as has been indicated before, studies have encompassed very few species, the most commonly used being *Saccharomyces cerevisiae*, *Neurospora crassa*, *Aspergillus nidulans* and *Penicillium chrysogenum*. This, together with the general employment of optimized conditions, makes for difficulty in extrapolating laboratory observations to natural situations. Major exceptions to this are found amongst investigations on symbionts of plants, especially mycorrhizal fungi, where nutrient transfer from host to symbiont has been studied.

Central aspects of nutrient acquisition, that is transport, the influx or efflux of a solute across the plasmalemma; uptake, the net gain of a solute within a cell in comparison with its external concentration; and assimilation, the metabolism of a solute within the cell, have been described regularly and in detail (see Scarborough, 1985; Jones & Gadd, 1990). In view of this, and the fact that fungal cells have much in common with those of other organisms, a necessarily brief account is given here.

Transport mechanisms

Mechanisms for transport of ions and of organic molecules across the plasmalemma are essentially similar, and three systems have been recognized: passive diffusion, facilitated diffusion and active transport (Fig. 2.9).

Passive diffusion of a solute can occur in either direction across the membrane in response to a difference in its concentration between the outside and inside of the cell. Rates of transport are intrinsically slow, due to the retarding effect of membrane lipids, but uptake will take place if the solute is assimilated within the cytoplasm. Quicker uptake occurs during facilitated diffusion, in which specific protein carriers or permeases bear solutes from the outer to the inner membrane surface. If the carrier has a lower affinity for a solute at the inner surface then this, together with solute assimilation within the cytoplasm, facilitates more rapid accumulation.

During active transport, energy is expended, which speeds uptake further via either direct or indirect coupling (Fig. 2.9). In direct coupling, ATP (or polyphosphate in the case of sugars) phosphorylates either the carrier or the solute, and brings about solute accumulation. Indirect coupling also requires ATP, but here it is used to drive a proton pump that excretes H^+ from the cytoplasm, thus reducing the pH of the external medium and creating an electrochemical gradient. This gradient enables solute uptake, often by means of proton cotransport with a carrier.

Membrane carrier proteins exhibit saturation kinetics at high nutrient concentrations, are affected by specific inhibitors, and can be affected by competition for binding sites with closely related molecules, these characteristics being held in common with enzyme proteins. Each carrier seems to have a unique range of properties. For

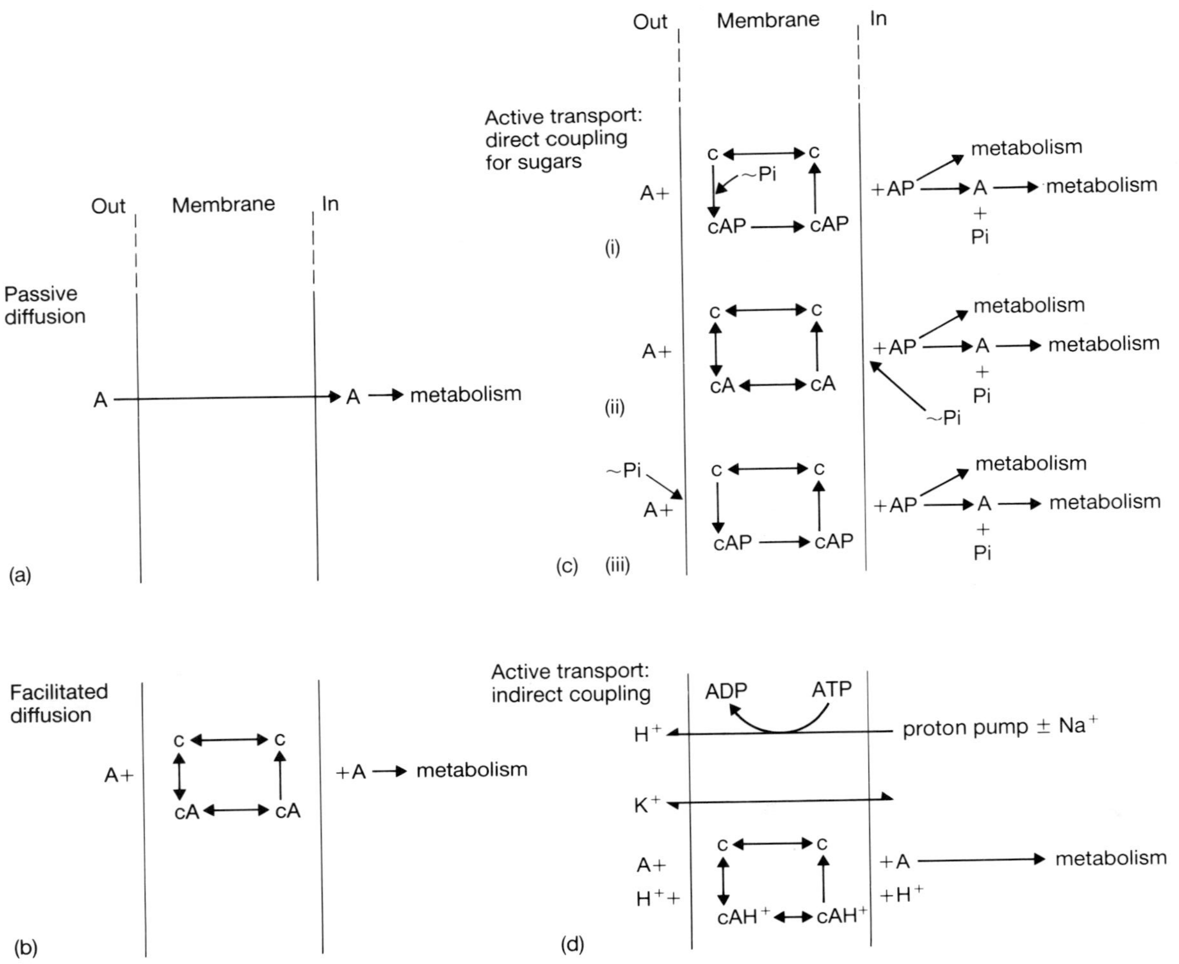

Fig. 2.9 Transport mechanism for a solute (A) passing through the cell membrane to enter metabolism. (a) Passive diffusion in either direction possible, determined by concentration differences inside and outside the membrane, predominantly inward if metabolized; (b) facilitated diffusion in either direction possible with solute attached to a protein carrier (c) or with permease aiding movement, movement inward if solute is metabolized or has a lower affinity with the carrier at the inner membrane surface; (c) active transport using energy and predominantly one way, showing direct coupling (group transfer) of a sugar. The sugar is phosphorylated and transferred with high energy phosphate (~Pi). The phosphorylation site is not clear, (i), (ii) and (iii) indicating the possibility of this occurring within the membrane or on its inner or outer surface. (d) Active transport with indirect coupling. An ATP-driven proton pump reduces pH of the external medium creating a proton or Na^+ gradient which drives solute transport through co-transport of H^+ via a carrier. Electroneutrality may be maintained by uptake of K^+ or other cations.

example, in *Saccharomyces cerevisiae*, glucose uptake is by means of facilitated diffusion, and on starvation, glucose uptake is decreased. Similarly, in *Neurospora crassa*, uptake is via facilitated diffusion, but under starvation conditions an active mechanism for glucose uptake is induced (Barnett & Sims, 1976).

Transport of organic nutrients

Monosaccharides can be transported by both facilitated diffusion and active transport, with the different carrier systems being either specific for only one sugar or having an ability to transport several. Some disaccharides, for example maltose, are actively transported intact, but others, notably sucrose, are first hydrolysed externally by most fungi, although some may take sucrose up directly (Sheard & Farrar, 1987). For most sugars, uptake is followed by rapid assimilation, but accumulation without further metabolism may occur when fungi are osmotically stressed (Eddy, 1982).

Uptake of amino and organic acids is via mechanisms similar to those for sugars. Most fungi, with the exception of some yeasts, require aerobic conditions for maximum amino acid uptake, and very high accumulation ratios (concentration inside : concentration outside) of 100 : 1 to 1500 : 1 have been observed. Fungi can have specific amino acid requirements for growth. For instance, some require amino acids as nitrogen sources for mycelial growth whilst others require them as sulphur sources.

Many fungi are prototrophic for vitamins, sterols and other essential organic growth factors (Table 2.8), that is they can synthesize them sufficient for their needs. It is, however, common for fungi to be auxotrophic and to require an exogenous supply of one or more of these compounds. In some cases synthesis by a prototrophic species may be insufficient to satisfy demand and must be supplemented exogenously. Whilst their importance in fungal metabolism is manifest, transport and uptake of vitamins and other factors have been little studied, although for *Saccharomyces cerevisiae* there is some evidence for the existence of a range of carriers (Shavlovsky & Sibirny, 1985).

Ion transport

Cation transport is an active process. Uptake of K^+, for instance, requires an external fermentable substrate in the absence of which, in non-growing cells, there is slow K^+ leakage irrespective of endogenous respiration. Transport of K^+ can take place against a concentration gradient of 5000 : 1, with transport of other monovalent, alkali cations being in the order $K^+ > Rb^+ > Cs^+ > Na^+ > Li^+$. Potassium ion transport inwards occurs at a similar rate to that of H^+ inwards, but transport outwards favours H^+ or Na^+ rather than that of K^+. Divalent cations have a transport affinity in the order $Mg^{2+} > Zn^{2+} > Mn^{2+} >$

Table 2.8 Roles of some vitamins and other factors required by fungi.

Compound	Function
p-aminobenzoic acid	Forms tetrahydrofolic acid used as a co-enzyme in one-C transfers
Biotin (vitamin B_7)	Co-enzyme for carboxylations
Cyanoco-balamin (vitamin B_{12})	Forms co-balamin derivatives used as co-enzymes for methyl transfers
Nicotinic acid (vitamin B_3) and nicotinamide (niacin)	Form nicotinamide adenine dinucleotide (NAD) and nicotinamide adenine dinucleotide phosphate (NADP) used as co-enzymes for dehydrogenases
Pantothenic acid (vitamin B_5)	Forms co-enzyme A used in two-C transfers
Pyridoxine (vitamin B_6)	Forms pyridoxyl phosphate and pyridoxamine phosphate used as co-enzymes for transaminations
Riboflavin (vitamin B_2)	Forms flavin mononucleotide (FMN) and flavin adenine dinucleotide (FAD) used as co-enzymes for dehydrogenases
Inositol and cholesterol*	Membrane components

* Although not co-enzymes, some fungi have an absolute requirement for an exogenous supply of these at micromolar concentrations.

$Ca^{2+} > Sr^{2+}$. Glucose, phosphate and K^+ all stimulate divalent cation uptake. In *Neurospora crassa* the activity of the electrogenic proton pump appears to be linked to a Ca^{2+}–$2H^+$ electroneutral exchange, which would serve to expel Ca^{2+} from the cell. In other species, K^+–H^+ exchanges may be more important. However, the systems involved in maintenance of ionic levels within the cytoplasm and vacuoles, and in controlling cation exchange between the external medium and the internal compartments, are far from clear.

Iron transport in conditions of high availability is not active and does not require carriers or membrane receptors, although hydroxy-acids can dissolve adsorbed ferric polymers. With low levels of iron availability, which commonly occur in nature, transport of Fe^{3+} is mediated by siderophores (or siderochromes), a group of specifically induced chelating compounds (Fig. 2.10). Amongst fungi, the most common of these are the ferrichrome or hydroxamate types, but others also occur and, for example, *Candida albicans* can produce phenolate types as well. A single species may produce several different siderophores as a response to a variety of environmental conditions (Winkelmann, 1986). Their structure confers a very high affinity for iron, and their release into the immediate environment increases

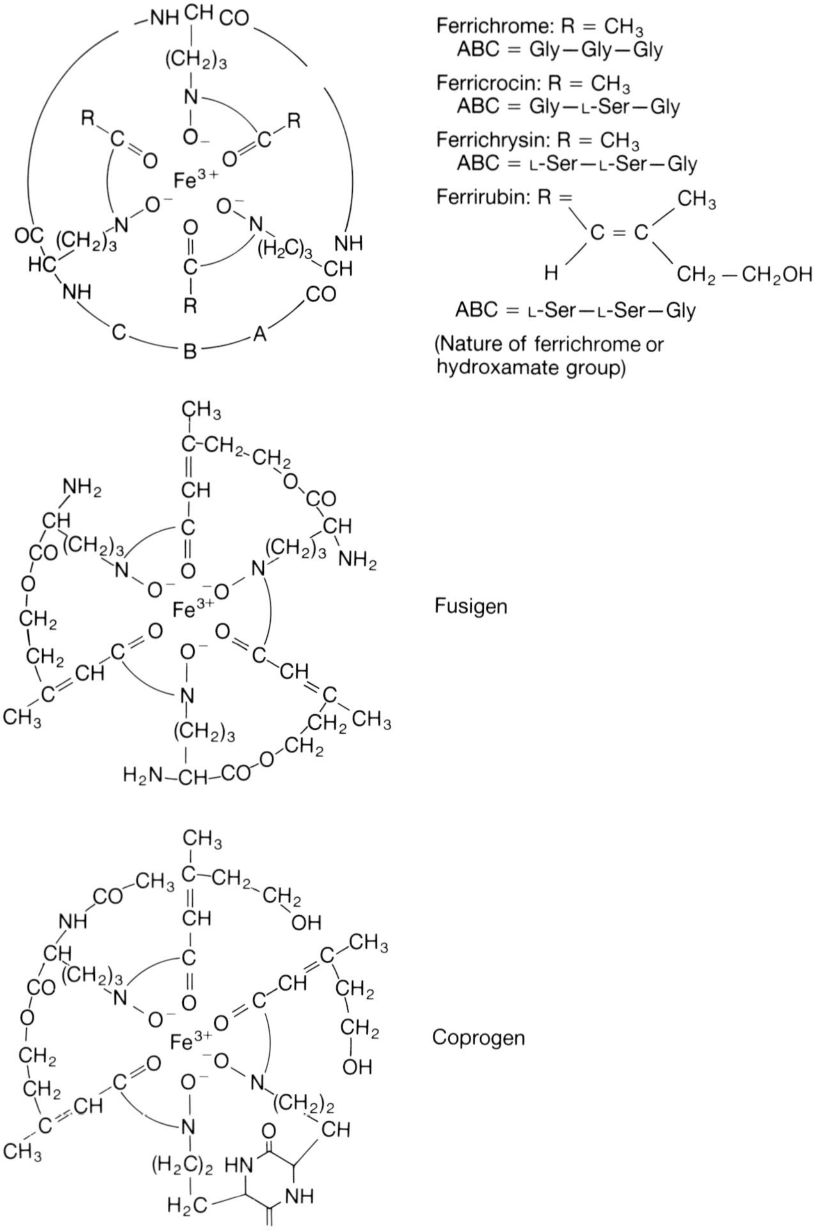

Fig. 2.10 Structure of some fungal siderophores showing ligation position of Fe^{3+} (from Wiebe & Winkelmann, 1975, © American Society for Microbiology).

the availability of iron in solution for transport. Siderophore–Fe^{3+} complexes such as coprogen and ferrichrome (*Neurospora crassa*) can pass directly through membranes via carrier systems and accumulate within the cell. Others, for instance ferrichrome A (*Ustilago sphaerogena*) and rhodotorulic acid (*Rhodotorula pilimanae*) remain outside, with only the iron entering. In *Neurospora crassa*, transport involves a single carrier that has distinct recognition sites for different siderophores, whereas in *Penicillium parvum* only ferrichrome-specific receptors are present. Optimum transport of Fe^{3+}–siderophore complexes takes place at low pH, which implies a need for protonation of the amino acid moieties in the siderophores for either recognition or transport. In *Saccharomyces cerevisiae*, which does not secrete siderophores, iron must be reduced before uptake, after which it is stored in vacuoles, as are most other divalent cations (Lesuisse & Labbe, 1989). More commonly ferric iron, once inside the cell, may enter metabolism directly or be stored within iron-rich proteins. In *Neurospora crassa* and *Saccharomyces cerevisiae*, copper is stored similarly in cysteine-rich proteins (metallothioneins) and these may play a role in detoxification and intracellular regulation of metal ion concentrations in general (Gadd, 1986).

In nature, inorganic nitrogen is commonly available as either NH_4^+ or NO_3^-. Ammonium is actively transported and is utilized preferentially over NO_3^- in ion mixtures, during which counterion exchange with H^+ results in a decrease in external pH. Nitrate appears to enter cells by diffusion as a result of nitrate reductase activity. In general, NH_4^+ carriers have low affinity in high nitrogen conditions, but high affinity systems become induced during starvation. Inability to ulilize NO_3^- is common amongst Basidiomycotina, Chytridiomycetes and Oomycetes, which seem to be deficient in nitrate reductase (Plassard *et al.*, 1986).

Phosphate is actively transported against concentration gradients in excess of 100 : 1, and both high and low affinity systems occur. In *Saccharomyces cerevisiae*, aerobic conditions and a fermentable substrate are necessary, but in mycorrhizas transport is independent of exogenous carbohydrate. During transport, external pH falls whilst internal pH rises, which suggests that $H_2PO_4^-$ uptake is accompanied by simultaneous uptake of protons. Phosphate uptake by *Saccharomyces cerevisiae* is stimulated by K^+, and Na^+ has the same effect in marine fungi (Eddy, 1982). In mycorrhizas in particular, very high accumulation ratios may be achieved due to the additional factor of polyphosphate synthesis. This occurs within the vacuole, effectively removing phosphate from the cytoplasm where it would otherwise interfere with metabolism, and allows considerable amounts of phosphate to be stored, often as insoluble metachromatic granules

(Gianinazzi-Pearson & Gianinazzi, 1986). Divalent cations, especially Ca^{2+}, aid this process via the formation, for example, of calcium polyphosphate.

Translocation

Water and solutes may travel considerable distances within mycelia, but translocation in this sense should be distinguished from mass migration of cytoplasm (cyclosis or transmigration) which is evidenced in many fungi by movement of nuclei, mitochondria and storage granules. Cytoplasmic flow is driven metabolically, probably by microtubule action, at rates of up to $25\,cm\,h^{-1}$ but is unlikely to be the main translocatory mechanism in hyphae (Eamus & Jennings, 1986a).

Translocation may be effected in several ways, the most common of which involves bulk flow of solutions (Fig. 2.11). For example, in aerial sporophores transpiration may induce water flow from the supporting mycelium and surrounding substratum. However, within mycelia, bulk flow is most frequently driven by a hydrostatic pressure source–sink system (Jennings, 1987). In *Serpula lacrimans*, glucose enters the hypha by active transport, being then converted to trehalose, which generates hydrostatic pressure at the source. The resultant pressure-driven flow translocates trehalose to the hyphal apex where it is either used for metabolism or is converted to arabitol. This serves to maintain a carbon sink and to create an osmotic gradient which further aids bulk flow. The internal hydrostatic pressure causes exudation of liquid droplets at the hyphal surface, a phenomenon commonly seen during the organogenesis of fruitbodies, sclerotia and rhizomorphs into which rapid translocation takes place.

A further mechanism involves ionic movement (Fig. 2.11). In older parts of a hypha, ATPases or ATP-driven H^+ (Na^+)–K^+ exchangers in the plasmalemma may decrease H^+ (Na^+) concentrations and increase that of K^+. However, at the extending hyphal tip ATPases are absent from the plasmalemma, and so K^+ travels down a concentration gradient towards the tip, at the same time inducing movement of water and vesicles in the same direction. This process might also serve in the maintenance of polarity of apical growth (Gow, 1989).

In situations where mycelia emerge from a nutrient-rich resource to travel across or within a nutrient-poor substratum, translocation will be predominantly in one direction. It is, however, probable that bidirectional flow is more commonplace; this is conspicuous in mycorrhizal species in the hyphae of which there is simultaneous movement of water and minerals towards the host and of carbon compounds away from it.

In some Basidiomycotina, translocatory ability is greatly enhanced by the aggregation of hyphae into cords or rhizomorphs which may exhibit a high degree of internal organization. For example, in the

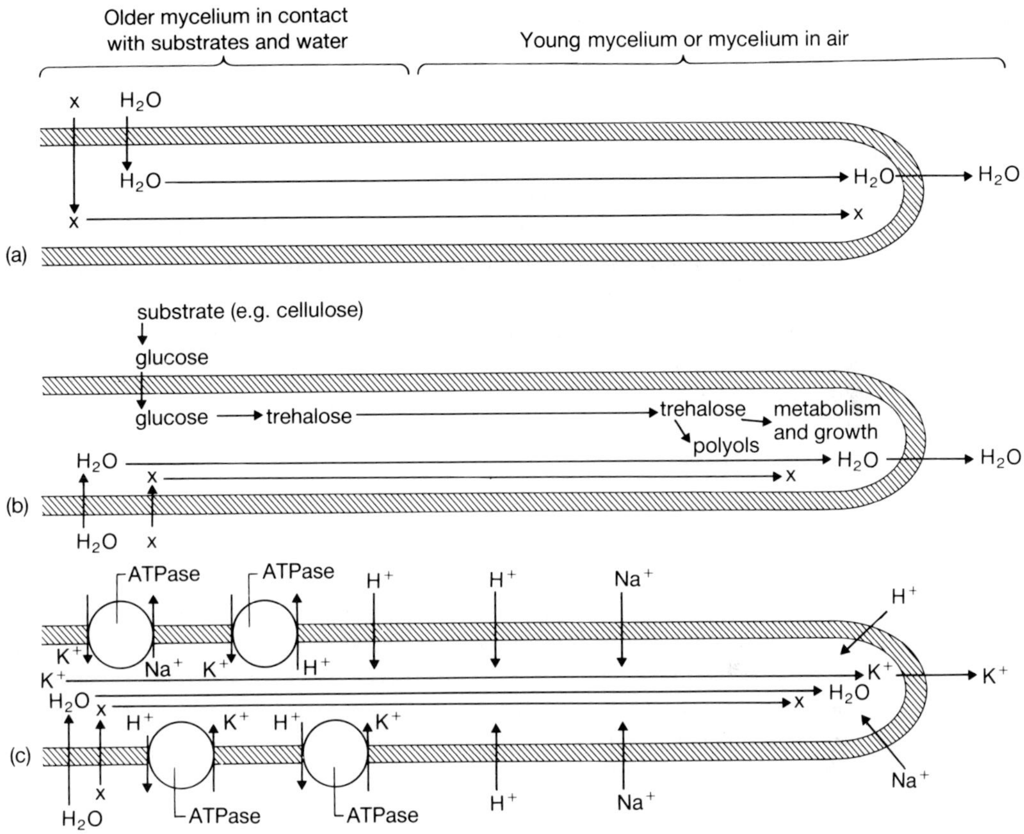

Fig. 2.11 Possible mechanisms for translocation of a nutrient (x) along a hypha. (a) Driven by evaporation of water. (b) Driven by a hydrostatic pressure source–sink system. Both systems involve bulk flow of solution. (c) Driven by a concentration gradient of K^+ created by ATPase activity present only in older parts of the mycelium. In all three cases water enters the hypha and moves along the water potential gradient accompanied by the nutrient in solution (from Jennings, 1979, 1984a; Lysek, 1984, © Cambridge University Press).

ectomycorrhizal species *Suillus bovinus*, strands are differentiated into central, wide, cytoplasm-free vessel hyphae surrounded by cytoplasm-rich sheathing hyphae (Brownlee *et al.*, 1983). If the walls of such vessel hyphae were relatively impermeable, then different materials could travel independently within adjacent hyphae, and in opposite directions, by bulk flow (Jennings, 1987).

Translocation has several important implications for growth and survival, some of which will be expanded upon in subsequent chapters. As has already been noted, it permits fungi to travel over or through nutrient-poor substrata and thus to explore and exploit new territory. By moving trehalose to hyphal tips for subsequent conversion to polyols (sugar alcohols), turgor pressure, and so growth, can be maintained under conditions of low water availability. Polyols may also have a role in control of proton availability, and consequently on overall ionic balance, in addition to activity of proton pumps in the plasmalemma, via the reaction:

$$\text{Sugar} + \text{NAD(P)H} + \text{H}^+ \rightleftharpoons \text{polyol} + \text{NAD(P)}^+$$

Continual bulk flow towards the hyphal tip can, however, create problems in that such an excess of some nutrients may occur as to require detoxification. This difficulty may be overcome by mass migration of cytoplasm in the opposite direction, by excretion of materials in solution at the hyphal tips, or by the formation of excretory crystals on the exterior of the hypha (Jennings, 1987).

Metabolism

Primary metabolism is maintained throughout the life cycle of fungi but is generally linked to growth. There are very few intermediates at the monomer level and they rarely accumulate. By contrast, secondary metabolism is largely associated with postgrowth or differentiation phases. Secondary metabolites have great structural diversity, are produced specifically by various taxonomic groupings of fungi and, although they have no apparent direct function in growth, they may have importance for survival. Selected, salient features only of metabolism are delineated here; consequences for natural behaviour, if and where possible, will be brought out in later chapters.

Primary metabolism

Glycolysis, fermentation and respiration. Glucose can be metabolized to pyruvate by three different, but closely connected, routes; the Embden–Meyerhof–Parnas (EMP) pathway, the Entner–Doudoroff (ED) pathway and the hexose monophosphate or pentose phosphate (HMP) pathway. Several other sugars and some amino acids can also feed into these routes. Essentially the EMP pathway produces pyruvate, some ATP, and reducing power in the form of NADH. After conversion to acetyl CoA, the pyruvate is used in the tricarboxylic acid (TCA) cycle and in fatty acid, isoprenoid, and aromatic synthetic pathways. Pyruvate, phosphoenolpyruvate and 3-phosphoglycerate are also used for synthesis of some amino acids. In those fungi that can develop anaerobically, NAD is regenerated by fermentation either to lactate or ethanol. The EDP and HMP pathways generate NADPH which is utilized in several syntheses, including those for polyols, fatty acids and cell wall components, as well as providing a pentose source for the manufacture of nucleic acids, co-enzymes and some amino acids.

Respiration is the outcome of three linked processes: the TCA cycle, electron transport, and oxidative phosphorylation, all of which take place in the mitochondrion. The overall reaction of the TCA cycle is:

$$CH_3COOH + 2H_2O \rightleftharpoons 2CO_2 + 8H$$

Three pairs of electrons (equivalent to the hydrogen atoms) become associated with NADH, and two pairs with FADH, from the oxidation of succinate. The TCA cycle can accept inputs derived from lipids, purines, pyrimidines and amino acids but, similarly, is also a supplier of amino acid precursors in the form of oxaloacetate and α-ketoglutarate as well as providing carbon skeletons for the biosynthesis of other compounds; these latter functions can drain the cycle. However, a series of anapleurotic reactions which produce TCA cycle intermediates help to maintain the cycle at maximum efficiency. In fungi that contain glyoxysomes, the glyoxalate cycle operates, which enables 2-carbon units entering metabolism from acetyl-CoA to feed into the TCA cycle without immediate loss as CO_2. This may be particularly important in germinating spores, where lipids are used to synthesize carbohydrates. Considerable feedback mechanisms operate on the TCA cycle, as well as on glycolysis, which control required energy levels and amounts of intermediates. It is only when substrate supply becomes limiting, or environmental effects perturb normal function of primary metabolic routes, that secondary metabolic shunts are activated.

The hydrogen atoms formed, or rather the equivalent electrons, are transferred to an electron transport chain. They pass down this in a series of oxido-reduction reactions to finally combine with oxygen to form water. Paired electrons pass to co-enzyme Q but subsequently, since each cytochrome contains only one iron atom, they move singly between the cytochromes, necessitating two cycles of oxido-reduction for each pair. The chain also accepts electrons from sources other than the TCA cycle. As the electrons progress, oxidative phosophorylation occurs at specific positions along the chain, so producing ATP.

Functioning of the normal electron transport chain has been elucidated largely by using either uncouplers, for instance 2,4-dinitrophenol, which prevent phosphorylation but not electron transport, or blockers, which prevent transport at specific positions along the chain. However, some fungi possess another, cyanide-resistant electron transport path which is inhibited by hydroxamate, but no oxidative phosphorylation occurs along this route.

Nitrogen, phosphorus and sulphur metabolism. Most fungi can utilize a variety of nitrogen sources, but when supplied with NO_3^- or NO_2^-, N must be converted to NH_4^+ before it can enter metabolism. This conversion involves a series of electron transfer reactions which employ NADPH as an electron donor (Table 2.9). Initially NO_3^- is reduced to NO_2^- in a two-electron step via nitrate reductase, the subsequent six-electron step from NO_2^- to NH_4^+ being catalysed by nitrite reductase. Ammonium ion enters metabolism through incorporation into glutamate or, with further addition of NH_4^+, glutamine. Nitrate reductase is induced by NO_3^- but repressed by NH_4^+, even with NO_3^- present in the medium. High NH_4^+ levels therefore depress NO_3^- assimilation. In low nitrogen conditions endogenous proteins may break down, with their constituent amino acids becoming deaminated to yield NH_4^+.

Phosphate enters metabolism rapidly, forming phosphorylated compounds through the action of ATP. In conditions of good phosphate supply polyphosphate is often synthesized (Table 2.10). This obviates osmotic and buffering problems that might otherwise result from high endogenous phosphate levels, as well as providing for phosphate storage. In times of shortage polyphosphate is quickly degraded by a range of enzymes (Capaccio & Callow, 1982).

Table 2.9 Incorporation of inorganic nitrogen into metabolism.

Enzyme	Reaction
Nitrate reductase	$NO_3^- + 2e + 2H^+ \rightarrow NO_2^- + H_2O$
Nitrite reductase	$NO_2^- + 6e + 8H^+ \rightarrow NH_4^+ + 2H_2O$
Glutamate dehydrogenase*	α-ketoglutarate + NH_4^+ + NAD(P)H $\rightleftharpoons$ $NAD(P)^+$ + glutamate + H_2O
Glutamine synthetase	Glutamate + NH_4^+ + ATP $\overset{Mg^{2+}}{\rightleftharpoons}$ glutamine + ADP + Pi + H^+

* This enzyme uses NADPH when forming glutamate and NAD when forming α-ketoglutarate.

Table 2.10 Formation and metabolism of polyphosphate.

Process	Enzymes	Reaction
Synthesis	Polyphosphate kinase	$(PolyP)_n + ATP \rightleftharpoons (PolyP)_{n+1} + ADP$
Degradation (specific)	Exopolyphosphatase	$(PolyP)_n + H_2O \rightarrow (PolyP)_{n-1} + Pi$
	Endopolyphosphatase	$(PolyP)_{n+m} + H_2O \rightarrow (PolyP)_n + (PolyP)_m$
	Polyphosphate glucokinase	$(PolyP)_n$ + glucose $\rightarrow$ glucose-6-P + $(PolyP)_{n-1}$
Degradation (non-specific)	Acid phosphatase	$(PolyP)_n + H_2O \rightarrow (PolyP)_{n-1} + Pi$

Sulphur is generally taken up as sulphate and is then activated by ATP to form adenosine phosphosulphate (APS). Subsequently, APS is phosphorylated and reduced to sulphite and sulphide in a series of steps whilst still bound to protein. The reduced sulphur is then transferred to serine to produce cysteine, or to *O*-acetyl homocysteine to produce methionine, from which it may be further metabolized.

Biosynthesis
Amino acids, nucleotides, nucleic acids and proteins

Amino acids are synthesized from a relatively small number of precursors, in the main siphoned off from the TCA cycle and EMP pathways. Of possible evolutionary significance is the synthesis of lysine from oxaloacetate through aspartate (DAP pathway) by Hyphochytriomycetes and Oomycetes, and from α-ketoglutarate (AAA pathway) by other fungi (see Table 1.2).

Nucleotide precursors consist of a purine or pyrimidine base glycosidically linked to ribose or deoxyribose phosphates. Formation of both types of precursors has a common activation step in which ribose-5-phosphate, from the HMP pathway, is phosphorylated with ATP to yield phosphoribosyl pyrophosphate. Further lengthy steps are required to produce active nucleosides.

In fungi, processes involved in DNA replication and regulation of transcription are relatively poorly understood but are thought to be analogous with those in other eukaryotes. Transcription of DNA is brought about by three RNA polymerases apparently specific for the type of RNA produced. Type I results in ribosomal RNA (rRNA) and accounts for 50–70% of RNA synthesis. Type II produces messenger RNA (mRNA) and constitutes 20–40% of synthesis. Type III, 10% of RNA synthesis, yields small nuclear RNA and transfer RNAs (tRNAs). Mitochondria have a different RNA polymerase with similar properties to bacterial RNA polymerase. The RNAs produced often undergo post-transcriptional modifications before becoming fully active. A range of RNAases are known which may help to control cellular activities.

Protein synthesis proceeds with activation of amino acids through coupling to specific tRNAs utilizing ATP. The tRNA–amino acid complexes so formed then associate with mRNA, rRNA and accompanying proteins, the new protein synthesized being based upon the mRNA code.

Reserve carbohydrates and wall polysaccharides

The major reserve carbohydrates of fungi are polyols (with exceptions shown in Table 1.1), trehalose, glycogen and, rarely, starch (Blumental, 1976; Jennings, 1984b). These can be derived from glucose or fructose, but synthesis requires initial phosphorylation and either input of reducing power in the form of NADPH or further activation using sugar nucleotides (Fig. 2.12). Polyol formation is closely linked to the

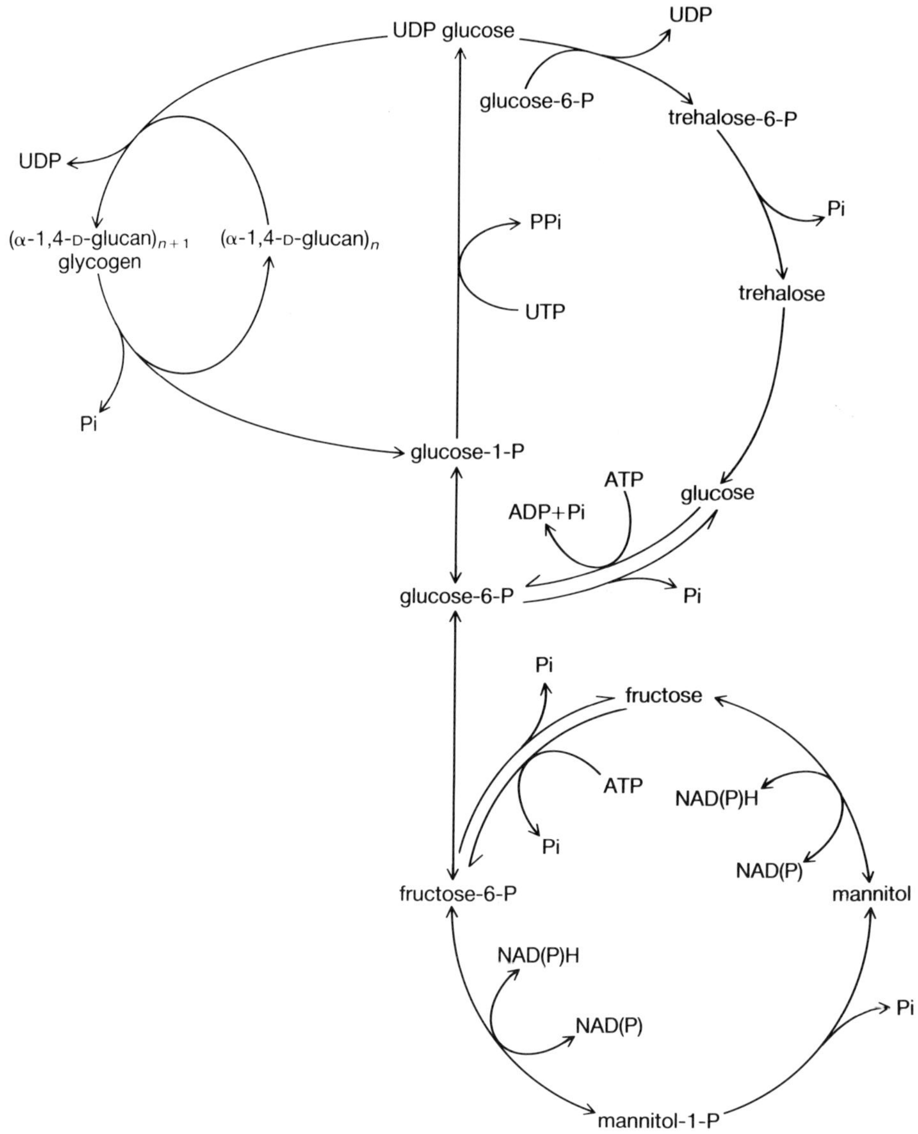

Fig. 2.12 Interconversions of sugars involved in the formation and degradation processes of some reserve carbohydrates.

HMP pathway, and arabitol and several other polyols are thought to be formed from sugars derived from this via enzyme systems similar to those for mannitol synthesis. Interestingly, reserve carbohydrates are generally synthesized by one pathway and degraded by another, giving particularly good control of these processes. Which pathway operates is dependent on the growth stage, environmental conditions and overall nutrition of the fungus. As has already been mentioned, polyols may also play a dynamic role in controlling proton availability.

Synthesis of cell wall polysaccharides generally proceeds by addition of a sugar, derived from a sugar nucleotide, to a pre-existing primer of the appropriate polysaccharide. In *Saccharomyces cerevisiae*, a lipid carrier is involved in mannan synthesis, which takes place in the endoplasmic reticulum. Synthesis of glucans, chitin and other fibrillar polymers occurs extracellularly, with the necessary enzymes being transported, in an inactive form, to the cell wall by means of vesicles. Enzyme activation then requires the action of protease within the wall.

Secondary metabolism

Biochemical differentiation, in the form of secondary metabolism, commonly accompanies morphological differentiation, such as occurs during reproduction, when primary metabolism and growth cease. On the basis of their precursors and synthetic pathways, secondary metabolites can be separated into five main groups: those derived from sugars and amino acids, and those arising from the shikimic acid, mevalonic acid and acetate–malonate pathways (Fig. 2.13). Precursors

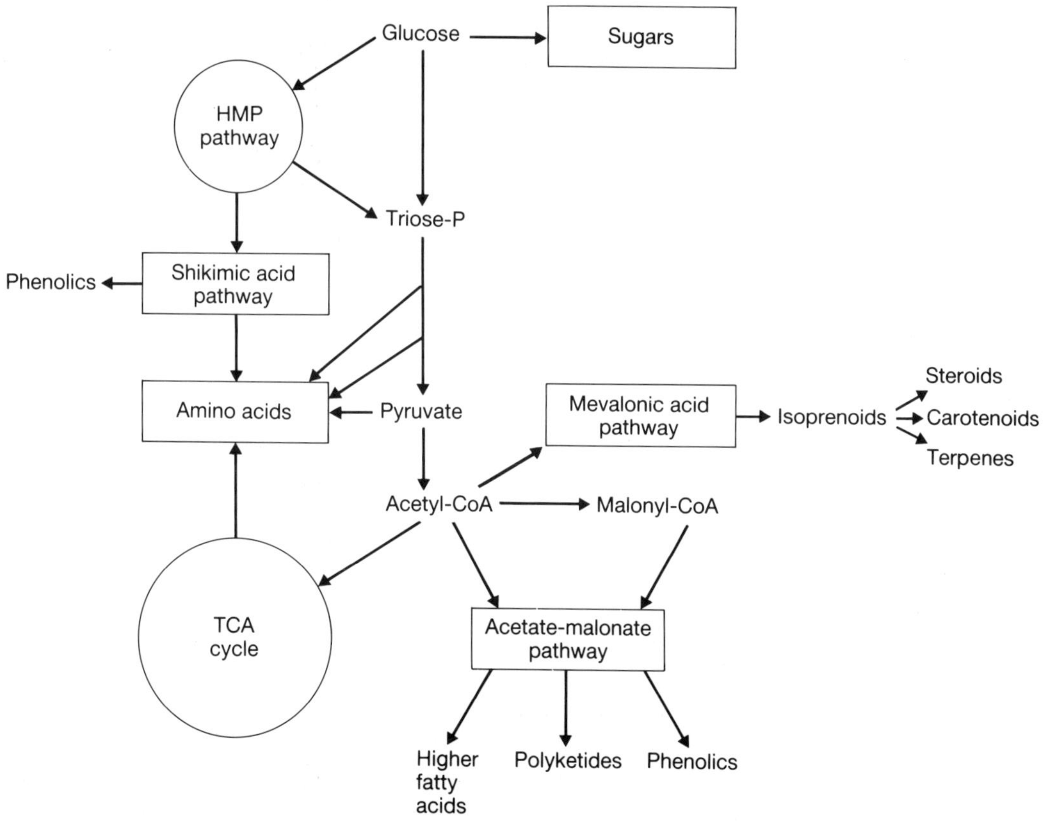

Fig. 2.13 Major pathways leading to secondary metabolite groups.

all emerge from primary metabolism. Secondary metabolites derived from acetate–malonate, mevalonate and, perhaps, amino acids, are the most widespread as a whole. However, in lichenized fungi and some Deuteromycotina the acetate–malonate pathway is predominant (Bennett & Ciegler, 1983; Jong & Donovick, 1989).

Although the roles of secondary metabolites are not yet fully understood, a range of possible functions can be ascribed to them. They may have no explicit role, merely reflecting an extension of failing, uncontrolled primary metabolism, or they may serve to maintain primary metabolism coupled to growth for as long as possible, perhaps by removal of toxic intermediate metabolites. Possession of a range of such flexible shunt pathways, not all of which need be active at any one time, would be of considerable benefit for continued vegetative growth. Production of veratryl alcohol and siderophores during lignocellulolysis under iron-limiting conditions are cases in point.

Secondary metabolism may also act to generate compounds such as melanin and sporopollenin which may enhance survival. These extremely resistant materials, laid down in or on hyphal and spore walls, may protect fungi from damaging radiation, lytic enzymes and desiccation. Many secondary metabolites are diffusible and antimicrobial, whilst some act as animal antifeedants. Some may also have a role in fungal communication and interactions by controlling cellular differentiation. For example, antheridiol produced by *Achlya ambisexualis*, and trisporic acids synthesized by some Zygomycotina, have become essential components of the reproductive process. Similarly, it may be no coincidence that lichens, with their high degree of organization, are a rich source of secondary metabolites.

3 Growth dynamics and transformations

Most fungi are of potentially indeterminate vegetative growth so that, with some notable exceptions, limitations to growth are largely imposed by environmental conditions. The effects of many exogenous factors, both singly and in combination, can be investigated relatively easily in the laboratory. However, in nature, environmental fluctuations and interactions may be so complex as to prevent accurate assessment of their influence. The aim here is to present a broad picture of major modes of vegetative growth and development as these occur under optimal laboratory conditions. This will be followed, in the next chapter, by a consideration of how growth patterns are affected by the environment.

Cell structure and growth mechanisms

Fungal cell walls are composed of polysaccharides, varying amounts of proteins and lipids, and lesser components such as melanins and other pigments. A fundamental feature of walls is their microfibrillar structure, which enables the moulding and weaving of materials to produce varied and often complex cell architecture (Bartnicki-Garcia, 1987). Mention has already been made of the constancy of cell wall composition within different Subdivisions and Classes of fungi, in which possession of chitin or cellulose is almost mutually exclusive (see Chapter 1, Table 1.1). However, other cell wall polysaccharides, for example β-glucans, are far more evenly dispersed. Furthermore, even with the few fungi so far studied in detail, it is clear that composition may differ widely between Orders within the same Subdivision. This suggests that the range of options available to fungi with respect to cell wall chemistry has been exploited to the full, largely to produce filamentous growth. It might be noted here that cell walls of yeasts, and yeast forms of normally hyphal fungi, are enriched with mannan.

Present understanding of hyphal growth mechanisms is based on a handful of fungi, predominantly *Aspergillus nidulans* and *Neurospora crassa*, but there is little doubt that these exhibit basic behaviour patterns with which filamentous fungi at large comply. In general, the apical compartment of a septate hypha has five intergrading regions (Fig. 3.1). Distally, there is an extension zone consisting of a primary fibrillar chitinous wall coated with amorphous protein or other material, the plasticity of the well gradually decreasing with distance from the tip. The cytoplasm of this zone is non-vacuolate, but it is rich in vesicles which may be aggregated to form a distinct, compact structure, the Spitzenkörper (Fig. 3.2). These vesicles are generated at some distance from the tip and then migrate to it, to finally join with the plasmalemma and release their contents into the cell wall

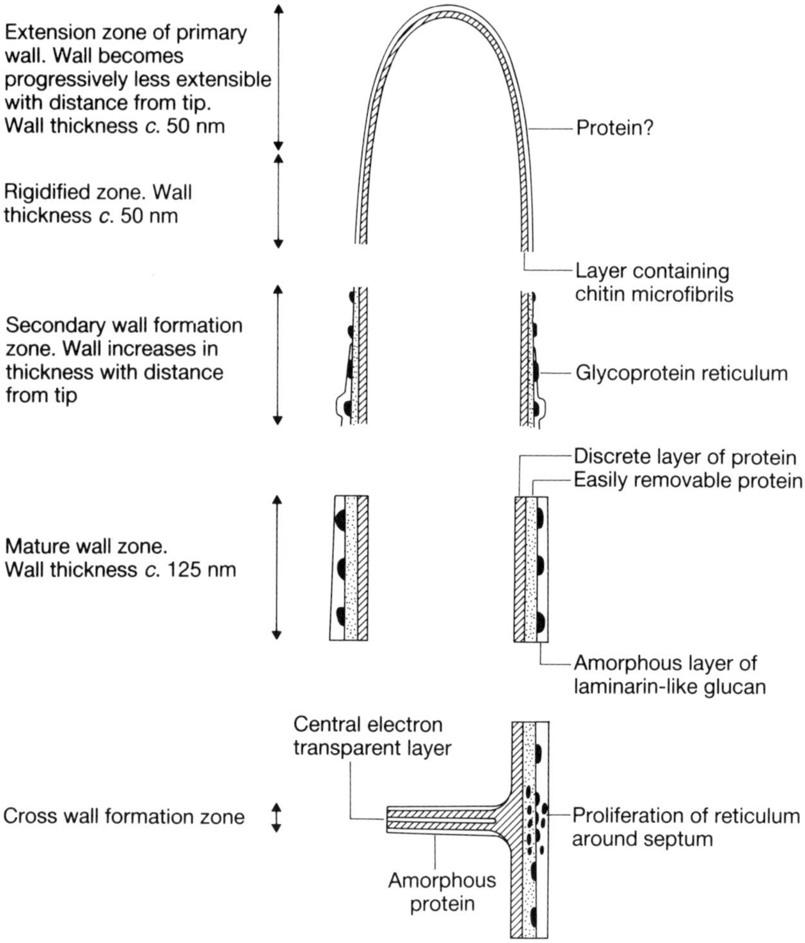

Fig. 3.1 Cell wall zonation in the apical compartment of a hypha of *Neurospora crassa* (after Hunsley & Gooday, 1974, © Springer-Verlag; Trinci & Collinge, 1975; Hunsley & Kay, 1976, © Society for General Microbiology; Trinci, 1978, by permission of Oxford University Press).

(Fig. 3.2). In this way the lytic and synthetic enzymes necessary for extension growth are delivered to the continually expanding cell wall. It is envisaged that a unit event in this process first requires a vesicle to discharge lytic enzymes, and so cause breakage of a number of wall microfibrils (Bartnicki-Garcia, 1990). Turgor pressure then stretches the wall at that point, and a second vesicle discharges wall-synthesizing enzymes into the affected area (Fig. 3.3). These, together with wall precursors passing into the wall from the cytoplasm, bring about repair of the microfibrils. Since the amorphous outer layer of the wall must also expand, it is probable that a third kind of vesicle containing this material is also involved.

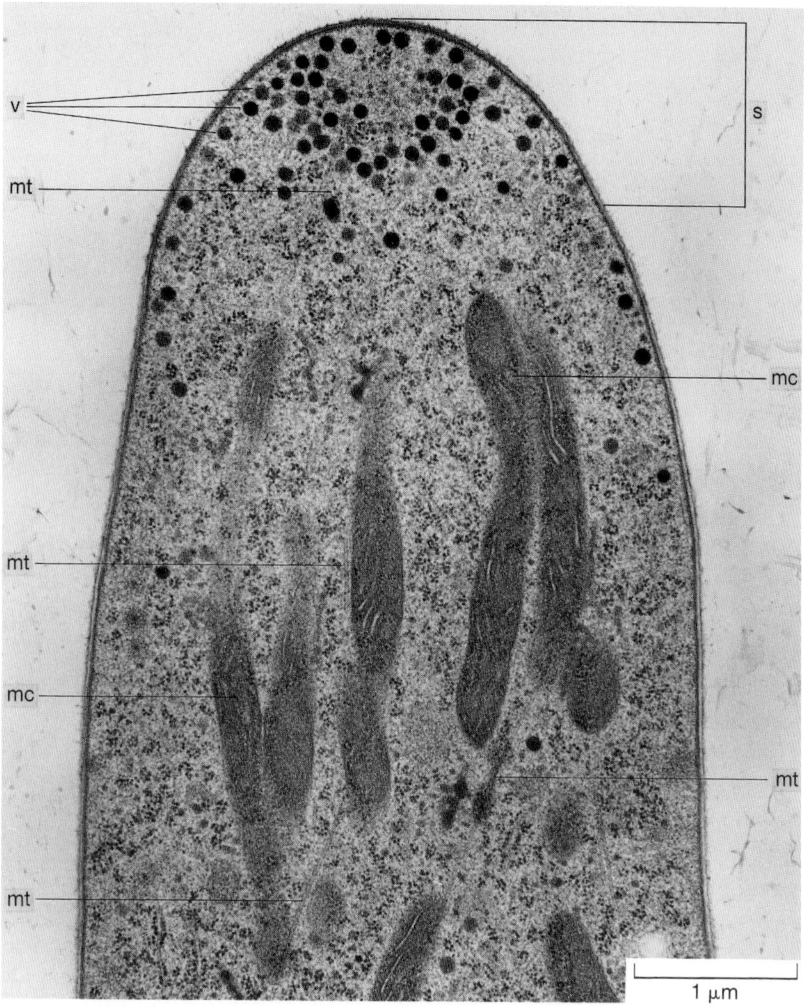

Fig. 3.2 Median section through the hyphal tip of *Fusarium acuminatum*. The Spitzenkörper region (s) with large numbers of vesicles (v) lies within a region of microfilaments at the tip. Below it there is a longitudinal arrangement of mitochondria (mc) and microtubules (mt) (from Howard & Aist, 1980, *Journal of Cell Biology*, **87**, © Rockefeller University Press).

Regular extension rates, and maintenance of the shape of the apical compartment, are dependent on a fine balance between wall lysis and synthesis, and on an ordered decrease in the numbers of vesicles fusing with the plasmalemma with distance from the tip (Fig. 3.4). Direction and flow of vesicles may be controlled by ion movement to the tip in conjunction with microtubule guidance (Derksen & Emons, 1990; Harold & Caldwell, 1990; Steer, 1990). Adequate Ca^{2+} is required to maintain cell wall rigidity and consistency of actin gels in the cyto-

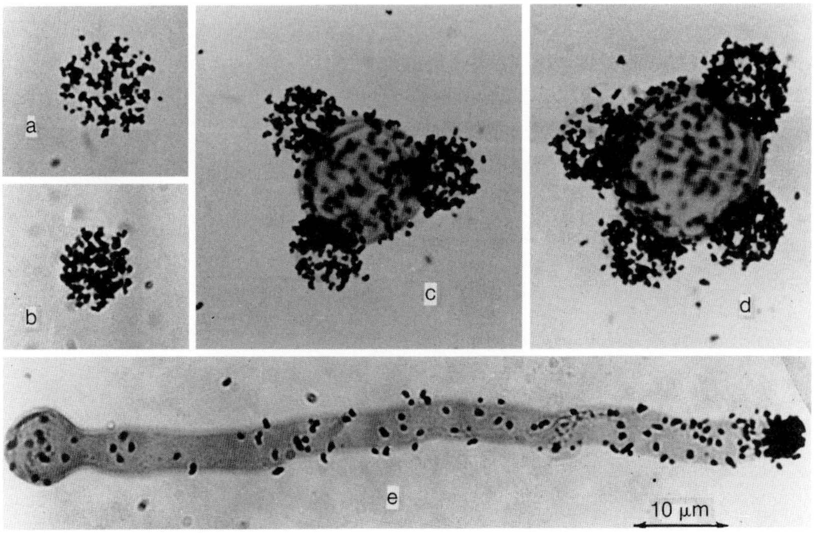

Fig. 3.3 Location of cell wall synthesis in yeast and hyphal forms of *Mucor rouxii*. Cells were exposed to tritiated *N*-acetyl-D-glucosamine and autoradiographed. Black silver grains indicate sites of chitin synthesis. (a) Germinating sporangiospores prior to budding; (b) germinating sporangiospores prior to germ tube emergence; (c, d) yeast cells with buds; (e) young hypha growing from a sporangiospore (from Bartnicki-Garcia & Lippman, 1969, © AAAS).

skeleton. Additionally, a constant turgor pressure must be maintained if morphological stability is to be preserved. That hyphal tips retain their form in the face of fluctuating cultural conditions attests to their adaptability in these respects. Recent mathematical models simulating hyphal growth require only that a sustained supply of vesicles migrate randomly to the wall from a continually displaced vesicle supply centre, which may correspond to the Spitzenkörper (Bartnicki-Garcia *et al.*, 1989). If the vesicle supply centre is stationary, the cell grows as a sphere, and any displacement of the centre distorts the spherical shape, and thus accounts for germ tube initiation and yeast phase : mycelium phase morphic switches.

The extension zone merges into a rigidified region having the same primary wall structure, which then gives rise to an area of secondary wall formation. Here, a glycoprotein reticulum appears upon the protein layer and, as the cell wall matures further, this becomes overlaid with some kind of amorphous polysaccharide, for instance glucan. Finally, the base of the compartment generates a septum, which then remains behind as the tip continues its extension. The pores in the cross walls so formed remain open for some time, so that the first few subapical cells communicate both with each other and the hyphal tip. Metabolites and organelles originating in these cells are

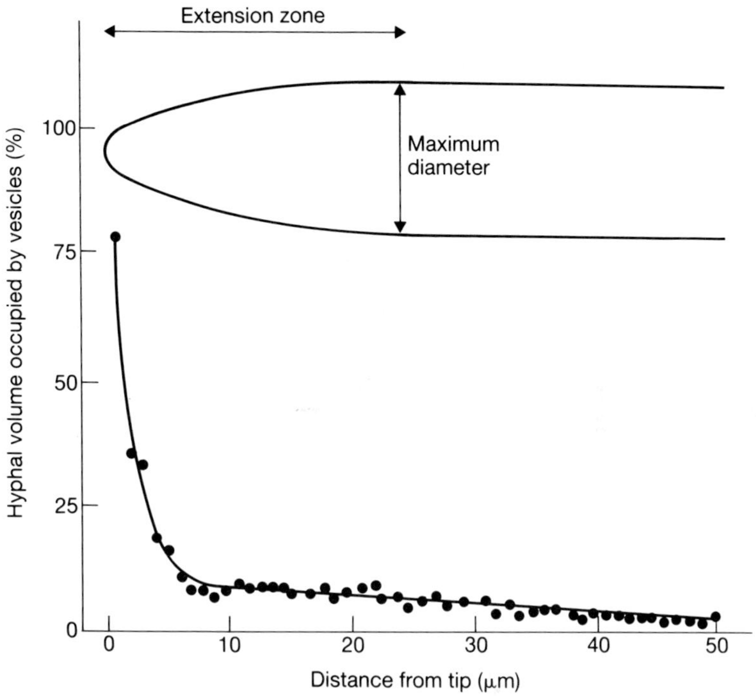

Fig. 3.4 Vesicle distribution within the hyphal apex of *Neurospora crassa* (from Collinge & Trinci, 1974, © Springer-Verlag; Trinci, 1978, by permission of Oxford University Press).

then translocated to the apical compartment. In addition, as subapical cells age, their vacuolar volume increases so as to displace their cytoplasm towards the apex. This makes an essential contribution to growth, the rate of protein synthesis within the apical compartment alone being insufficient to maintain its extension. The vacuoles also play an important role in metabolite storage, cytosolic ion and pH homeostasis, and macromolecular degradation (Klionsky *et al.*, 1990). Eventually the septal pores become plugged, and the resulting autonomous units, whilst being capable of differentiation and secondary metabolism, cease to contribute to the extension process (Fig. 3.5).

Growth characteristics

Theoretically any fungus growing in an 'ideal' environment that provides unlimited nutrients and space would remain undifferentiated and exhibit an exponential increase in biomass. When growing with excess, but not unlimited, resources, exponential growth might still be expected to occur, but would be terminated, at some point, by a deceleration phase followed by a period during which biomass either remained constant or fell because of autolysis (Fig. 3.6). During exponential growth the specific growth rate, that is the amount of

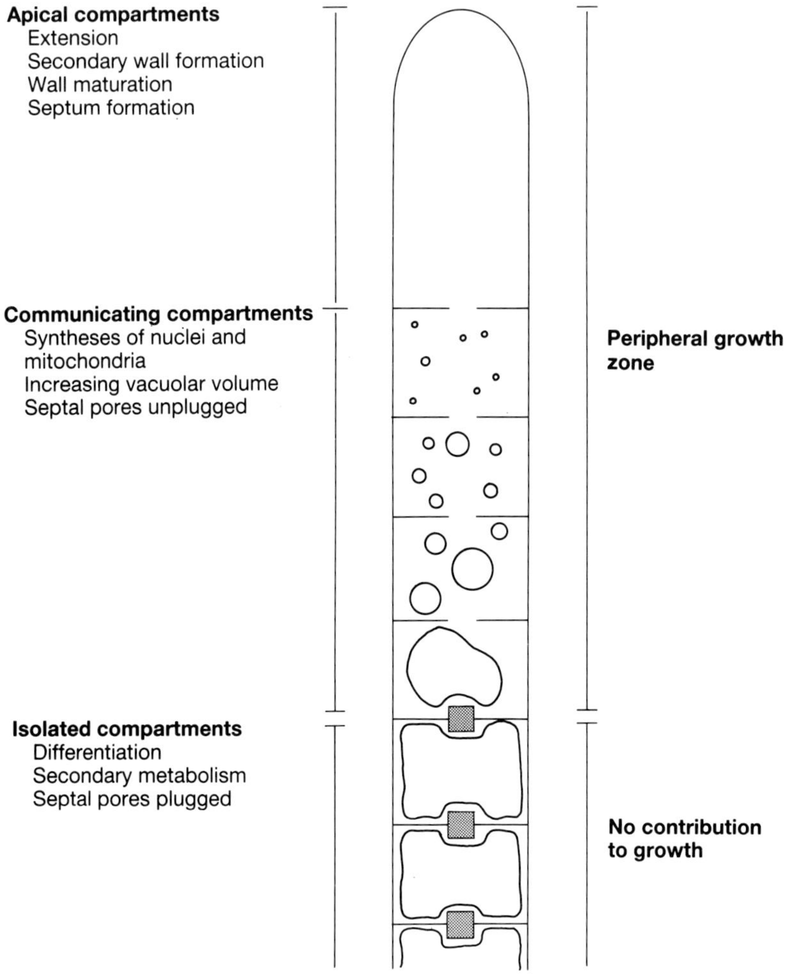

Fig. 3.5 Diagrammatic representation of the various hyphal compartments and their contribution to hyphal function.

biomass produced by a unit of biomass in unit time, is constant for any particular fungus in specified environmental conditions. It is defined as:

$$\frac{dx}{dt} = \mu x$$

where μ is the specific growth rate constant, x is biomass and t is time. Then, by integration:

$$x_t = x_0 e^{\mu t}$$

where x_t is biomass at any time t after the establishment of exponen-

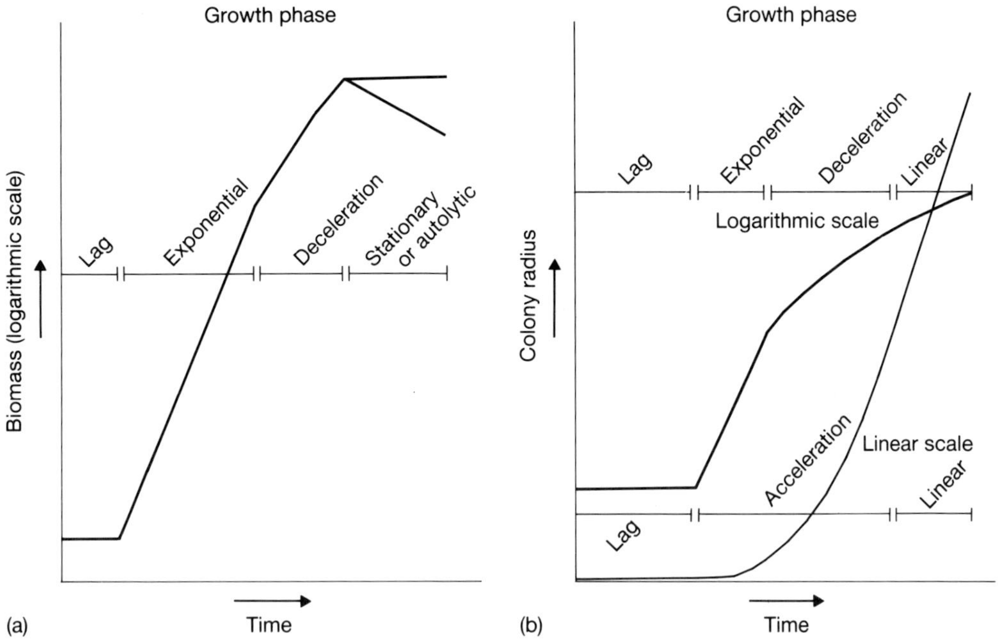

Fig. 3.6 Models of fungal growth under idealized conditions. (a) Mycelial biomass production in a liquid medium with an adequate, but limited, nutrient supply; (b) colony extension on a solid medium (after Trinci, 1969, 1971, © Society for General Microbiology).

tial growth, and x_0 is the starting biomass. This simplifies to:

$$\ln x_t = \mu t + \ln x_0$$

Hence, when biomass is plotted using a logarithmic scale, its increase during exponential growth is linear (Fig. 3.6).

Extension growth, as for instance measured on solid media, similarly has an exponential phase. However, within the colony centre, nutrients become depleted, O_2 availability decreases, and there may be adverse changes in pH and accumulation of secondary metabolites. Consequently only an annulus, consisting of the colony margin, is involved in extension. Its growth can be expressed as:

$$\frac{dx}{dt} = \mu x_a$$

where μ is again the specific growth rate constant and x_a is the mass of the annulus. The value of x_a approximates to $2\pi rHad$, where r is colony radius, H is the height of the colony, d is its density, and a is the depth of the annulus. Since a is small in relation to r:

$$\frac{dr}{dt} = \mu a$$

and by integration:

$$r_t = \mu a t + r_0$$

where r_t is colony radius at time t and r_0 its initial radius. As μ and a are constants, and even though the annulus is growing exponentially, the rate of increase in colony radius is linear, and so can be plotted graphically using a linear scale (Fig. 3.6).

Annulus width is equal to the average length of that terminal part of the individual hypha that contributes material for apical growth, which equates to the width of the peripheral growth zone. Since colony radius increases linearly, its width cannot significantly increase or decrease with time. There is thus a correlation between the width of the peripheral growth zone and linear extension rate such that:

$$d = \frac{K_r}{w}$$

where K_r is a constant for radial increase per unit time, w is the width of the peripheral growth zone and d is the specific growth rate constant: that is the increase in hyphal length per unit time in relation to the width of the zone involved in supporting that increase. It follows that:

$$K_r = wd$$

so that the wider the peripheral growth zone the greater the rate of colony increase per unit time.

Exponential growth requires that the quantity of macromolecules and organelles, and primary metabolic rate, all increase in parallel. Thus mycelial growth can be considered as being analogous to that of uninucleate unicells in terms of the duplication of a hypothetical growth unit equivalent to the peripheral growth zone. This unit consists of a hyphal tip and a strain-specific length of hypha of constant internal diameter (Bull & Trinci, 1977). Marginal hyphae of an expanding colony are differentiated into leader and branch hyphae which can have an apical compartment up to 2 mm in length and an intercalary compartment up to 250 μm in length. Leader hyphae extend at a constant rate and have apical and intercalary compartments of constant mean length. They are oriented radially outwards from the colony centre and remain a fixed distance apart, so that each grows into an area of unexploited resource. Branch hyphae generally have a smaller diameter, compartment length, and extension zone, but with time their extension rate increases, as does the mean length of their apical and intercalary compartments, until they attain the characteristics of leader hyphae. Hyphal density at the colony edge is constant, irrespective of colony size, so that generation of new apices via branching must therefore occur at a rate which is proportional to the number already present. Hence, the number of apices produced by branching of leader hyphae will increase exponentially as the colony expands, just as the number of yeast cells increases in a culture by budding or fission. In some filamentous fungi, the hyphal tip also rotates, involving either unequal stretching under the influence of turgor pressure or the presence of spiral structures in the hyphal tip wall. On solid surfaces, friction between the rotating tip and the substratum causes the tip to roll over, producing a spiral growth form (Madelin *et al.*, 1978).

Branching is probably initiated by a cytoplasmic event, possibly a localized alteration in electrical current, leading to a change in the pattern of vesicle migration and actin distribution (Gow, 1989; Harold & Caldwell, 1990; Steer, 1990). Differences in Ca^{2+} and cyclic AMP concentration along hyphae may also be involved (Pall & Robertson, 1986). Localized accumulation of vesicles and actin is associated with branching in filamentous fungi and bud initiation in yeasts. In the latter the newly forming cell wall is derived from the inner layers of the parent cell wall, the bud wall expanding until the daughter cell is approximately half the diameter of the mother cell; a septum is then formed centripetally by invagination of the plasmalemma. This becomes a double wall both anchoring and separating bud and parent cell until abscission.

In some filamentous fungi, septa may impede vesicle transport, which might explain the common occurrence of branching just below the most recently formed septum behind the apical compartment. Perhaps significantly, branches arising from older, intercalary compartments are less regularly associated with septa. In coenocytic fungi, localized vesicle accumulation may occur if their production rate exceeds their rate of utilization for apical growth, so initiating branching.

Senescence and autolysis

Under normal circumstances, vegetative growth ceases at some point, and hyphae then senesce and autolyse. For most fungi this is occasioned by exhaustion of exogenous nutrients; a fact which has resulted in the tacit assumption that fungi require considerable concentrations of energy-yielding materials in order to maintain growth. However, there is increasing evidence that, in nature, many fungi are adapted to oligotrophic conditions and can maintain sparse, but rapid, hyphal development at low concentrations of energy-yielding substrates. Some are chemolithoheterotrophic, and so are able to gain additional energy through oxidation of reduced forms of certain elements (Wainwright, 1988). Whatever the basis for oligotrophy may be, starvation growth of fungi on carbon-deficient or carbon-free laboratory media is a common occurrence. During this, cytoplasm is confined to the apical compartment and a very few intercalary compartments and, as extension proceeds, is continually relocated within these metabolically active cells. Empty compartments then become sealed by plugging of their septal pores. In laboratory culture, the effects of nutrient exhaustion are usually compounded by accumulation of secondary metabolites and drastic changes in pH; in nature, competition between fungi for diminishing resources may similarly exacerbate nutrient stress.

The onset of senescence is signalled by increased septum frequency, retarded nuclear division, and pronounced vacuolation. These are followed by cessation of RNA and protein synthesis, nucleic acid breakdown and plasmalemma leakage. Loss of control over degradative enzymes then results in autolysis. When developing in ideal cultural conditions, some fungi exhibit senescence long before nutrients become limiting or secondary metabolites reach a harmful level. This is due to the expression of cytoplasmically inherited lethal genes. The ascomycete *Podospora anserina* possesses a ring plasmid which, in juvenile hyphae, is incorporated into the mitochondrial genome. Cell ageing coincides with self-replication of the plasmid, and senescence and death rapidly ensue (Esser *et al.*, 1984). The plasmid may be a transposable element capable of movement from one location to another within the fungal genome, there to bring about sudden and drastic physiological events (Wright & Cummings, 1983). The latter may then be manifested in rapid sensescence and death or, in parasites, changes in pathogenicity. In other cases, presence of plasmids may simply result in decreased growth. Such plasmid-induced changes are discussed further in Chapter 4.

Differentiation and morphogenesis

In a strict sense, the only parts of the mycelium that remain undifferentiated are the terminal portions of apical hyphal compartments, their role being confined to that of cell wall expansion. Secondary wall synthesis and septum formation are then the initial steps in a process of increasingly complex differentiation and elaborate morphogenesis. The key to this is septum formation, which produces compartments that, whilst they may still remain in communication, can often act with a high degree of autonomy. This facilitates the generation of cellular systems with a capacity for division of labour and concomitant differentiation into a variety of pseudotissues, each having a distinct function. Septa also confer on hyphae the ability to anastomose, a process that opens up possibilities for both communication and organogenesis. Septa are absent from, or occur rarely in, the vegetative mycelia of Oomycetes and Zygomycotina; hyphal fusions do not take place except between specialized gametangia which differentiate after being delimited by septa. Therefore, septum formation in both unspecialized and specialized hyphae confers an ability to recognize other hyphae with which fusion may be possible.

Individual hyphae

Normally, branching in filamentous fungi is monopodial, although hyphae may respond to environmental stress by branching subapically, dichotomously or sympodially. The relationship between branch length and surface area exposed in the monopodial system imparts maximum efficiency in exploring a substratum, and draining resources from it, with minimum expenditure on biomass – a highly cost-effective system. During colony growth, hyphae of the peripheral growth zone deplete, but do not exhaust, nutrients as they advance through new territory. Any residual nutrients are taken up by those hyphae immediately behind the growth zone. Here branches are numerous, but become progressively narrower as branch systems develop, and do not exhibit consistent orientation towards the colony margin. Mathematical models have recently substantiated these concepts (Ritz & Crawford, 1990).

In many fungi there are considerable morphological distinctions between exploring and exploiting hyphae (Boddy & Rayner, 1983). For

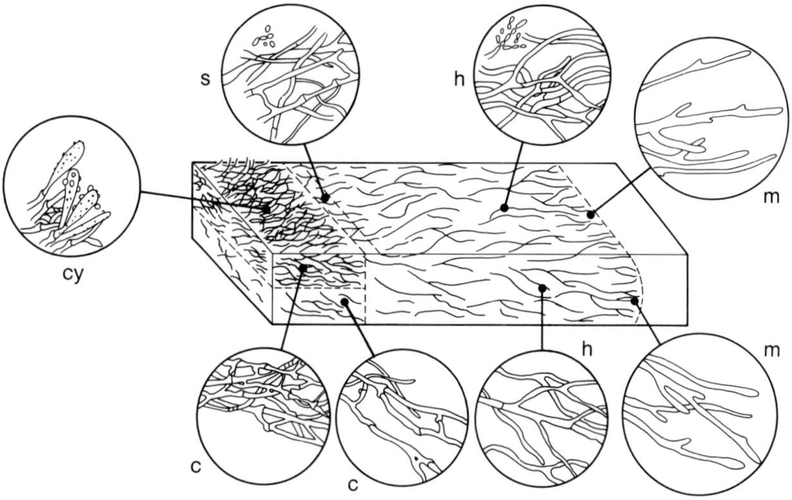

Fig. 3.7 Characteristics of a dikaryotic colony of *Phlebia radiata* growing through malt agar. Direction of growth is from left to right, with the upper row of diagrams representing surface hyphae and the lower row submerged hyphae: m, wide coenocytic marginal hyphae; h, heterogeneous hyphae, mostly coenocytic with surface hyphae producing oidia; s, heterogeneous hyphae with septa, pseudoclamps and oidia; c, hyphae with clamp connections and anastomoses, closely packed above, more loosely below; cy, hyphae with clamp connections and cystidioles (from Cooke & Rayner, 1984, © Longman).

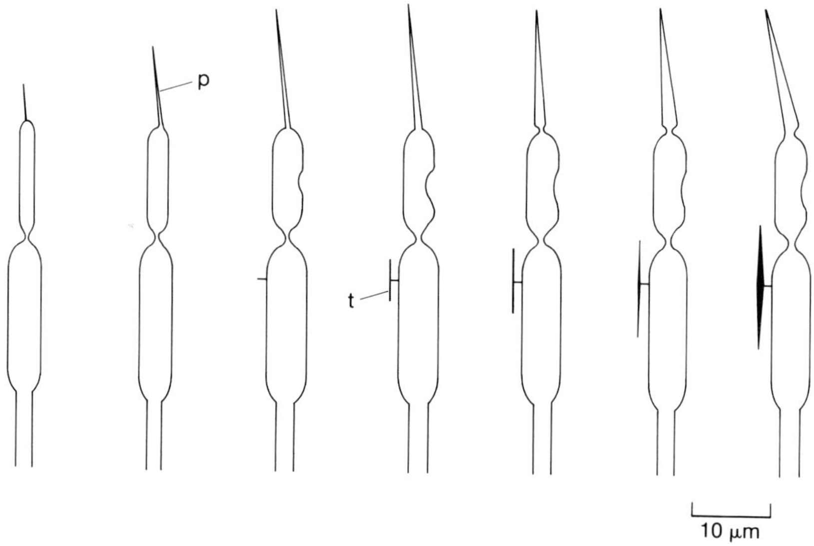

Fig. 3.8 Longitudinal section view of *Phialophora hoffmannii* growing through birch fibres showing development of a proboscis hypha (p) and a T-branch hypha (t) by the fungus (based on photographs in Hale & Eaton, 1985, © British Mycological Society).

instance, the marginal hyphae of agar-grown colonies of the basidiomycete *Phlebia radiata* are coenocytic, do not anastomose, and are not antagonistic with respect to other fungi (Fig. 3.7). In older parts of the colony, hyphae become septate, form numerous fusions, and are combative towards hyphae of fungi which they encounter. Even more extreme differences occur in other fungi. When *Lecythophora (Phialophora) hoffmannii*, an ascomycetous soft-rot species, penetrates cell walls in wood, it produces a fine hypha which passes through the wall and inflates to normal size on the opposite side. Alternatively, whilst still within the wall, cycles of extension growth and cavity formation occur. Subsequently, new hyphae may form at right angles or apically, producing characteristic T-branch hyphae, or fine proboscis hyphae (Fig. 3.8). Perhaps uniquely, growth of hyphae within cavities may be non-apical, as they appear to expand from parent hyphae of a more normal diameter. Such marked differentiation of individual hyphae is widespread amongst pathogens; notable examples are the penetrating and eroding hyphae of some Dermatophytes, the travelling and penetration hyphae of some root parasites, and the appressoria, infection cushions and penetration pegs of stem- and leaf-infecting fungi.

Morphic switches

The mycelium has been compared with an embryo of indeterminate development, so that, particularly when occupying complex niches, it must be an interactive and changeable entity capable of responding to varied, and often contradictory, environmental conditions (Rayner & Coates, 1987). Such reactions may occur with respect to either short-term adaptation or long-term innovation, and are manifested in switches between functionally and morphologically distinct modes. Mycelia can exhibit considerable developmental plasticity by growth into new structural types but, although the adaptive significance of particular modes adopted in nature is often self-evident, mostly the underlying mechanisms are poorly understood. Evidence is, however, accumulating that mode switches may perhaps be regulated by cytoplasmically mobile factors, the intervention of which exposes differentiation pathways that are not otherwise expressed (Coates & Rayner, 1985a–c). Although some fungi, for instance *Fusarium* species, are highly polymorphic and have several variants, a more common occurrence appears to be that of dimorphism, in which there is an option between two developmental modes.

Yeast phase: mycelial phase dimorphism

Increasing numbers of fungi are being found to have an ability for alternation between unicellular (determinate) growth and mycelial (indeterminate) growth. Thus, at different stages of their life cycle they are able to exploit the ecological advantages of either mode. Adoption of the yeast form confers stress tolerance, and facilitates dispersion in fluids; the mycelial form enables penetration of solid substrata and colonization of a fixed spatial domain (Cooke & Rayner, 1984; Cooke & Whipps, 1987). Many pathogens of arthropods, mammals and vascular plants colonize their hosts in the yeast phase, commonly via the blood or sap stream, transforming to the mycelial phase when the host becomes sufficiently debilitated so as to alleviate stress (Table 3.1). Some ecologically important mutualistic associates of wood-boring insects are also transmitted in the yeast form, the unicellular state being imposed on them by fungistatic host secretions.

This kind of dimorphism appears to be rare among asymbiotic saprotrophs in nature, although this may simply reflect the much greater research emphasis towards symbiotic fungi. However, there are hints that, as a group, mucoralean species have this propensity. An interesting example is *Mycotypha microspora* which adopts the yeast phase when physically and chemically stressed, and also does so in the presence of the biotrophic mycoparasite *Piptocephalis fimbriata*. This prevents the release of growth-directing compounds which normally guide *Piptocephalis* hyphae towards those of *Mycotypha*, so reducing the chance of infection (Evans *et al.*, 1978; see also Chapter 10).

As with all morphic switches, wall morphogenesis has an obvious central role. But a barrier to elucidating control mechanisms is the multiplicity of environmental factors that can effect change, perhaps

Table 3.1 Examples of fungi exhibiting yeast phase:mycelial phase dimorphism with, where known, some major environmental factors involved in the switch.

Symbiotic state	Species	Habitat	Major factors involved
Symbionts of insects	*Beauveria bassiana*	All tissues, but initially blood	Nature of available C, N and P
	Entomophthora spp.	All tissues, but initially blood	
	Ambrosia fungi	Mycetangia of xyleborine beetles; tunnel systems in wood	Proline and glutamic acid levels
	Amylostereum and *Stereum* spp.	Ovipositor sacs of siricid wasps; decaying wood	
Symbionts of endothermic vertebrates	*Blastomyces dermatiditis*	Lungs, skin and bone	Temperature
	Candida albicans	Mucosa, but other tissues also	Concentration of glucose and/or cysteine and proline, temperature, pH
	Cladosporium wernekii	Skin	pCO_2, temperature
	Coccidioides immitis	Skin, bone, internal organs	pCO_2
	Cryptococcus neoformans	Bone, meninges	
	Geotrichum candidum	Mouth, intestines, lungs	
	Histoplasma capsulatum	Blood cells, lungs, intestines	Temperature
	Mucor spp.	Blood vessels	
	Paracoccidioides brasiliensis	Skin, lymph nodes, other organs	Temperature
	Phialophora dermatiditis	Skin	
	Sporothrix schenckii	Skin, lymph nodes	
Symbionts of plants	*Aureobasidium pullùlans*	Leaves, stems	Inoculum density, nature of available C and N, pH, age
	Ophiostoma ulmi	Vascular tract, bark	Nature of available N
	Verticillium dahliae	Xylem elements	
Asymbiotic saprotrophs	*Mucor rouxii*	Soil, litter	pO_2, pCO_2, sugar concentration
	Mycotypha microspora	Soil, litter	pO_2, pCO_2, sugar concentration, pH
	Mycotypha africana	Soil, litter	

via their modulation of endogenous ionic fluxes and Ca^{2+}–calmodulin interactions (Muthukumar & Nickerson, 1984; Stewart *et al.*, 1988). There is considerable evidence for the involvement of cyclic AMP, and the suggestion is that environmental changes may alter the activity of adenyl cyclase and cAMP phosphodiesterase, which are enzymes respectively concerned with synthesis and degradation of cyclic AMP either directly or via effects on ionic fluxes (Stewart & Rogers, 1983; Maresca & Kobayashi, 1989; Orlowski, 1991). Cyclic AMP may act on gene transcription by directly or indirectly controlling those products that bear on cell wall synthesis. For instance, there could be activation of cyclic AMP-dependent protein kinases concerned with synthesis, interaction with microtubules and microfilaments and, hence, regulation of the flow of precursors to sites of synthesis. There would also be effects on nitrogen metabolism and polyamine synthesis that affect fungal growth.

Slow-dense: fast-effuse dimorphism

When draining substrata of resources, mycelia may make subtle adjustments of two basic kinds to their morphology. The first involves changing internode length, so altering the balance between extension growth and branching frequency. The second involves varying branching angle, which may give rise to two distinct types of growth, slow-dense and fast-effuse. The slow-dense form is characterized by low extension rates and short internode length, the resultant closely spaced branches diverging widely from their parent hyphae. In the fast-effuse form there is rapid extension coupled with long internode length, the more widely spaced branches having a relatively small branching angle, thus being directed in approximately the same direction as their parent hyphae. In ecological terms fast-effuse growth facilitates exploration and initial occupation of domain, whilst slow-dense growth aids consolidation within captured territory and exploitation of its energy resources. These differing roles are exemplified by some Basidiomycotina in which the primary homokaryotic mycelium, associated with establishment, shows slow-dense growth, whilst the secondary heterokaryotic mycelium, associated with exploration, has fast-effuse growth (Rayner & Coates, 1987).

Aerial growth: appressed growth dimorphism

Mycelia commonly produce aerial hyphae which, since they lack a major trophic function, divert resources from those within the substratum. They do, however, act as a means of escape from the substratum, especially under conditions of secondary metabolite accumulation or poor aeration. Aerial and non-aerial (appressed) growth may, at two extremes, occur simultaneously or disjunctly, the latter leading to distinctive sequential growth phases. They may also be only partially uncoupled, in which case rhythmic morphic switches will result.

This kind of dimorphism is expressed by Deuteromycotina, Ascomycotina and Basidiomycotina. The wood-decaying basidiomycete *Hymenochaete corrugata* has two interconvertible colony types with similar extension rates: one aerial and white, the other appressed and yellow-brown. On wood the appressed type is associated with regions of greater decay, and only this type possesses tyrosinase and laccase, enzymes associated with ligninolytic activity (Sharland *et al.*, 1986; Rayner & Coates, 1987). In *Schizophyllum commune*, cysteine-rich proteins accumulate at high concentration at the time of aerial hypha formation and may be responsible for suppression of appressed growth (Wessels *et al.*, 1991).

Aggregation and organogenesis

It has already been pointed out that the ability of septate hyphae to group and anastomose opens up the possibility of organogenesis. The latter begins when normally divergent hyphal growth is replaced by convergent growth. Aggregation and fusion can then generate plectenchymatous, vegetative and reproductive structures of numerous kinds. The basic architecture of some of these forms is outlined here; problems surrounding the initiation and developmental control of reproductive structures are discussed in the appropriate place in subsequent chapters.

Cords and rhizomorphs. Cords and rhizomorphs are linear vegetative organs, produced largely by Basidiomycotina, formed from the compaction of collaterally aligned, outwardly extending hyphae (Rayner *et al.*, 1985). Two basic types of cord morphogenesis have been distinguished (Fig. 3.9). The first occurs in the violet root rot pathogen of beet, *Helicobasidium purpureum*. As its mycelium expands, separate, fine hyphae come into contact to form leader groups, which then aggregate into a cord that is widest at its base and narrows towards its tip. There is considerable interweaving and anastomosis of hyphae within the cord, which itself may branch or join with other cords. The

Fig. 3.9 Types of cord morphogenesis. (a–c) *Helicobasidium purpureum*, with direction of growth from left to right; (a) young, robust hypha extending from food base; (b) loose aggregation of hyphae and contact between established cords; (c) older region of cord with anastomoses (from Valder, 1958, © British Mycological Society); (d, e) *Serpula lacrimans*; (d) fourth node behind the tip of a young leader hypha showing branching to produce primary (p) and secondary (s) tendril hyphae; (e) main leader hypha becoming surrounded by branch hyphae (from Butler, 1958, © Academic Press).

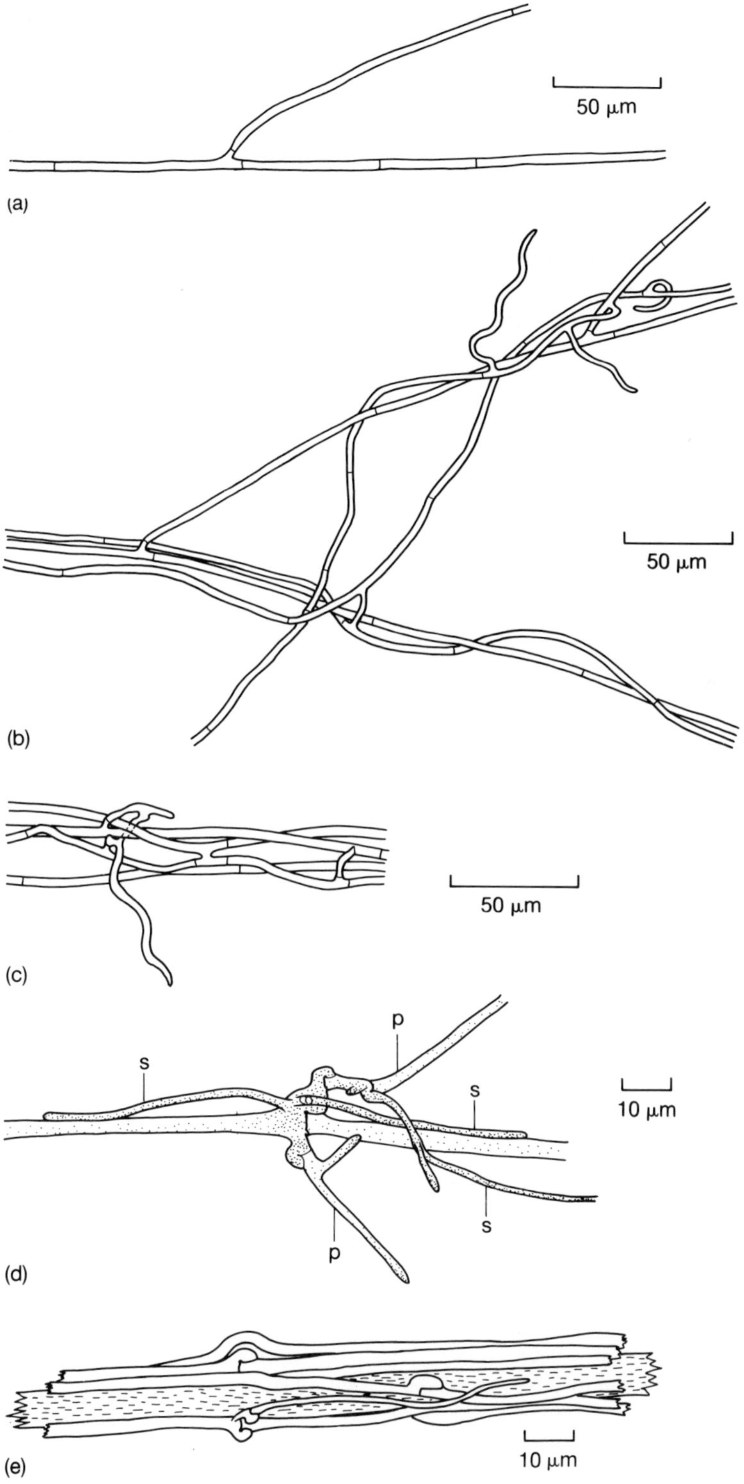
50 μm
(a)
50 μm
(b)
50 μm
(c)
p
s
s
10 μm
s
p
(d)
10 μm
(e)

second, and most common, type of cord arises from individual, wide leader hyphae that become ensheathed by their own branches. In *Serpula lacrimans* the primary branches which surround the leaders become bound together by much narrower tendril hyphae (Fig. 3.9). Abundant glycogen reserves are deposited within the ensheathing hyphae and an extrahyphal matrix is produced, the leaders losing their cytoplasm and becoming transformed to empty vessel hyphae. Some surrounding hyphae then autolyse, and thickened fibre hyphae differentiate, so that the mature cord is essentially a series of hollow tubes surrounded by fibrous material, all of which is embedded in a non-cellular matrix (Fox, 1987).

By contrast with cords, rhizomorphs are generated from distinct, multicellular, apices. In the ascomycete *Sphaerostilbe repens*, the apex is made up of parallel hyphae aligned with the long axis of the rhizomorph (Fig. 3.10). From these, inward cell production and differentiation lead to the formation of a clearly defined cortex, medulla and central lacuna. The meristematic zone in *Armillaria* species hitherto has been thought to consist of an extensive subapical dome. If this were the case, then creation of rhizomorph tissues, including an apical

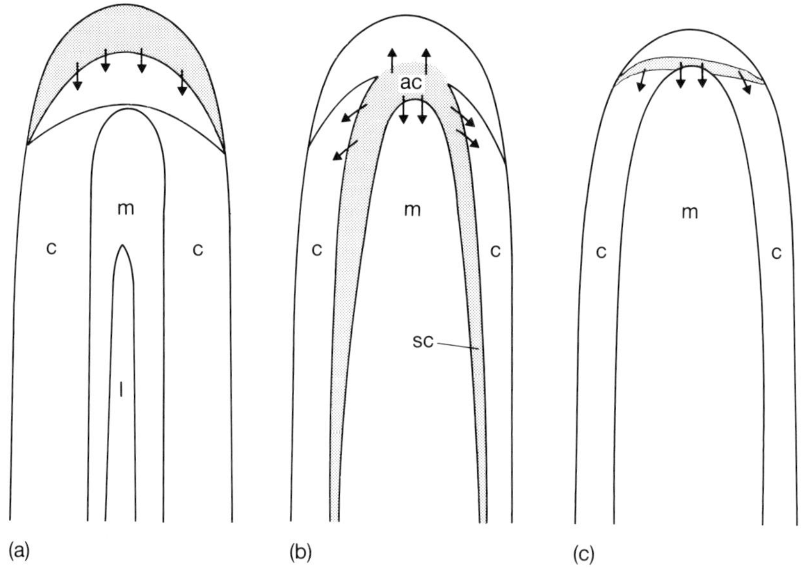

Fig. 3.10 Apical structure of rhizomorphs. (a) *Sphaerostilbe repens* (from Botton & Dexheimer, 1977, © Gustav Fischer Verlag); (b) *Armillaria mellea* and *Armillaria tabescens* (from Motta, 1969; Motta & Peabody, 1982, *Mycologia*, **74**, © 1982 New York Botanical Garden); (c) *Armillaria bulbosa* (from Rayner *et al.*, 1985, © Cambridge University Press). Meristematic regions (shaded) produce cells predominantly in the direction indicated by arrows: c, cortex; m, medulla; l, lacuna; ac, apical centre; sc, subcortex.

cap, would involve cell production simultaneously inwards, outwards and laterally (Fig. 3.10). Recent evidence suggests that the situation may be much simpler and that there is a true apical meristem. This is composed of a close-knit mass of apically extending hyphae which repeatedly form septa, branches and anastomoses. Further back, cells are arranged in parallel, so giving rise to a cortex and medulla (Rayner *et al.*, 1985). Rhizomorph extension would then take place by means of a balanced pressure–lysis mechanism similar to that operating in individual hyphae. Plasticity of the tip would be aided by mucilage production, rigidification achieved by melanization and compaction of the outer tissues, and forward pressure generated by osmotically driven water flow (see Chapter 2).

Mature rhizomorphs are differentiated into an outer rind or crust of thickened, laterally connected, melanized hyphae surrounding a cortex and medulla of less compacted cells held together by an extracellular matrix. The cells of the outer medulla may be concentrically thickened, but the inner medulla is made up of more loosely woven hyphae which eventually collapse to form the central lacuna. The latter facilitates O_2 supply to both the inner medulla, which is strengthened by longitudinal fibre cells, and the meristematic region (Cairney *et al.*, 1989; Cairney, 1990).

Cords and rhizomorphs are common amongst wood-decomposing fungi, and the ability to produce them confers obvious ecological advantages on species that must search for, and exploit, discrete resource units. However, whilst high nutrient levels inhibit cord formation, they promote rhizomorph initiation, which indicates differing ontogenies for the two types of organ. Putting this aside, aggregations of these kinds additionally provide protection against environmental extremes and antagonistic microorganisms, facilitate conservation and re-utilization of nitrogenous metabolites through lysis of redundant tissues, promote reallocation of phosphorus between resource units, and may increase sensitivity to external stimuli over a much greater distance than that shown by less organized mycelium (Thompson, 1984; Dowson *et al.*, 1986; Wells *et al.*, 1990).

Although cords and rhizomorphs are generally found in soil or leaf litter, aerial rhizomorphs, such as those of *Marasmius* species, appear to be particularly common in the canopy of tropical rain forest. They attach to stems and leaves by forming specialized hyphal adhesion zones, and by then producing net-like aggregates may trap and exploit falling leaf litter and small branches (Hedger, 1990).

Sclerotia. Major morphological features of sclerotia remain more or less constant for any particular species but, taken at large, they form a diverse array of hyphal aggregates serving as persistent resting stages in adverse conditions. When they are able to germinate they do so by producing either vegetative hyphae or sexual or asexual sporophores. Sclerotia most commonly arise from localized proliferation and compaction of hyphal branches, formed after a switch to either slow-dense or aerial growth; although some, for instance those of *Claviceps* and *Cordyceps* species, result from differentiation of either extensive, pre-existing mycelia or of an asexual fruiting structure. Indeed, some may be degenerate or much-modified sexual or asexual fruitbodies, or at least have close structural and developmental affinities with them (Moore, 1981; Backhouse & Stewart, 1988).

Those not formed from pre-existing mycelia originate from one of four distinct kinds of primordium: loose, terminal, lateral or strand (Table 3.2). Mature sclerotia may be divided additionally into the majority which have a rind of flattened, dead and usually melanized cells, and the minority – for example those of *Rhizoctonia solani* and *Verticillium dahliae* – which do not. Interior to the rind, tissues consist largely of storage cells embedded in a mucilaginous matrix, major storage compounds being lipids, glycerol, polyols and glucans. Several sclerotium-specific polypeptides may also occur (Newsted & Huner, 1988).

A characteristic of sclerotium formation is copious exudation of fluid from initials, which probably results from generation of high internal hydrostatic pressure forcing water to the sclerotium surface. At first this is associated with rapid translocation of materials into the growing initials, later it brings about the reduced tissue hydration levels that are necessary for maturation (Cooke & Al-Hamdani, 1986; Al-Hamdani & Cooke, 1987).

Coremia. Organization of conidiophores into discrete structures is frequent amongst ascomycetous fungi, some of the most distinctive forms being coremia or synnemata. There is first a switch to aerial growth, with subsequent aggregation of hyphal tips to form a single advancing apex, sporulation then occurs on this compound conidiophore. Developing coremia of *Penicillium* species may require unidirectional light in order to maintain aligned growth (Bennink, 1972).

Ascomata. Three basic structures are known: the closed cleistothecium; the flask-shaped ostiolate perithecium, either formed within a stroma or freely exposed on the substratum; and the cup-shaped apothecium, bearing asci and paraphyses on its exposed upper surface (Fig. 3.11). Under appropriate environmental conditions, development begins either from somatic, monokaryotic hyphae in a preformed

Table 3.2 Types of development and differentiation of sclerotia in septate fungi.

Subdivision	Loose (irregular branching of vegetative hyphae with intercalary septum formation)	Terminal (branching of single or several hyphae at their tips)	Lateral (intercalary branching of a single main hypha)	Strand (branching of several hyphal strands)
Ascomycotina	*Mycosphaerella ligulicola*	*Pyronema domesticum* *Sclerotinia fructicola* *S. sclerotiorum* *S. trifoliorum*	*Sclerotinia gladioli* *S. minor*	
Deuteromycotina		*Botrytis cinerea* *B. allii* *Sclerotium cepivorum*		*Aspergillus alliaceus* *Colletotrichum lindemuthianum* *C. coccodes* *Macrophomina phaseolina* *Phymatotrichum omnivorum* *Verticillium dahliae*
Basidiomycotina	*Rhizoctonia solani*			*Sclerotium rolfsii* *Typhula* spp.

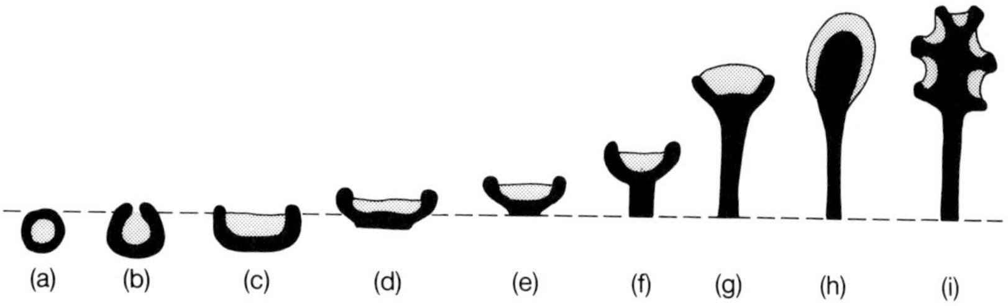

Fig. 3.11 Diagrams of a range of ascoma types. (a) Cleistothecium; (b) perithecium; (c–i) apothecia. Shaded areas represent the ascogenous region, with the dotted line indicating the substratum surface (from Hawksworth, 1987, © Cambridge University Press).

stroma before archicarp formation or, in the absence of a stroma, from stalk cells of the archicarp.

In *Sordaria humana* the perithecium is initiated by an ascogonial coil arising as a side branch from a vegetative hypha. The coil becomes enveloped in sterile hyphae which form a spherical protoperithecium. Subsequently, the perithecium develops radial symmetry along its vertical axis with a polarized distribution of cellular types. Ascogenous hyphae, asci and hymenial paraphyses are located at the bottom, and are surrounded by a rigid protective peridium containing thick-walled cells from which other hyphae and hyphal-like elements extend (Figs 3.12 & 3.13). Development involves transitions from free, hyphal-like structures to coherent regions where hyphal-like elements become firmly fixed to each other. For example, some periphyses become part of neck peridium, and some enveloping hyphae become basal peridium. Similarly, transitions occur in the other direction. For instance, basal peridium can give rise to fringe hyphae, and neck peridium to neck hyphae. Cells in the non-ascogenous regions may be formed from hyphae derived either from vegetative cells or from the ascogenous coil, suggesting that differentiation of perithecial hyphae is dependent upon their final position rather than their origin (Jensen, 1983; Read, N.D., 1983).

Considerable amounts of mucilage surround the hyphal-like elements within the perithecium, and this may serve to lubricate their longitudinal growth. It may also facilitate elongation and retraction of asci through the ostiolar canal during ascospore release. Here, elongating asci are under strain due to internally generated hydrostatic pressure necessary for ascospore discharge.

Basidiomata. Basidiomata are produced when suitable combinations of endogenous and exogenous conditions occur; thus a specific period

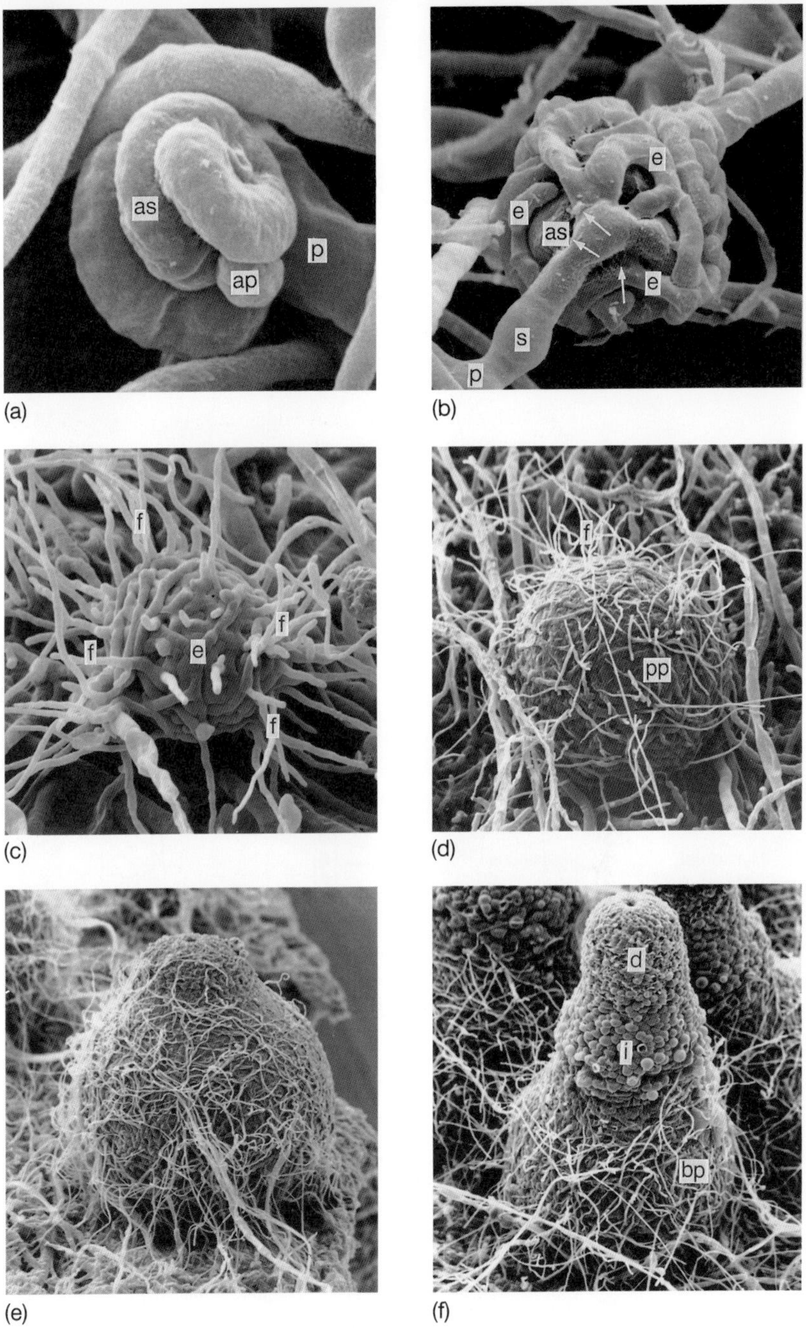

Fig. 3.12 Scanning electron micrographs of different stages during perithecium development in *Sordaria humana*. (a) Ascogonial coil (as) with a tapering apex (ap) which has curled upon itself, with parent hypha (p); (b) enveloping hyphae (e) surrounding the ascogonial coil. These have arisen from the swollen coil base (s) and extracellular material is visible (arrowed); (c) protoperithecium with interwoven enveloping hyphae and fringe hyphae (f); (d) old protoperithecium (pp); (e) young ostiolate perithecium; (f) mature perithecium with basal peridium (bp), inflated neck peridial cells (i) and distorted neck peridial cells (d) (from Read, N.D., 1983, © National Research Council of Canada).

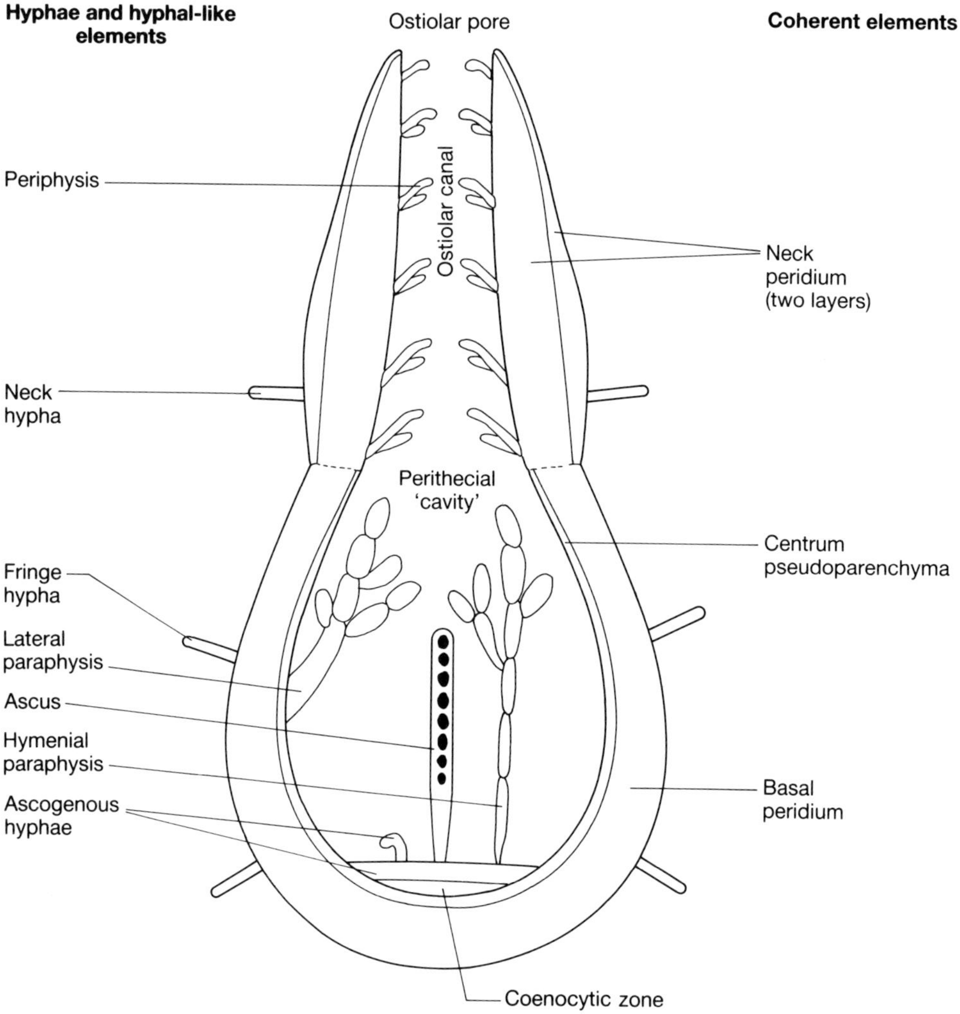

Fig. 3.13 Structure of the mature perithecium of *Sordaria humana* with its elements classified as either hyphal or coherent via cohesion of adjacent hyphae (from Read & Beckett, 1985, © National Research Council of Canada).

of vegetative growth may be necessary before the mycelium achieves competence to initiate their formation (Ross, 1985). Commonly, light and nutrient depletion act as environmental triggers, development then requiring low pCO_2. In agarics, rapid differentiation of the primordium establishes basic basidioma shape which then becomes enlarged to final form via cell expansion, division and further differentiation (Fig. 3.14). In *Agaricus bisporus*, continuous breakage and re-formation of weak hydrogen bonds among the glucan chains, and passive reorientation of the glucosaminoglycan chains in a transverse

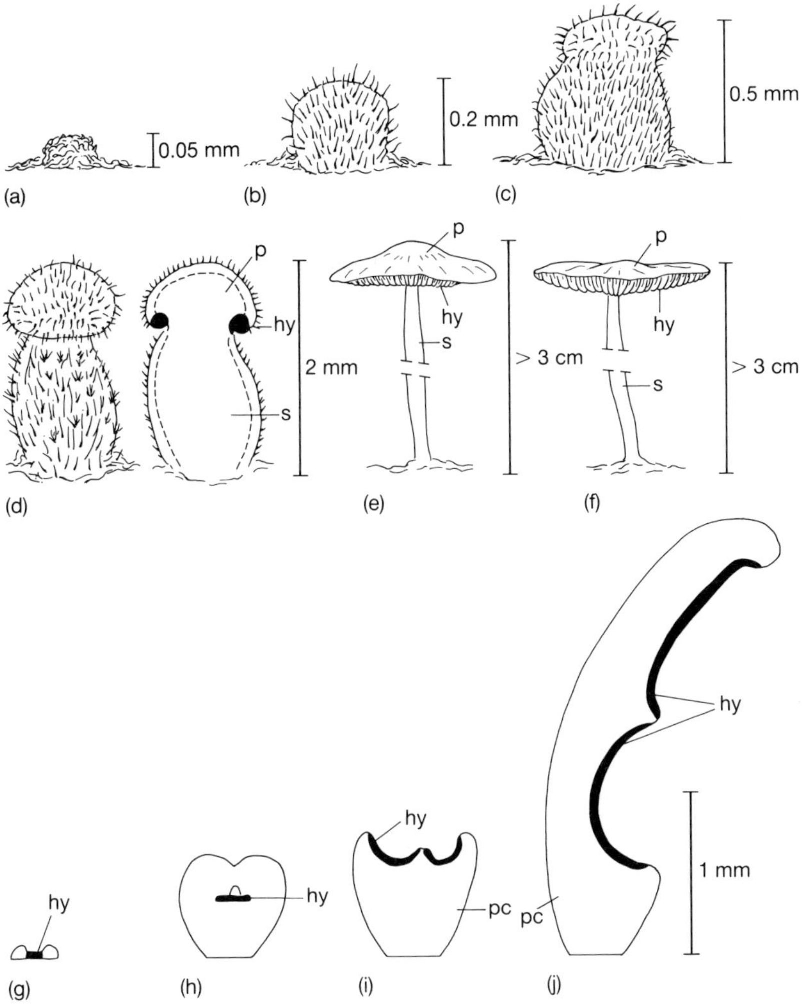

Fig. 3.14 Development of basidiomata. (a–f) Exemplified by the agaric *Flammulina velutipes*; (g–j) a member of Aphyllophorales, *Schizophyllum commune*. (a–d) Development from initial aggregation of hyphae to a primordium, and organization of tissue into stipe (s), pileus (p) and hymenium (hy); (e, f) expansion to the mature fruitbody with little further differentiation; (g, h) development and differentiation of the hymenium from vegetative hyphae; (i, j) continued growth and differentiation of the primordium cup (pc) margin producing further hymenial surfaces (from Williams *et al.*, 1985; Wessels *et al.*, 1985, © Cambridge University Press).

direction under the influence of turgor, allows diffuse extension over the whole wall surface of hyphae in the stipe. This cannot occur in substratum mycelium, as the hydrogen bonds between glucan chains in the hyphal walls are too strong (Mol *et al.*, 1990; Mol & Wessels, 1990). By contrast in Aphyllophorales, for example *Schizophyllum commune*, development is through continued hyphal proliferation at the margin of a cup-shaped hymenium (Fig. 3.14).

Formation of the agaric basidioma involves a change from a homogeneous aggregation of hyphae in the initial to a well-defined structure, the organization of which takes place along two axes: from base to apex, and from inner surface to outer surface. The impetus for this differentiation may derive from a diffusion gradient of a specific chemical, but transfer of the latter may be slow because of the relatively low frequency of cytoplasmic connections between adjacent basidioma hyphae. It would then necessarily have to cross the extracellular matrix, cell wall and plasmalemma (Rosin *et al.*, 1985).

Basidiomata are differentiated internally into several kinds of physically and physiologically distinct hyphae. For instance in trimitic polypores, such as *Coriolus versicolor*, there are thin-walled generative hyphae with dense cytoplasm and clamp connections, and these give rise both to basidia and other types of hyphae. The latter have a largely mechanical function and consist of skeletal hyphae, which are unbranched and thick walled, and binding hyphae which are also thick walled but highly branched.

4 Constraints, limitations and extreme environments

In nature, fungi rarely encounter conditions which allow exponential growth. In general, this is prevented by nutritional and other abiotic environmental constraints, or by the effects of competitive interactions. Rarely, some fungi exhibit cytoplasmically controlled self-limited growth in the laboratory, but how widespread this phenomenon might be in natural habitats is not clear.

The fundamental niche of any organism is determined by the range of environmental conditions under which it can survive and reproduce. Major factors include water availability, temperature, pH, O_2 requirements and CO_2 tolerance, as well as nutrient availability. In some extreme or specialized habitats, heavy metals, salts and pressure can also be important. The capability of a fungus to grow under specific laboratory conditions may explain, in part, how it can occupy a particular realized niche under the influence of competition from other fungi in similar niches. However, it must be remembered that in nature, at any one time, the mycelium may exist in several discrete microsites, each influenced by different biotic and abiotic factors. Similarly, environmental conditions, such as temperature and water availability, may vary both spatially and temporally. By contrast, much laboratory experimentation has centred on the effects of unvarying factors on growth of the hyphal tip, and thus exclusively on mycelial extension.

Environmental factors

Water availability

Access to water is probably the single most important environmental factor affecting growth. Within the cytoplasm and vacuole, ions, organic acids, sugars and enzymes are all either in solution or in colloidal suspension, and consequently the free energy of cellular water determines its participation in all metabolic activity. The term water potential is used to describe availability of water, and may be defined as the free energy of water in a system, related to the free energy of a reference of pure, free water having a specified mass or volume (Papendick & Mulla, 1986). Hence, as water flows naturally from high to low (more negative) potentials, water availability decreases as water potential is lowered. Correspondingly, energy must be expended to raise water potential and thus make water more available.

Water potential (ψ) has been defined mathematically as energy per unit volume of water:

$$\psi = \frac{(\mu_w - \mu_w^*)}{V_w}$$

where μ_w is the chemical potential of the water ($J\,mol^{-1}$), μ_w^* is the chemical potential of water under reference conditions ($J\,mol^{-1}$)

and V_w is the partial molar volume of the water ($m^3 mol^{-1}$). Water potential thus has the dimensions of $J m^{-3}$, which is equivalent to units of pressure in pascals (Pa; 1 MPa = 10 bars = $10^6 J m^{-3}$; 1 kPa = 10 mbar). The total water potential is due to several different components:

$$\psi = \psi_\pi + \psi_m + \psi_p + \psi_g$$

where ψ_π is the osmotic or solute potential, ψ_m is the matric potential, ψ_p is the turgor or pressure potential and ψ_g is the gravitational potential. Because energy is used to move water from the cell to the reference state, ψ_π and ψ_m are always negative, whilst ψ_p is always positive or zero, and ψ_g is either negative, zero or positive, depending on the choice of reference level.

Another term used in describing water relations, particularly in relation to the spoilage of foodstuffs, is water activity (a_w) which is defined:

$$a_w = \frac{P}{P_0}$$

where P is the vapour pressure of water in solution and P_0 is the vapour pressure of pure water. There is a further relationship between relative humidity (RH) of the atmosphere in equilibrium with any solution:

$$\text{RH} = 100 \times \frac{P}{P_0}$$

Water potential and water activity are related as:

$$\psi_\pi + \psi_m = \left(\frac{RT}{V_w}\right) \ln a_w$$

where R is the universal gas constant and T is the temperature (K). But a_w, in contrast to ψ, does not take account of pressure or gravitational potentials, and consequently does not consider the turgor pressure of cells. Hence, a_w has only limited application for describing water relations between cells and their external environment.

Mycelial extension has a characteristic relationship to solute potential (Fig. 4.1). There is no growth at high ψ_π because solutes, and therefore nutrients, are absent. Subsequently, as nutrients are increased, with corresponding decrease in ψ_π, extension rates rise, but with further decreases in ψ_π extension rates decline. By contrast, extension declines continuously as matric water potential, ψ_m, decreases. Fungi may be grouped according to their ability to grow over any particular range of water potentials, the total range lying between just less than zero to −70 MPa. *Monoascus bisporus* is the most

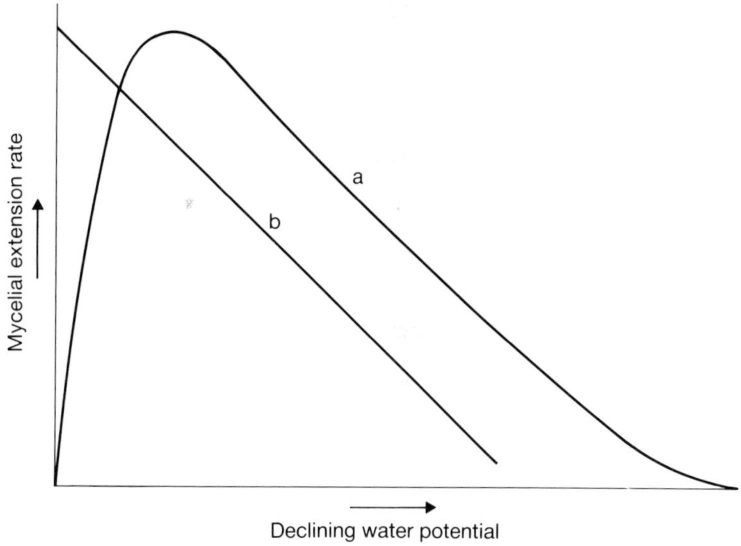

Fig. 4.1 Idealized relationships between mycelial extension and water potential. (a) Solute potential; (b) matric potential (from Griffin, 1981, © Plenum Press).

extreme example, its aleuriospores germinating at −69 MPa (Table 4.1). Although groups overlap, it is clear that some, for instance species of Mastigomycotina, can only grow at high water potentials whilst others, termed xerophiles, are restricted to low potentials. Xerotolerant fungi are those still able to grow at low water potentials, even though their optimum growth rate may occur at much greater potentials.

Xeric environments can be categorized into two groups, depending upon which factors contribute to prevailing low water potentials (Griffin, 1981). The first is characterized by high ψ_m but low ψ_π and includes syrups and preservative brines, where yeast forms predominate. The second, characterized by low ψ_m and variable ψ_π, includes soils of low water content, and stored food products. Here mainly filamentous fungi, including *Aspergillus*, *Chrysosporium*, *Eremascus*, *Penicillium* and *Wallemia* species, are found. In xeric environments provided by solid substrata, growth of unicellular organisms is constrained by lack of suitable water-filled pathways in which ψ_m is high, whereas filamentous fungi are able to bridge air-filled pores and to penetrate the solid phase, where water potentials may be more favourable.

Growth responses to water potential are modified by other environmental conditions. For example, in general, tolerance to low water potential is greatest at the optimum temperature for spore germination and growth, and is reduced on either side of this. Isopleths obtained

Table 4.1 Fungi grouped in relation to water potential (Ψ) tolerance (after Griffin, 1981).

Characteristics	Optimum Ψ for growth (MPa)	Minimum Ψ for growth (MPa)	Examples
Extremely sensitive to low Ψ	−0.1	−2.0	Wood decay fungi, soil Basidiomycotina
Sensitive to low Ψ	−1.0	−5.0	Mastigomycotina Zygomycotina Some Ascomycotina and Basidiomycotina, especially some coprophilous fungi
Moderately sensitive to low Ψ	−1.0	−10 to −15	Some Ascomycotina and Basidiomycotina
Xerotolerant (tolerant of low Ψ)	−5.0	−20 to −50	Ascomycetous yeasts, e.g. *Debaryomyces hansenii*, *Saccharomyces rouxii* Ascomycetous filamentous fungi, e.g. *Aspergillus*, *Eremascus*, *Eurotium*, *Penicillium*, *Wallemia sebi*
Xerophilic (require low Ψ)	Fail to grow above −4.0	−40 or less	*Aspergillus restrictus* *Chrysosporium fastidium* *Monoascus bisporus*

from studies on ψ–temperature interactions effectively represent two dimensions of the fundamental niche of a fungus, given favourable states of other environmental variables (Fig. 4.2). The nutritional, and sometimes toxic, effects of compounds used to induce water stress in the laboratory can also markedly influence growth responses (Wheeler *et al.*, 1988a).

In natural substrata, most fungi do not grow over the full range of water potential that allows them to do so on laboratory media. Competition during colonization, and many other factors, are undoubtedly involved in narrowing the water potential range. However, at water potentials lower than −20 MPa, the ability to grow in axenic culture is a good indicator of competence in natural habitats.

Not only the amounts but also the distribution of water can have a profound effect on the fungal flora of natural habitats. For example, in forestry plantations both *Mycena galopus* and *Marasmius androsaceous* grow under Sitka spruce, but the fruiting mycelium of the latter is restricted to the surface 3 mm of the L horizon, and that of the former to the lower, F_1 horizon (Newell, 1984a,b). The dominant mycophagous collembolan of spruce litter grazes preferentially on *Marasmius androsaceous* but is drought sensitive. *Marasmius androsaceous* is drought tolerant, being capable of growth at −6.1 MPa,

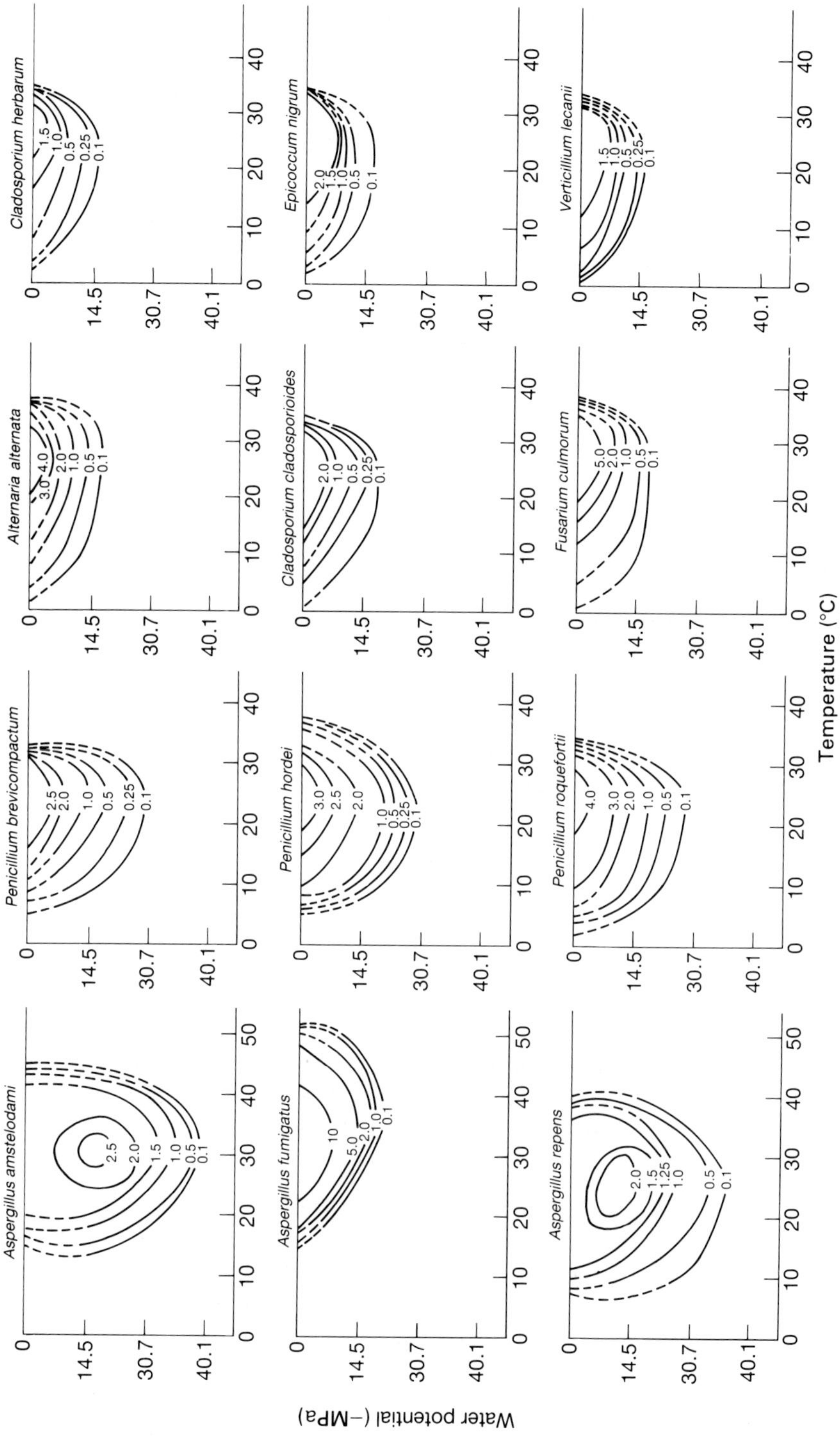

Fig. 4.2 Effect of water potential and temperature on growth rates of *Aspergillus* and *Penicillium* spp. from stored grain, and a range of fungi from the field. Numbers on the isopleths are extension rates of colonies in mm/day (from Magan & Lacey, 1984a, © British Mycological Society).

whereas *Mycena galopus* ceases growth at water potentials below −2.8 MPa (Dix & Frankland, 1988). Thus, grazing and drought tolerance are the major factors controlling distribution of these two agarics. Similarly, variations in water potentials in rabbit faeces determine not only the time of appearance of coprophilous fungi fruiting on them, but also which species appear (Kuthubutheen & Webster, 1986a,b). Decomposition of leaves and wood by early colonizers may result in increased water potentials within such substrata, so encouraging further decomposition by late colonizers that might have greater sensitivity to water stress (Dix, 1985).

Periodicity in water supply can influence establishment and growth in exposed habitats. For example, microcolonial species of *Lichenothelia* are common on southern faces of rocks in hot arid and semiarid regions with less than 50 cm annual rainfall. Lichens are unable to tolerate such conditions, but newly exposed rocks are colonized rapidly by these fungi, although their growth is slow and colonies rarely exceed 100 μm in diameter. The most favourable conditions occur during spring and autumn, when moisture is frequently available and insolation maintains surface temperature above that of the atmosphere. An added advantage to microcolonial fungi *vis-à-vis* lichens is the ability to grow in the dark. This assumes greatest importance in summer when daytime temperatures become too high to permit growth (Palmer *et al.*, 1990).

Seasonal variations also affect fungi in soil. In southwestern Australia optimum conditions for rhizomorph growth in *Armillaria luteobalina* only occur for limited periods at 12 cm depth during mid to late autumn and early summer. From late autumn to mid spring, soil is at or near field moisture capacity, extension decreasing by up to half, and in summer it is too dry for any significant growth. This situation is compounded further by the occurrence of optimum temperatures during summer, during which period they can have no favourable impact on rhizomorph development (Pearce & Malajczuk, 1990).

Rhizomorphs and mycelial cords, for example those of the dry-rot fungus *Serpula lacrimans*, facilitate growth over substrata at water potentials much lower than those which limit extension of unaggregated mycelia (Clarke *et al.*, 1980). Water is moved through the internal cord hyphae in response to endogenously generated turgor pressure at the growing point (see Chapter 2).

Although substratum water potential may be important in influencing mycelial growth, humidity may have equally strong effects on fruiting. Thus, for commercial production of *Agaricus bisporus* and *Pleurotus* fruitbodies an RH between 70 and 90% gives optimum yields. Relative humidities below 60% are likely to cause drying of the substratum and thus inhibit basidioma development.

Temperature Fungi vary widely in their temperature requirements. However, in the main they are mesophilic, having an optimum for mycelial extension or biomass production between 15 and 40°C, whilst still having some capacity for growth around 0–5°C (Table 4.2). Psychrophiles have a range from 0°C or less to 20°C or less, with an optimum of 15°C or less. By contrast, thermophilic fungi are unable to grow below 20°C, have optima for growth in excess of 35°C, and are often capable of growth above 50°C. Other environmental variables, principally nutrient availability and water potential, can influence these cardinal points considerably.

Many fungi survive extreme temperatures even if incapable of growth at them. In this regard, it might be noted that 80% of the Earth's biosphere is permanently cold, that is at 5°C or below, and that a large proportion of the remainder is seasonally cold. When temperatures in arctic and antarctic terrestrial regions ameliorate in the short summer season, fungal growth, including that of lichens, increases, whereas during the rest of the year growth is minimal, being restricted to that of psychrophiles. At temperatures above 65°C the membranes of most eukaryote cells are irreparably damaged. Nevertheless, some fungi can grow at temperatures near to this. For instance, *Dactylaria gallopava*, a deuteromycetous species from thermal spring outlets, has a maximum growth temperature of 61.5°C (Tansey & Brock, 1973). Ascospores of the food spoilage organisms *Neosartorya fischeri* var. *glaber* and *Talaromyces flavus* can survive heat treatments in excess of 80°C for an hour (Conner *et al.*, 1987; Beuchat, 1988). Thermophilic fungi are commonly isolated from compost heaps, grain stores, coal refuse piles, hot springs and soils receiving high insolation. Here, temperatures are maintained for

Table 4.2 Fungi grouped according to their temperature requirements.

Group	Temperature range for growth (°C)	Optimal temperature for growth (°C)	Examples
Psychrophilic	<0–20	0–17	*Mucor psychrophilus* *M. strictus* *Sclerotinia borealis*
Mesophilic	0–50	15–40	Most fungi
Thermotolerant	0–>50	15–40	*Aspergillus candidus* *A. fumigatus*
Thermophilic	20–>50	>35	*Mucor miehei* *Rhizomucor pusillus* *Sporotrichum thermophile* *Thermomyces lanuginosus*

some time above the normal range for growth of mesophilic fungi, and consequently thermophiles come to form part of the dominant microflora. The marine environment also provides examples of temperature-determined fungal distribution. *Asteromyces cruciatus*, *Sigmoidea marina* and *Varicosporina ramulosa* grow *in vitro* at temperatures of 10–30°C, 20–30°C and 20–40°C respectively. *Asteromyces cruciatus* is a temperate-water fungus occurring in the North Atlantic, and North and South Pacific. *Sigmoidea marina* is also a temperate-water species, occurring over the same range as *Asteromyces cruciatus*, but additionally is found in subtropical regions of the West Atlantic. *Varicosporina ramulosa* is found in the tropical–subtropical regions of the Atlantic and Pacific Oceans (Boyd & Kohlmeyer, 1982).

Temperature may also have a controlling effect on mesophilic populations if competition is also involved. For example, in the organic matter-rich surface layers of a spruce forest soil, different species of *Trichoderma* showed different seasonal patterns of abundance (Widden & Abitbol, 1980; Widden & Hsu, 1984). Subsequent one-to-one competition assessments on spruce needles in the laboratory showed that *Trichoderma polysporum* and *Trichoderma viride* competed best at 5–10°C, whilst *Trichoderma koningii* and *Trichoderma hamatum* did so at 15–25°C. This was in agreement with the known temperature preferences of these fungi in their natural habitats (Widden, 1984). However, one *Trichoderma* isolate found abundantly in the summer season was not a good competitor at the higher temperatures against either *Trichoderma koningii* or *Trichoderma hamatum*. Temperature–competition effects therefore may not be always simply defined.

Aeration In nature, many fungi must rarely encounter ideal aeration conditions during their life cycle. Oxygen depletion is a common occurrence in aquatic habitats and in closed environments such as wood, with the effects of depletion often being compounded by increased CO_2 levels. Although CO_2 has direct effects on growth and morphogenesis, indirect effects may also occur as it dissolves, so changing the pH of the milieu:

$$CO_2 + H_2O \rightleftharpoons H_2CO_3$$
$$H_2CO_3 \rightleftharpoons H^+ + HCO_3^-$$
$$HCO_3^- \rightleftharpoons H^+ + CO_3^{2-}$$

Bicarbonate ion is the most important one in the pH range 7–10, but with increased CO_2 concentration the equilibrium changes towards bicarbonate thus lowering pH. At values below pH 5, ionization of carbonic acid becomes insignificant and only dissolved CO_2 is of importance.

Under some conditions growth may require CO_2, as it can maintain the functioning of several metabolic pathways, including the tricarboxylic acid cycle, purine and pyrimidine biosynthesis, and the formation of malonyl-CoA. For example, *Verticillium alboatrum* requires CO_2 when glycerol or glucose are sole carbon sources (Hartman *et al.*, 1972). However, the majority of fungi appear to be strict aerobes, and growth decreases at lowered CO_2 concentrations. This is understandable, as anaerobic respiration is energetically less effective than aerobic respiration, and large quantities of carbon-rich substrates would be required for continued growth in O_2-poor conditions. If other moieties, for instance nitrate and nitrite, could act as terminal electron acceptors in anaerobic conditions, then growth would occur where carbon substrates were limiting (Wainwright, 1988). Some species characteristic of stagnant water, such as *Blastocladiella pringsheimii*, are microaerobic, being able to grow at O_2 concentrations as low as 0.2% but being unable to do so in the absence of O_2.

Many other fungi, for example *Saccharomyces cerevisiae* and some Chytridiomycetes and Oomycetes, are facultative anaerobes. Of these, *Blastocladiella ramosa* lacks typical mitochondria and has only traces of b-type cytochromes, whilst *Aqualinderella fermentans* lacks both of these (Held *et al.*, 1969; Held, 1970). To grow anaerobically, several facultative anaerobes, including *Aqualinderella fermentans*, *Fusarium moniliforme*, *Fusarium solani*, *Mucor rouxii*, *Rhizopus* species and *Saccharomyces cerevisiae*, may need exogenous fatty acids, sterols or vitamins (Bull & Bushell, 1976; Gibb & Walsh, 1980). Consequently, for long periods fungi may be unable to grow unless these nutrients are supplied. In soil, where aerobic and anaerobic conditions alternate, it is likely that such essential nutrients are formed during aerobic periods, to then become available when conditions subsequently become anaerobic. Rhizomorphs enable *Armillaria* species to grow, albeit at reduced rates, through near-saturated soil with low O_2 and high CO_2 concentrations. The unique structure of the rhizomorph facilitates internal transport of O_2 to the growing apex (Smith & Griffin, 1971; Pearce & Malajczuk, 1990). Additionally, this O_2 transport system permits exploitation of new woody substrata, which may be low in available O_2, to take place before other organisms can colonize them.

Absence of O_2 seems to be an essential factor for full development of the Chytridiomycetes *Neocallimastix frontalis*, *Piromonas communis* and *Sphaeromonas communis*, which are generally confined to the herbivore rumen and caecum, although some are aerotolerant for short periods in faeces and saliva (Milne *et al.*, 1989). Other conditions optimizing growth of these fungi include a pH of 6.5, a temperature of 39°C, a cellulosic substrate, and the presence of CO_2. *Blastocladiella ramosa* and *Aqualinderella fermentans* also require CO_2 in concentrations of 5–20% for growth.

High CO_2 concentrations are generally inhibitory to growth, frequently causing morphic switches, but observations can be complicated by associated pH effects and interactions with other environmental variables (Magan & Lacey, 1984b; Ho & Smith, 1986). Relative sensitivity to CO_2 can have ecological impact. For example, basidiomycetes growing in leaf litter, which would not normally encounter high CO_2 concentrations, are inhibited by a pCO_2 of 10 kPa. However, some wood-inhabiting Basidiomycotina are adapted to such conditions and will grow at a pCO_2 above 30 kPa (Hintikka & Korhonen, 1970). Similarly, isolates of *Rhizoctonia solani* from foliage, crown and stem rots, and root rots show 80, 52 and 31% inhibition in radial growth rates respectively in 20% CO_2, reflecting the natural frequency of exposure to elevated CO_2 concentrations (Durbin, 1959).

pH, metals and salts

Because of its complex interaction with other factors, and its variation between local microsites, the precise effect of pH on fungi is difficult to evaluate. Nevertheless, it is clearly of key importance in fungal ecology. Fungi have the ability to alter the pH of their surroundings during growth (see Chapter 2), but in the laboratory usually grow best over a pH range of 5–7. Some are able to grow slowly at pH extremes. For example, *Candida krusei*, *Rhodotorula mucilaginosa* and *Saccharomyces exigua* can do so at pH 1.5, and *Saccharomyces fragilis* at pH 9 (Recca & Mrak, 1952; Battley & Bartlett, 1966). This ability is probably related to maintenance of a suitable internal pH and ionic balance, and will be discussed later. Turning to behaviour in nature, ericoid mycorrhizal fungi develop symbiotically in soils of pH 3, some wood-decaying basidiomycetes commonly tolerate similar pH values in the heartwood of some hardwood trees, and some spoilage fungi can grow in vinegar, but in general, examples of free-living fungi adapted to growth at pH extremes are rare. It is surprisingly difficult to correlate the occurrence of particular fungi in nature with prevailing pH, even in situations where it can be assumed that pH is a dominant ecological factor. It is known that soils and streams of different pH contain different fungal assemblages, but this does not always directly relate to the pH growth responses of representative species of these groups in the laboratory (Griffin, 1972; Rosset & Bärlocher, 1985). Interestingly, a relationship exists between the incidence of equine phycomycosis, caused by a *Pythium* species, and cyclical changes in the pH values of water in the lakes of north Queensland, Australia

(Shipton, 1983). In the rainy season, the risk to horses of contracting disease by drinking swamp water contaminated with large numbers of *Pythium* propagules increases as pH decreases to near the optimum for growth and sporulation of the pathogen.

The effects of pH on amounts of dissolved CO_2 and its associated ionic forms have already been mentioned. However, pH also affects the availability and ionic forms of nutrients, the solubility of many heavy metal ions being drastically altered. In low pH soils, Al^{3+}, Cu^{2+}, and Fe^{3+} are all freely available. At high concentrations, Al^{3+}, Cd^{+}, Cu^{2+} and Zn^{2+}, together with other heavy metal ions, can be toxic to fungi. Divalent transition elements and other metals forming insoluble sulphides, such as Ag, Mo and Sn, are poisonous as a result of their reactivity with enzymes; Cd, Cu, Hg and Pb react with membranes, so affecting their permeability (Ross, 1975). By contrast, at high pH many metal ions become insoluble. Lack of some metals for use as co-factors affects the functioning of many metabolic pathways and may inhibit growth. For example: aldolase, RNA polymerase and superoxide dismutase require Zn^{2+}; nitrate reductase contains Mo^{2+}; Fe^{2+} functions as an enzyme activator and is a component of porphyrins required for electron transfer reactions; Cu^{2+} activates laccase, tyrosinase and uridine nucleosidase; Mg^{2+} and Mn^{2+} bind to ATP and function in reactions involving transfer of phosphate groups; Mg^{2+} also activates enolase and isocitrate dehydrogenase, and is important in maintaining membrane structure and function.

Marine fungi are adapted to growth in seawater, which contains approximately 35% NaCl. Some marine fungi, for instance *Lulworthia floridana* and *Corollospora maritima*, grow optimally in seawater, whereas others, such as *Dendryphiella salina*, grow best with only 10% seawater (Jones & Byrne, 1976). Generally, Saprolegniaceae are not tolerant of saline conditions or low water potentials although some, such as an *Aphanomyces* species causing ulcerative mycosis of the salt water Atlantic menhaden fish (*Brevoortia tyrannus*), have a degree of salinity tolerance, with zoospores able to germinate during salinity stress provided that nutrients are present (Hearth & Padgett, 1990). However, the presence of salt *per se* may not be the overriding factor determining which species are native to the sea, as the growth of many terrestrial Deuteromycotina is enhanced on media containing seawater. Presumably seawater removes some limitation to growth, either by a change of ionic conditions or through providing extra nutrients. Nevertheless, *Basipetospora halophila*, a fungus isolated from dried salt fish, requires NaCl for optimum growth over the entire range of water potentials which support growth, and may be an obligate halophile in nature (Wheeler *et al.*, 1988b). An estuarine chytrid, *Phlyctochytrium*, requires higher concentrations of Na, K and

Ca than those found in freshwater, but lower than for seawater. Similarly, it has a broad optimum range for growth of pH 5–8, which is characteristic of ranges found in estuaries (Amon & Arthur, 1981).

Wallemia sebi (*Sporendonema sebi*) commonly occurs as a spoilage fungus in dried salted fish, and part of its success in this habitat is related to its ability to produce large numbers of small conidia on simple conidiophores within 24 hours of germination, even at water potentials of −22.4 MPa. Other xerophilic fungi, such as *Penicillium janczewskii*, take up to 14 days to produce conidia on more complex conidiophores. Ascospore production by another xerophile, *Eurotium chevalieri*, is an even slower process, and whereas ascomata can form at −22.4 MPa on media containing high concentrations of carbohydrates, they do not do so at −11.4 MPa in NaCl-based media (Hocking, 1986). *Wallemia sebi* retains intracellular glycerol, which is used to maintain turgor, much more efficiently than other xerophilic fungi investigated, and it may be thus more energetically efficient during growth and reproduction under saline conditions. High maintenance energy requirements of other xerophiles may limit their growth to energy-rich habitats such as dried fruits and confectionery.

Adaptations to extreme environments

An extreme environment is one in which, for most fungi, biomass production is restricted by the continuous imposition of stress. Simply, this may be due to the action of a single stress factor, such as lack or excess of a specific nutrient, an extreme temperature or pH, or low water availability. However, in nature, stress more commonly arises from a combination of several of these. In many situations growth will be prevented, but severe stresses may be tolerated by fungi which either possess appropriate physiological characteristics, or can adapt through a temporary alteration in their developmental pattern. Here, consideration is given to general mechanisms and adaptations exhibited in extreme environments. These have a considerable bearing on the natural development of both free-living fungi and symbionts of animals and plants. More specific details of behaviour are discussed, where appropriate, in subsequent chapters.

Low water potential

Growth under stress requires maintenance of turgor for extension growth, preservation of cell division, and continued metabolic activity. Small changes in water availability within stressed cells affect turgor, enzyme activity and protein stability. Consequently, any stress-induced changes in water availability, for instance directly via osmotica or indirectly through effects of temperature on membrane structure (see later), will have drastic effects on growth.

In general, growth requires positive turgor potential, which is dependent on the plasmalemma facilitating a lowered internal protoplasmic solute potential compared with that of the external environment. In conditions of reduced water potential, fungi maintain low protoplasmic and vacuolar water potential through polyol or organic acid synthesis. As well as being osmoregulators, polyols – which commonly include glycerol, arabitol and mannitol – also act as reserve carbohydrates which can enter into metabolism rapidly according to the energy requirements of cells (see Chapter 2). Thus, they provide fungi with considerable physiological flexibility, and by acting as compatible solutes, which can provide a more suitable medium for enzyme activity than solutions of other solutes, have an essential role in low water potential conditions. Glycerol particularly may additionally protect hydrated biopolymers and ensure their structural integrity and activity at low water potentials (Griffin, 1981; Eamus & Jennings, 1986a,b). *Saccharomyces cerevisiae* responds to lowered external water potential by synthesizing glycerol, but loses a constant proportion of this to the exterior. The xerotolerant yeast *Zygosaccharomyces rouxii* maintains a constant rate of glycerol production and has a much lower leakage rate, with arabitol present only at low concentration (van Zyl & Prior, 1990). As the osmoregulatory system of this species is energetically more efficient, this may partly explain why it is more xerotolerant than *Saccharomyces cerevisiae*. However, polyol synthesis is energetically expensive, and probably contributes to the generally slow growth of fungi at reduced water potentials. In fungi lacking polyols, proline is the organic compound most commonly synthesized in response to lowered water potentials. *Mucor hiemalis*, *Phytophthora cinnamomi*, *Pythium debaryanum* and an unidentified member of the Saprolegniales all synthesized proline in media containing a range of different osmotica with lowered water potential (Luard, 1982b). Like polyols, this amino acid has little effect on enzyme activity.

For filamentous fungi, the osmoticum used in the laboratory can affect the type of compatible solute synthesized (Table 4.3). When NaCl, $MgCl_2$ and inositol are used as osmotica, *Dendryphiella salina* respectively accumulates predominantly glycerol, mannitol and inositol. Accumulation, within hyphae, of the solute used to control osmotic potential of the medium has also been found in the xerotolerant species *Penicillium chrysogenum* and *Chrysosporium fastidium*, as well as in *Phytophthora cinnamomi* which is sensitive to reduced water potential (Luard, 1982a,b). Changes in relative amounts of organic compounds in the mycelium also take place when water potential of the substratum decreases (Table 4.4). For example, sclerotia of *Sclerotinia sclerotiorum* contain mannitol as the major soluble carbohydrate and this, together with arabitol and glycerol, increases with a decrease in water potential to −5 MPa. Trehalose

Table 4.3 Percentage contribution to osmotic potential (Ψ_π) of major classes of solute within the protoplasm of *Dendryphiella salina* and *Thraustochytrium aureum*, together with values for osmotic potential calculated from concentrations of known solutes (from Eamus & Jennings, 1986b, © Cambridge University Press).

Osmoticum added to basal medium (water potential)	Contribution (% Ψ_π) Carbohydrate	α-amino nitrogen	Inorganic ions	Organic acid	Major organic solute	Major organic solute	Calculated Ψ_π (−MPa)
Dendryphiella salina							
NaCl (−0.96 MPa)	28	31	41	0	14	Glycerol	1.63
(−1.92 MPa)	27	14	59	0	15	Glycerol	3.30
$MgCl_2$ (−0.96 MPa)	29	20	51	0	19	Mannitol	1.88
(−1.92 MPa)	19	11	66	4	8	Mannitol	4.32
Inositol (−0.96 MPa)	26	34	32	8	14	Inositol	1.58
(−1.92 MPa)	33	26	33	8	23	Inositol	2.30
Thraustochytrium aureum							
Seawater 50% (−1.24 MPa)	1	9	90	0	0.5	Proline	1.36
Seawater 75% (−1.85 MPa)	1	11	88	0	0.5	Proline	1.78
Seawater 100% (−2.46 MPa)	2	15	83	0	6	Proline	2.45

Table 4.4 Amounts of soluble carbohydrates in sclerotia of *Sclerotinia sclerotiorum* formed at different substratum water potentials. Values in parentheses are percentages of total (from Al-Hamdani & Cooke, 1987, © British Mycological Society).

Substratum	Soluble carbohydrates and glycerol (mg g^{-1} dry weight)					
Ψ (−MPa)	Glucose	Trehalose	Mannitol	Arabitol	Glycerol	Total
0.1	—	12.6 (20.3)	42.6 (68.9)	1.2 (1.9)	5.4 (8.8)	61.8
1.0	—	18.8 (19.5)	65.4 (67.7)	2.4 (2.5)	10.0 (10.4)	96.6
2.0	8.9 (6.6)	3.1 (2.3)	99.0 (73.5)	4.9 (3.6)	18.8 (14.4)	134.7
3.0	22.4 (10.2)	—	162.9 (74.7)	5.1 (2.3)	28.5 (13.0)	218.9
4.0	37.5 (14.0)	—	186.6 (69.7)	7.7 (2.9)	35.9 (13.4)	267.7
5.0	42.7 (14.7)	—	200.4 (69.0)	8.4 (2.9)	39.0 (13.4)	290.5

occurs at high potentials but disappears below −3 MPa. Conversely, glucose occurs only at −2 MPa and below (Al-Hamdani & Cooke, 1987). This process allows both water influx and translocation of nutrients for sclerotium formation to occur over a wide range of substratum water potentials.

It has been proposed that lichens have become highly adapted to growth under fluctuating conditions of water availability, and that water is the predominant factor controlling their growth (Smith, 1979). Their drought survival mechanisms again involve polyols. The photobiont passes most of its photosynthetic carbohydrate to the mycobiont, within which it is rapidly converted to polyols. This transfer is in excess of that necessary to support the slow rate of thallus growth, and polyols appear to serve as physiological buffers which enable lichens to survive the often extreme fluctuations in environmental conditions to which they are commonly subjected. Polyols not only aid survival but also have a function during rewetting. At this time, a short period of resaturation respiration occurs whilst membrane barriers are restored and organic solutes and mineral ions are being lost. The large store of polyols enables the thallus to survive these regular depletions of its material.

High and low temperatures

There is considerable evidence which suggests that membrane composition determines the ability of fungi to grow over specific temperature ranges. Ultrastructural observations on mesophilic species held at supraoptimal temperatures show that damage to mitochondrial and nuclear membranes, and endoplasmic reticulum, occurs rapidly and also, later, to the plasmalemma. In psychrophilic yeasts, for example species of *Candida*, *Leucosporidium* and *Torulopsis*, constituent fatty acids are more unsaturated than those of mesophiles, and depressed incubation temperatures increase this degree of unsaturation (Kerekes & Nagy, 1980). A similar pattern of change has also been found for some *Mucor* species (Dexter & Cooke, 1984a,b). However, with other *Mucor* species, membrane phospholipids rather than general fatty acids (which can include storage lipids) differ between psychrophiles, mesophiles and thermophiles (Hammonds & Smith, 1986). Here, levels of membrane phospholipid unsaturation decrease from psychrophile to mesophile to thermophile. In addition, the thermophile *Mucor pusillus* has membrane phospholipids of greater chain length compared with related mesophiles and psychrophiles.

The structure and composition of membranes is likely to affect the temperature at which their properties change from an inactive, thermotropic gel phase to an active, liquid crystalline phase. For instance, since membrane phospholipids of *Mucor pusillus* show little change in composition with temperature, membrane activity in this species may cease at even median temperatures as its phospholipids transform to an inert, rigid, tightly packed gel phase. On the other hand, temperatures above those necessary for the fluid state to form will produce increased thermal motion, with concomitant loss of permeability control and associated protein functioning resulting in cell death. In this regard, the psychrophile *Mucor strictus* shows the greatest potassium leakage in comparison with a range of closely related mesophiles at 25°C, but the smallest loss at 0°C (Dexter & Cooke, 1985).

Proteins and sterols within membranes can also influence their stability (Dexter & Cooke, 1984a). For instance, new proteins appearing in ascospores of *Neosartorya fischeri* are associated with an increase in heat tolerance which develops with age (Conner *et al.*, 1987). Glycerol and mannitol may also increase so as to maintain turgor pressure against heat-mediated decreases in external water potential. Furthermore trehalose, which is known to stabilize membranes during dehydration (which may result from both heat and cold stress), increases in these heat-tolerant ascospores. This disaccharide is also synthesized in response to nutrient deprivation and exposure to toxic chemicals, and appears to be a general stress protectant in the cytosol (Wiemken, 1990). Similarly, a few specific proteins are synthesized rapidly and to high concentration when cells are exposed to supraoptimal, but non-lethal, temperatures. Anaerobiosis, ethanol, heavy metals and certain antibiotics also elicit this response. Following such exposure, cells become resistant to otherwise lethally high temperatures, and to other physical and chemical stresses. These heat-shock proteins have highly conserved amino acid sequences, and some may be involved in maintaining protein transport across membranes (Plesofsky-Vig & Brambl, 1985; Craig *et al.*, 1990).

Enzyme thermostability is also an important feature of heat tolerance. In many cases, enzymes extracted from thermophiles show higher optimum and maximum temperatures for activity in comparison with those extracted from mesophiles, but there is little evidence that thermophilic fungi possess enzymes of exceptional thermostability (Anderson & Smith, 1976; Woodin & Wang, 1989).

pH and toxic levels of metals and salts

Ericoid mycorrhizal fungi exhibit important physiological characteristics associated with growth of their hosts in soils of low pH. In these environments, mineralization processes are slow, and proteins may constitute a significant proportion of the available nitrogen pool. *Hymenoscyphus ericae* secretes extremely stable proteinases with a pH optimum of 2–3, and these enzymes provide access to a source of nitrogen otherwise unavailable to the host, *Calluna vulgaris* (Leake & Read, 1990). Interestingly, the endophyte from the calcicolous alpine shrub *Rhodothamnus chamaecistus* also produces a proteinase with a similar pH for optimum activity, but this enzyme retains activity at much higher pH values than that from *Hymenoscyphus ericae*. With the alpine endophyte, proteinase activity at high pH values would be advantageous, as thermal limitations inhibit nutrient mineralization by many free-living microbes.

Mycorrhizal fungi have been widely studied in relation to heavy metal tolerance, their role in establishing and maintaining plant growth in metal-polluted soil having an obvious ecological significance. Both endo- and ectomycorrhizal fungi are involved. For example, clover plants infected with the vesicular–arbuscular fungus *Glomus mosseae* have been recovered from Zn- and Cd-contaminated soils (Gildon & Tinker, 1981). The endophyte strain from these plants was able to infect clover at soil concentrations of Zn and Cd that severely restricted infection by a strain obtained from agricultural soil. As plants infected with either endophyte contained similar quantities of Zn and lower quantities of Cd in comparison with non-mycorrhizal plants, the metal-tolerant strain did not transport additional amounts of metals to the plant, even when these were present in the soil in high concentrations. In soybean, vesicular–arbuscular mycorrhizal infection similarly resulted in reduced accumulation of heavy metals in the foliage of plants growing in soils heavily contaminated with Cd, Mn and Zn but, in contrast, enhanced foliar accumulation of these metals from soils with low heavy metal concentrations, which suggests a possible scavenging effect of infection under the latter conditions (Heggo *et al.*, 1990).

The presence of *Hymenoscyphus ericae* in roots of *Calluna vulgaris* on metal-contaminated soil enabled host plants to grow following transfer to sand cultures containing up to $75\,mg\,l^{-1}$ Cu and $150\,mg\,l^{-1}$

Zn, in which non-mycorrhizal plants failed (Bradley *et al.*, 1981). Here, non-mycorrhizal plants contained significantly more Zn and Cu overall compared with non-mycorrhizal plants (Fig. 4.3). Non-mycorrhizal plants also had high metal contents in their roots relative to their shoots. This suggests that metal exclusion, and hence toxicity avoidance, was the mechanism involved. Ascomycotina related to *Hymenoscyphus* are known to bind metal cations to their cell walls, and extensive mycelial growth in and around roots would provide a large volume for metal ion retention. The extramatrical mycelium of the ectomycorrhizal basidiomycete *Paxillus involutus* may function

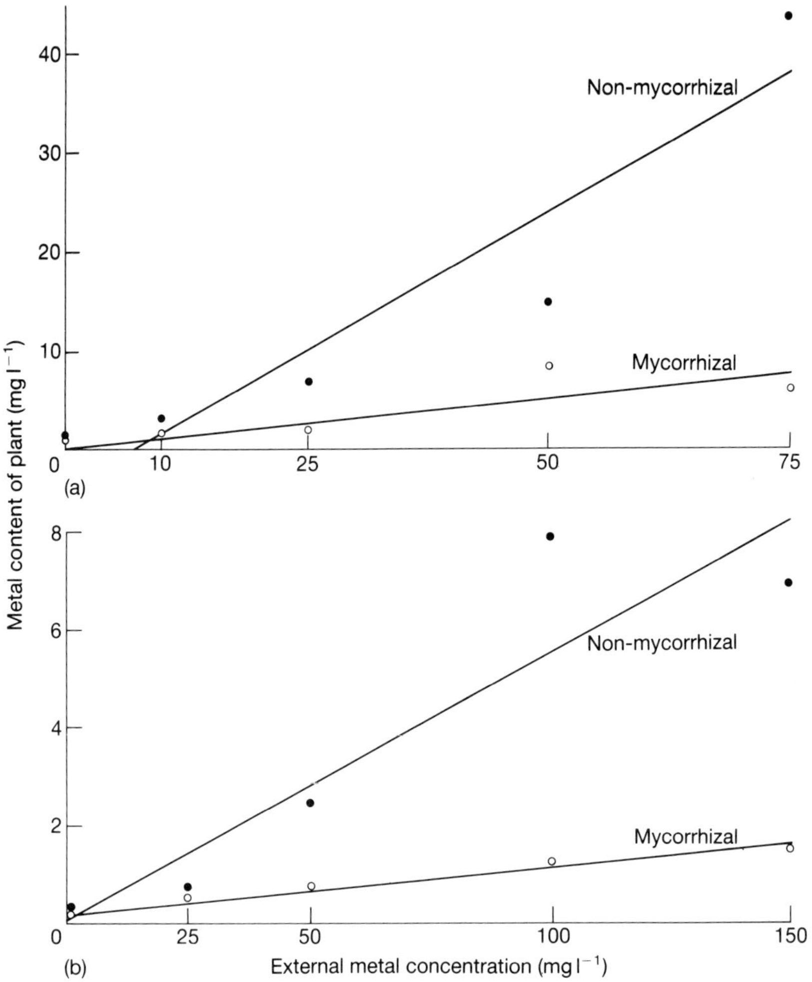

Fig. 4.3 (a) Copper; (b) zinc contents of mycorrhizal plants of *Calluna vulgaris* from metal-contaminated soil grown in sand culture containing various concentrations of copper and zinc (from Bradley *et al.*, 1981, © MacMillan Journals).

similarly with *Betula* species, Zn^{2+} ions having been found to bind to electronegative sites in the hyphal wall and to extrahyphal polysaccharides (Denny & Wilkins, 1987). Such metal-binding systems in *Suillus luteus* may also be involved in the reduction of leaf concentrations of Cd, Ni and Pb in ectomycorrhizal *Quercus rubra* growing in heavy metal-contaminated soil (Dixon, 1988).

In these symbioses, the ability of mycorrhizal fungi to survive toxic environments may depend on carbon supply from the host. Because the fungus receives adequate carbon, it may be able to withstand toxic metals as well as to expend energy on detoxification processes. Free-living saprotrophic fungi could not do so in soils of low carbon availability. However, in the laboratory, Cu-tolerant strains of *Aureobasidium pullulans* and *Saccharomyces cerevisiae* ameliorated the effects of high Cu levels by restricting uptake in comparison with Cu-susceptible strains (Gadd & Griffiths, 1980; Ross & Walsh, 1981). The Cu-tolerant strain of *Saccharomyces cerevisiae* still accumulated large quantities of Cu, but this was bound to specific proteins (metallothioneins) and was effectively immobilized. *Saccharomyces cerevisiae* and several filamentous fungi have also been shown to deposit Cu as sulphide at the cell wall, with *Cunninghamella blakesleeana* binding Cu and Co within the cell wall in hydroxyproline-rich and citrulline- and cystathione-rich proteins respectively (Venkateswerlu & Stotzky, 1989). Similarly, wood-rotting fungi such as *Poria* species, *Serpula lacrimans* and *Coriolus palustris*, and others such as *Penicillium ochro-chloron* (isolated from copper-plating solutions) when growing on Cu-rich substrata, produce copper oxalate crystals which are deposited on the cell wall (Gadd, 1986). Calcium oxalate crystals are commonly formed on hyphae and may serve to detoxify oxalate from within the mycelium or, more likely, to remove excess Ca from the environment adjacent to it (Whitney & Arnott, 1987, 1988). Besides improving the activity of some extracellular enzymes which are inhibited by high Ca levels, the crystals could provide a hydrophobic coating protective against bacterial and fungal attack, and may also act as a physical deterrent to grazing by soil microfauna. Mercury resistance in fungi may occur by intracellular detoxification due to the presence of thiol compounds, or via methylation, or because of changes in membrane permeability that decrease uptake. Another possible detoxification mechanism is via intracellular storage of Mn^{2+} and Sr^{2+} in the vacuole as low molecular weight polyphosphate complexes.

The ability of marine and terrestrial fungi to grow in potentially toxic saline environments depends upon the maintenance of the lowest ionic balance within the mycelium ($K^+ > Na^+$) and a water potential sufficient to maintain turgor. Any deviation from these will affect hyphal growth (Eamus & Jennings, 1986b). The marine yeast *Debaryomyces hansenii* grows better in high Na^+ concentrations and in alkaline media than does *Saccharomyes cerevisiae.* It shows optimum Mg^{2+}-ATPase activity at pH 8 when grown in media containing 1.6 mol l^{-1} NaCl, whereas that of *Saccharomyces cerevisiae* is optimum at pH 6.5–7 (Comerford *et al.*, 1985). Under alkaline, saline conditions, pH of the cytoplasm of *Debaryomyces hansenii* increases, hence decreasing competition between protons and Na^+ for exchange with K^+ across the plasma membrane via Mg^{2+}-ATPase. A bicarbonate pump may also be present at the plasma membrane, but not in *Saccharomyces cerevisiae*, which would enhance this beneficial effect (Hobot & Jennings, 1981). Thus, in seawater, *Debaryomyces hansenii* is able to maintain the correct levels of K^+ within its cells, whereas *Saccharomyces cerevisiae*, which lacks active Mg^{2+}-ATPase at alkaline pH, cannot do so. In the salt-marsh fungi *Alternaria chlamydospora*, *Alternaria phragmospora* and *Ulocladium chartarum*, the major fatty acid (oleic 18:1) and most phospholipids decrease with increasing salinity, and this may be related to the maintenance of membrane function at high salt concentrations (Mulder *et al.*, 1989).

A response of many fungi to increased external salt concentration is polyol synthesis, notably production of glycerol. The marine species *Dendryphiella salina*, *Thraustochytrium aureum* and *Thraustochytrium roseum* may also synthesize organic and amino acids, principally proline, in response to increased external salt levels and, depending on the species, these may contribute up to 90% of cell solute water potential (Wethered & Jennings, 1985). However, internal ionic concentration may also increase, and the proportions of solutes contributing to external osmotic potential of the cytoplasm may vary with the external osmoticum (see Table 4.3). In *Debaryomyces hansenii*, up to 2.6 mmol Na g^{-1} cell dry weight has been found after growth has taken place in a medium containing 2.7 mol l^{-1} NaCl, although much of this was probably either bound to the cell wall or stored in the vacuole (Hobot & Jennings, 1981). In *Penicillium ochro-chloron*, an increase in intracellular glycerol concentration occurs in response to high extracellular concentration of both Na^+ and Cu^{2+}, reflecting a general response to decrease in water potential in the environment (Gadd *et al.*, 1984). In this fungus, resistance to high levels of Cu^{2+} must, in part, be coincidental with its ability to maintain turgor at low water potential.

High Na concentration can be expected to severely affect enzyme functioning but there is some evidence that, provided Mg^{2+} levels are adequate, enzymes of marine fungi may not be so affected, particularly as glycerol stabilizes protein structure and function. Marine isolates of some fungi, such as *Aureobasidium pullulans* and *Zalerion eistla*, require high Na for growth at elevated temperatures (Ritchie & Jacobsohn, 1963; Torzilli *et al.*, 1985). This is due to an osmotic requirement by *Zalerion eistla*, but the possibility remains that some marine fungi have a large demand for specific ions. For instance, in the marine bacterium *Vibrio marinus* inhibition of protein synthesis at high temperatures was relieved by high salination, which suggests that high salt concentrations stabilized the structure of thermally labile protein-synthesizing enzymes (Cooper & Morita, 1972).

Nutrients In the laboratory, fungi are routinely grown in conditions of nutrient excess. However, although nature also provides a wide range of nutrient-rich habitats, it is probable that, at large, a common problem faced by fungi is that of potential starvation. This is particularly acute in situations where there are only very low amounts of available carbon compounds, either because they are insufficient, as in some mineral soils, or because although organic substrates are present, they are refractory.

Refractory substrates. Numerous refractory substrates can be degraded by fungi (see Table 2.1). For example, chitinophiles, keratinophiles, and lignicoles utilize chitin, keratin and lignin respectively, and maintain slow growth on them. In soil, some fungi are able to use humic and fulvic acids, and sundry phenolic decomposition products which are not readily available to other microorganisms. Fungi utilizing such refractory substrates may be tolerant of fungistatic and fungitoxic aromatic compounds that also might be present. Possession of varied enzymic competence by different fungi partly explains the temporal changes in assemblages on complex substrata, such as wood, which contain a number of refractory components. Development of *Aspergillus fumigatus* and *Cladosporium resinae* in aviation fuel and the degradation of hydrocarbons by the arenicolous species *Corollospora maritima*, *Didymosphaeria enalia* and *Lulworthia lignoarenaria*, are perhaps extreme examples of the ability of fungi to use toxic refractory substrates for growth (Thomas & Hill, 1976; Kirk & Gordon, 1988). The arenicolous species are of potential importance in some coastal situations in decreasing the impact of oil spills.

Oligotrophy. Mineral soil in particular is generally regarded as a milieu poor in available nutrients, and in which fungi can grow only after an influx of organic materials. Using media containing low carbon levels, large numbers of fungi can be isolated from soil. These then show relatively high growth rates in culture at low levels of growth-yielding substrates (Wainwright, 1988). Such oligocarbotrophs have several characteristics which enable them to utilize low nutrient supplies efficiently, and to facilitate growth and sporulation without release of readily utilizable materials in large quantities (Table 4.5). Growth of only a few millimetres through soil may be sufficient for a hypha to be able to capture and utilize newly encountered resources. The ability of a fungus to exist or grow oligotrophically allows it to utilize new nutrient sources more effectively than one relying heavily on spores for survival. Mixed substrate utilization, together with the ability to scavenge low levels of nutrients, confers an obvious ecological advantage. Some soil fungi are able to utilize a range of carbon substrates, many simultaneously, including gases such as methane, *n*-butane, propane, ethene, ethane and carbon monoxide, which are commonly found at low concentrations in soil and in some refractory organic materials.

There have been several reports of fungal growth in the apparent absence of exogenous organic carbon (Wainwright, 1988). For example, some *Trichoderma* and *Fusarium* species produce fine hyphae, with many anastomoses, bearing large numbers of chlamydospores. This growth form has the effect of increasing the surface area : volume ratio of hyphae for nutrient uptake and provides for nutrient storage. Similarly, conidia of *Penicillium daleae* and *Penicillium billaii* when introduced into soil give rise to severely restricted growth, with conidia being formed on simplified penicilli at the germ tube tips (Sheehan & Gochenaur, 1984). Conidia and germ tubes formed during such iterative germination have a total volume of between 30 and 40% of those produced on nutrient-rich media. Available nutrients are thus partitioned such that survival is favoured at the expense of mycelial growth (see also Chapter 8).

Oligotrophic growth additionally may be supported by volatile carbon sources. For instance, culture flasks containing carbon-free

Table 4.5 Major characteristics of fungi adapted to oligotrophic conditions or capable of oligotrophic growth (after Wainwright, 1988).

Nutrient uptake
Capacity increased by high hyphal surface area : volume ratio or by high uptake site density per unit area
Uptake sites, with a high affinity for substrate molecules and low specificity transport systems
No requirement for substrate modification, for instance phosphorylation, as all available substrates utilizable
Other features
Simplified reproductive structures
Low capacity for antibiosis
High susceptibility to mycostasis

media and stoppered with cotton wool accumulated 0.5 $mg\,l^{-1}$ $week^{-1}$ of dissolved organic carbon (Geller, 1983). Fungi can also fix CO_2 which, in most cases, is associated with anapleurotic CO_2 reactions involving compensation for the removal of intermediates during metabolism (see Chapter 2). This could cause a net energy drain from the fungus. However, growth of *Cephalosporium* and *Fusarium* species in response to rising CO_2 or HCO_3^- concentration is linear in the absence of organic carbon in the medium (Mirocha & Devay, 1971). This suggests that oxidation of molecular hydrogen by molecular oxygen provides energy for substantial CO_2 fixation and growth. Another way open to fungi in soil is oxidation of reduced forms of elements such as S, N, Mn and Fe (Wainwright, 1988).

Fungi growing in conditions of low availability of essential heavy metals have developed systems for scavenging such ions. Of those systems examined, the production of siderophores for solubilization, transport and storage of iron (Fe^{3+}) has been the most thoroughly studied (Winkelman, 1986). Under low iron conditions, iron-free siderophore ligands are released by fungi, and the iron so sequestered is subsequently made available to the fungus. In *Ustilago sphaerogena* the siderophore, ferrichrome, is taken up complete, the iron-free ligand subsequently being excreted for a continuation of the cycle. However, another siderophore, ferrichrome A, does not enter the cell, but releases its iron to the external face of the cell membrane via reductive removal. In *Neurospora crassa*, the siderophore (coprogen) is also taken up complete with its iron, but as the iron is released only slowly, coprogen may also function as an iron storage compound. Importantly, *Hymenoscyphus ericae* may play a role in regulating iron uptake by ericaceous plants over a wide range of iron availabilities. With low iron concentrations, the fungus produces a hydroxamate siderophore which enhances iron uptake by roots of *Vaccinium macrocarpon* and *Calluna vulgaris*, whereas when iron is available at high concentration, mycorrhizal roots withhold the metal from transport to the shoots, thus maintaining the iron balance in the plant (Shaw *et al.*, 1990).

Self-limited growth

It seems to be assumed generally that fungi are capable of potentially indefinite growth if provided with an inexhaustible source of nutrients and unlimited space. This may be true, but there are some species which quickly senesce and die even when cultured in what might be assumed as being ideal laboratory conditions. Most notable are *Podospora anserina* and *Aspergillus glaucus*, in which growth rate gradually decreases and mycelia degenerate, with swelling and lysis of hyphal tips (Esser *et al.*, 1984). Similar morphological effects involving growth inhibition and sectoring of the colony margin, but without death, have been observed in the 'ragged' mutant of *Aspergillus amstelodami*, and 'poky' and 'stopper' mutants of *Neurospora crassa*. These types of behaviour are associated with effects on mitochondrial DNA produced by maternally inherited cytoplasmic infective factors.

Sectoring and decreased growth have also been reported for several plant pathogens, including *Endothia* (*Cryphonectria*) *parasitica*, *Gaeumannomyces graminis* var. *tritici*, *Helminthosporium victoriae*, *Ophiostoma ulmi* and *Rhizoctonia solani* (Hashiba, 1987; Nuss & Koltin, 1990). In general, this also is associated with cytoplasmically inherited traits which here involve virus-like double-stranded (ds) RNA or, rarely, DNA plasmids. The effects are more akin to disease than is the self-limited growth associated with mitochondrial dysfunction.

Mechanisms involving mitochondrial DNA. During onset of senescence in *Podospora anserina*, correlations have been found between failure of mitochondrial function and the appearance of self-replicating extrachromosomal fragments of circular DNA. Subsequent analysis has shown that the circular DNA, or plasmid, is derived from multiple copies of segments of mitochondrial DNA, and that during senescence large segments of mitochondrial DNA disappear. Two of these plasmids, carrying coding information for subunits I and III respectively of mitochondrial cytochrome oxidase, have been shown to transpose or integrate into the genome of senescing mycelium (Wright & Cummings, 1983). This action has two effects. First, as *Podospora anserina* is an obligate aerobe, the loss of a functional mitochondrial genome causes immediate breakdown of respiration. Second, by integration of 'alien' DNA into the genome, several gene sequences are likely to be rendered non-functional, so impairing normal cell activity. However, in the 'ragged' mutant of *Aspergillus amstelodami*, the wild-type mitochondrial DNA sequences are maintained, and only additional tandem repeats of mitochondrial DNA are produced. This could explain the ability of *Aspergillus amstelodami* to continue to grow, albeit more slowly, rather than to die (Lazarus *et al.*, 1980). The factors controlling the behaviour of mitochondrial DNA in these fungi are unknown.

Mechanisms involving dsRNA. Many fungi have been shown to contain virus-like particles of dsRNA, but there is much conflicting evidence as to whether these cause senescence *per se* or disease, both of which can be expressed via decreased or abnormal growth. In some cases, transmission of one or more dsRNA virus types can convert growth patterns from healthy to senescent. For example, transmission by hyphal anastomoses of one or two dsRNA viruses commonly found in apparently healthy 'carrier' isolates of *Helminthosporum victoriae* induced severe stunting in previously healthy isolates (Ghabrial, 1986). In others, multiple types may be required to induce abnormal growth. For instance, diseased or 'd' isolates of *Ophiostoma ulmi* are commonly infected with multiple types of dsRNA segments. In a d^2-infected isolate which contains 10 types of dsRNA segments, transmission of disease symptoms to healthy isolates is accompanied by transfer of all 10 types. If three specific dsRNA types, termed 4, 7 and 10, are lost, healthy isolates are obtained. Some infected isolates recover spontaneously and these are also found to have lost types 4, 7 and 10. It is concluded that the d^2 factor consists of one or more of these dsRNA types (Rogers *et al.*, 1986). Further, not all dsRNA types cause impaired growth.

There are also cases where senescence does not appear to be related to dsRNA. In some isolates of *Rhizoctonia solani* known to contain dsRNA, disease is apparently caused by small linear DNA plasmids in the cytoplasm rather than by dsRNA; and, like dsRNA, these can be transferred cytoplasmically (Hashiba, 1987). DNA plasmids have also been recovered from several other fungi, for example *Ascobolus immersus*, *Gaeumannomyces graminis*, *Kluyveromyces lactis*, *Morchella conica* and *Ophiostoma ulmi*, but are not necessarily associated with abnormal growth.

Perhaps the most obvious, but at present the least considered, possibility is that single gene mutations are involved with senescence. Forms of *Endothia* (*Cryphonectria*) *parasitica* have been obtained which produce slow colony growth and dense branching, and which segregate as if determined by single nuclear genes (Anagnostakis, 1984). Such simple genetic mutations are likely to be common in the field.

Ecological significance. In all cases of self-limited growth the outcome would appear to be deleterious, which raises the question of how such systems have evolved and survived and what, if any, is their ecological significance. Infections by dsRNA have been recorded for all major fungal groups, but only relatively rarely have they been associated with major deleterious effects on their carriers. Similarly, cases of cytoplasmically inherited senescence, such as that found in *Podospora anserina*, are comparatively rare. Consequently it is difficult to ascribe specific ecological significance to such effects. Most dsRNA viruses are probably benign, but where deleterious events occur these can be severe, as with dieback of *Agaricus bisporus* and the 'killer' phenomenon in *Saccharomyces cerevisae* and *Ustilago maydis* (Nuss & Koltin, 1990). In the 'killer' phenomenon, the dsRNA mycovirus codes for the production of a polypeptide toxin which kills susceptible fungal strains, but a nuclear gene maintains the dsRNA within the cell and confers resistance on the carrier. If there are no incompatibility barriers, spread of dsRNA from infected to non-infected individuals can occur. Strains of *Endothia* (*Cryphonectria*) *parasitica* carrying dsRNA are hypovirulent, and these have been used to control epidemics of chestnut blight caused by virulent strains by transferring the dsRNA (van Alfen *et al.*, 1975). By contrast, in *Rhizoctonia solani*, the presence of dsRNA within mycelium is associated with virulence, as strains lacking dsRNA do not cause disease and, in this case, have been used as prophylactic biological control agents, either preventing invasion of roots or inducing resistance in the plant (Koltin *et al.*, 1987).

One possible advantage of senescence in *Podospora anserina* relates to sexual reproduction. Strains can be maintained in a juvenile condition indefinitely by regular sexual propagation of them (Esser *et al.*, 1984). Senescence may be a natural mechanism for encouraging outbreeding, which would be of value in some ecological situations where stress is severe. The fact that the plasmid derived from the breakdown of mitochondrial DNA can integrate into host nuclear DNA, presents opportunities for its use as a vector in molecular biology in filamentous fungi.

PART THREE
Reproduction and Establishment

5 Induction and control of reproduction

At some time during trophic growth, or on or close to its cessation, reproduction is initiated, its onset being determined by the availability of sufficient somatic material of appropriate maturity to impart reproductive competence to the mycelium. In many instances, but by no means always, reproduction is favoured by conditions which are non-optimum for vegetative growth. It might, therefore, be expected that in situations favourable to vegetative expansion, resource capture would be the main function of the mycelium, whilst in unfavourable conditions propagule production would be promoted.

The time required to achieve competence varies widely, and depends on whether reproduction is vegetative, asexual or sexual, on ecological circumstances, and on the constitution of the fungus concerned. Thus for numerous conidial *R*-selected fungi, for instance species of *Aspergillus*, *Penicillium* and *Trichoderma*, young mycelia may reach competence within 24 hours of spore germination, whilst basidioma production by many *S*-selected fungi may occur only after many months, or even years, of vegetative growth. Several interacting factors are involved in mediating the switch from trophic growth to reproductive differentiation, and include various endogenous timing mechanisms, and a range of exogenous regulators. Generally, these act to ensure that expenditure of reproductive effort is concentrated at a time when environmental conditions are likely to favour, or demand, the release of spores, and either their subsequent germination or deposition in a dormant state. But no matter how favourable conditions might be for reproduction, the time at which it can take place will be determined first by the minimum period within which it is possible to achieve competence and, second, by the time required for the mycelium to respond to external stimuli which either encourage reproduction or are essential for triggering it. Furthermore, especially in lower fungi, sexual reproduction may be phased by the interplay of sequentially released morphogenetic compounds, which again places a restraint on the speed with which spores can be produced.

Autonomic controls

Autonomous mechanisms underlying reproduction have been most intensively researched in Chytridiomycetes, Oomycetes and Zygomycotina, although the number of species studied is very small. Here, during sexual reproduction, accurate placement of gametes is necessary for successful fertilization, and chemotaxis of flagellate cells, or directional growth of gametangia is brought about via sophisticated, and often complex, induction and guidance systems.

Detailed observations on chemotaxis of gametes have been confined to *Allomyces*, although there is good reason to suppose that such behaviour also occurs in other Chytridiomycetes. Haploid thalli release small, uniflagellate male cells which must fuse with larger, slower-moving females to form diploid zygotes. Female cells synthesize and release the bicyclic sesquiterpene diol, L-sirenin, which elicits a positive chemotactic response in males (Fig. 5.1). Its action is highly specific, other isomers and analogues having no effect, and also appears to be calcium dependent. Male gametes take up sirenin which is then inactivated, a process which possibly maintains their sensitivity to it, and so enables them to respond to a concentration gradient (van den Ende, 1983).

In those fungi where sexual reproduction involves fusion of non-motile cells, metabolite exchange may occur between sex organs of separate and physiologically distinct mycelia, or of the same mycelium. For example, female mycelia of the heterothallic Oomycete *Achlya ambisexualis* release constitutively produced steroid hor-

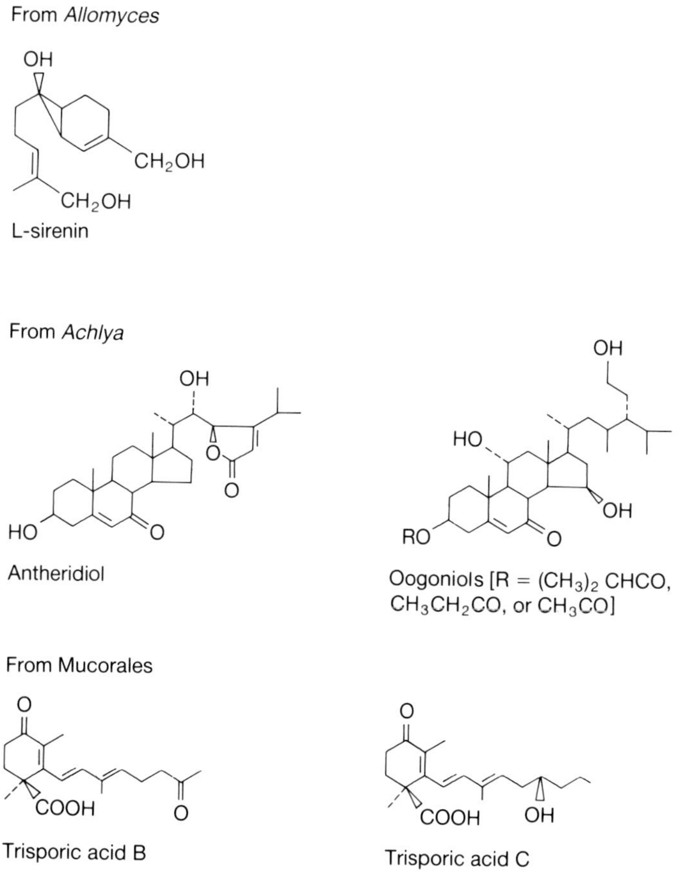

Fig. 5.1 Structures of some reproductive hormones.

mones that induce sexual differentiation in male mycelia; the resultant antheridial branches show directional growth towards oogonial initials on female mycelia. Sexual development of male hyphae, including their subsequent growth, is controlled by release of the steroid antheridiol from female thalli, and oogonium formation is then stimulated by antheridiol-induced production and release of oogoniols by male thalli (Fig. 5.1, Table 5.1). As is the case with sirenin, antheridiol activity is highly specific, and it is rapidly metabolized by recipient male cells (Musgrave & Nieuwenhuis, 1975). Similar control of sexual reproduction occurs in homothallic *Achlya* species.

In male mycelia, antheridiol induces changes in electrical currents associated with antheridial initiation, increases in rRNA, poly-(A^+)RNA (mRNA) and overall protein synthesis, as well as formation of specific proteins localized in distinct intracellular compartments (Brunt & Silver, 1987; Gow & Gooday, 1987; Horton & Horgen, 1989). Concomitant increases in cellulase activity may soften the walls of male hyphae, so enabling lateral branches to form. Antheridiol also induces a decrease in lipoxygenase activity in both *Achlya ambisexualis* and *Saprolegnia ferax*, which results in production of 20-carbon fatty acids (eicosanoids) that are known to regulate a variety of physiological processes in animals and plants (Herman & Luchini, 1989). Fucosterol, which is the precursor of antheridiol, is also active in stimulating oospore production in species of *Phytophthora* and *Pythium*, possibly by regulating cyclic AMP levels (Kerwin & Washino, 1984).

Directed growth of gametangia also occurs in heterothallic Mucorales, for instance *Blakeslea trispora*, *Mucor mucedo* and *Phycomyces blakesleeanus*, where zygophores of opposing (+) and (−) mating types arise some distance apart, grow towards one another, meet and fuse. In

Table 5.1 Effects of hormones on sexual interactions in *Achlya ambisexualis* (after Raper, 1939, © Botanical Society of America; van den Ende, 1983).

Hormone	Producer	Recipient	Result
Antheridiol	Vegetative female hyphae	Vegetative male hyphae	Induces formation of antheridial branches
Oogoniols	Antheridial branches	Vegetative female hyphae	Initiate formation of oogonial initials
Antheridiol?	Oogonial initials	Antheridial branches	Attracts antheridial branches, induces a thigmotropic response and antheridum delimitation
Oogoniols?	Antheridia	Oogonial initials	Induce delimitation of oogonium by basal septum formation

addition, just before meeting, each zygophore tip may give rise to an asymmetrical bulge or subapical peg through which contact is made. There is firm evidence that this mutual attraction is due to co-operatively produced prohormones, which are volatile precursors of trisporic acids, the latter being inducers of zygophore formation (Figs 5.1 & 5.2). Zygophores are formed only when two opposite mating types are in diffusion contact, as the mycelium of one mating type can only accomplish a limited part of the biosynthetic pathway from prohormones to trisporic acids. Trisporic acid synthesis is stimulated by diffusion of intermediates of this pathway between mating types. In addition, trisporic acid stimulates prohormone production, inducing a positive feedback on the trisporic acid synthesis pathway (Bu'Lock *et al.*, 1976; Sutter & Whitaker, 1981). Similar mechanisms probably act in homothallic species.

Although it has not been explored in such detail, directional growth during reproduction occurs amongst Ascomycotina and Basidiomycotina and is probably widespread in septate fungi. For example, in the coprophile *Ascobolus stercorarius* sexual reproduction is brought about by fusion of an oidium with a filamentous trichogyne arising

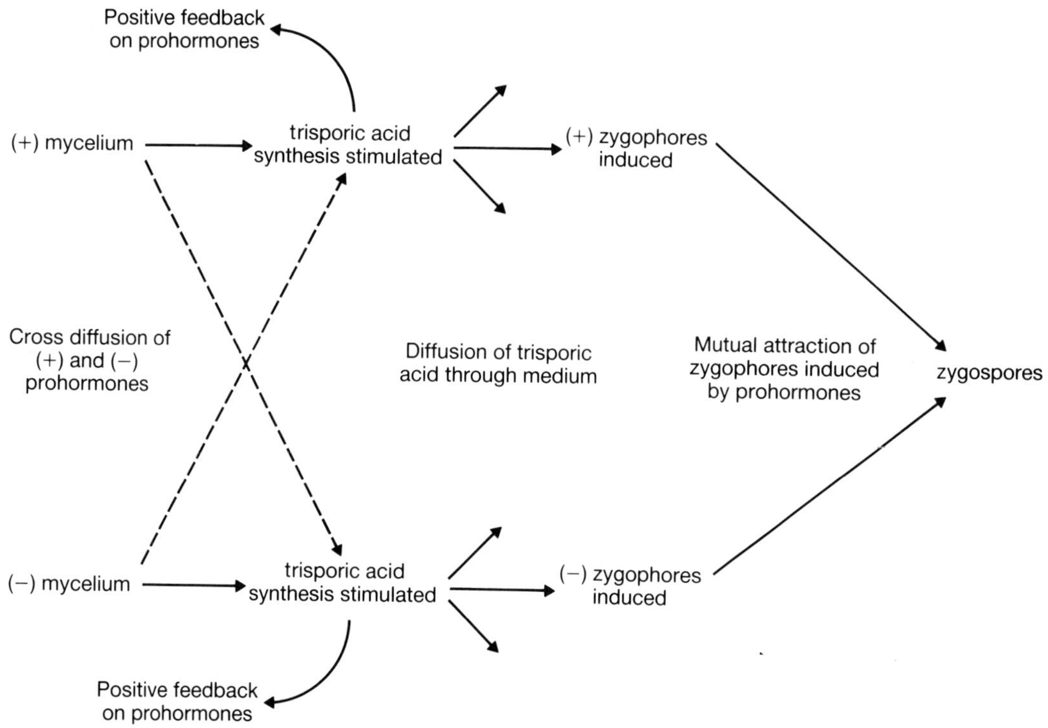

Fig. 5.2 Sequence of events during zygospore development in heterothallic Mucorales.

from an ascogonium. The trichogyne shows marked directional growth towards oidia, which only elicit this response after being activated in some way by diffusates from the female sex organ (Bistis, 1956, 1957). Biochemically controlled growth of trichogynes towards male cells has also been observed in *Bombardia lunata*, *Nectria haematocca*, *Neurospora crassa*, *Neurospora sitophila* and *Podospora anserina* (Bistis, 1983). Endogenous chemical factors may also stimulate sexual development in *Pyrenopeziza brassicae* (heterothallic) and *Aspergillus nidulans* (homothallic) (Ilott *et al.*, 1986; Champe *et al.*, 1987).

In many Basidiomycotina, dikaryotization is effected by fusion of uninucleate oidia with branches of a compatible homokaryon mycelium. Directional growth has been observed in species of *Clitocybe*, *Coprinus*, *Flammulina* and *Psathyrella*, and is probably an essential feature of the fusion process (Bistis, 1970; Kemp, 1970). Growth-directing factors released by oidia of *Coprinus* species are highly potent, and can elicit reorientation over distances in excess of 75 μm, with responses being apparent within 15 minutes of placing oidia in proximity to hyphal tips. A similar homing reaction has been observed between vegetative hyphae and basidiospores of *Laccaria laccata*, *Leccinum* species, *Polyporus dryophilus* and *Schizophyllum commune*, but for *Leccinum* details are more complex than for those occurring between hyphae and oidia in the other species (Fries, 1983; Voorhees & Peterson, 1986). Factors released by hyphae induce basidiospore germination, which results in the formation of a small germ vesicle; the activated spore then releases factors that guide hyphae towards the vesicle.

Directional growth responses are not confined to filamentous fungi, but are also found among ascomycetous and basidiomycetous yeasts, in which they facilitate conjugation (Crandall *et al.*, 1977). For example, adjacent sexually compatible cells of *Hansenula anomala*, *Kluyveromyces* species and *Saccharomyces cerevisiae* bud towards one another so that daughter cells are brought into contact. Alternatively, in *Hansenula* and *Saccharomyces*, one cell may bud preferentially towards its mate, which then ceases to bud but elongates, the long axis of the expanding cell being oriented in the direction of the approaching chain of bud cells (Yanashigima, 1988).

Amongst basidiomycetous yeast forms, orientation of conjugation tubes arising from sporidia occurs in *Rhodosporidium toruloides*, *Tremella mesenterica* and *Ustilago violacea* (Poon *et al.*, 1974; Abe *et al.*, 1975). In the first two species this is clearly due to chemically directed growth, and takes place even when cells are widely separated. In *Ustilago violacea* another mechanism may be involved. Conjugation takes place, for the most part, between sporidial cells that are in contact, and conjugation tubes are subsequently formed which push

From *Saccharomyces cerevisiae*

α1 NH_2–Trp–His–Trp–Leu–Gln–Leu–Lys–Pro–Gly–Gln–Pro–Met–Tyr–COOH
α2 NH_2–His–Trp–Leu–Gln–Leu–Lys–Pro–Gly–Gln–Pro–Met–Tyr–COOH
α3 NH_2–Trp–His–Trp–Leu–Gln–Leu–Lys–Pro–Gly–Gln–Pro–Met(SO)–Tyr–COOH
α4 NH_2–His–Trp–Leu–Gln–Leu–Lys–Pro–Gly–Gln–Pro–Met(SO)–Tyr–COOH

α factor

From *Rhodosporidium toruloides*

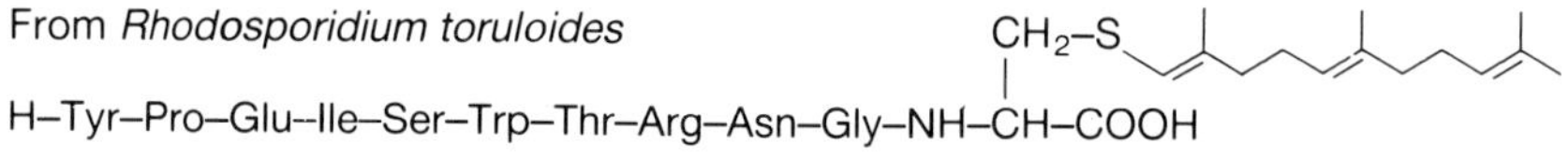

Rhodotorucin A

From *Tremella mesenterica*

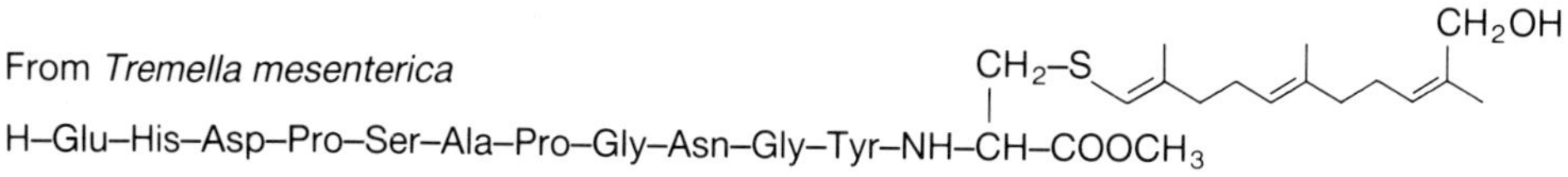

Tremellogen A–10

Fig. 5.3 Hormones from yeasts.

the cells apart, although conjugation tubes can bridge between cells separated by a distance of up to 20 μm. Sporidia bear fine fimbriae, about 10 μm long, those of adjacent cells joining end to end at the beginning of the mating process. Following this, conjugation tubes seem to grow along the lines of fimbrial connections, and may in some way be guided by them (Day, 1976). In the case of *Rhodosporidium*, *Saccharomyces* and *Tremella* there is an exchange of peptides between compatible mating strains (Fig. 5.3). Initially this arrests vegetative growth, inhibits DNA synthesis, and induces a change in the agglutinability of the cell surface, which is accompanied by either cell enlargement or oriented expansion (Sakagami *et al.*, 1979; Miyakawa *et al.*, 1986; Yanashigima, 1988).

Environmental controls

Nutrient quality

Following attainment of competence, often reproduction can take place only if particular organic or inorganic nutrients are available (Table 5.2). Many such effects are likely to involve alterations in those physiological processes directly associated with induction, but others may result from changes in normal metabolic pathways remote from induction itself. For example, calcium-mediated sexual reproduction in Oomycetes and *Rhodosporidium toruloides* involves uptake and binding to calmodulin, which is known to have secondary messenger

Table 5.2 Examples of nutrients required for reproduction.

Nutrients	Species	References
Organic		
Arginine	*Sordaria macrospora*	Bahn & Hock, 1973
Aspartate and phenylalanine	*Mucor miehei*, at 50°C but not at 35°C	Mehrotra & Mehrotra, 1980
Phosphoglycerate	*Chaetomium globosum*	Buston & Khan, 1956
Phospholipids	*Phytophthora cactorum*	Ko, 1985
Sterols	Some Peronosporales	Elliott, 1977
Vitamins: (biotin and thiamin)	*Aspergillus nidulans*, only in microaerobic conditions	Adler *et al.*, 1981
	Chaetomium convolutum	Lilly & Barnett, 1949
	Ophiostoma spp.	Turian, 1978
	Saccharomyces cerevisiae, aerobic and anaerobic conditions	Adler *et al.*, 1981
	Sordaria fimicola	Barnett & Lilly, 1947
	Sordaria macrospora	Hock *et al.*, 1978
Inorganic		
B	*Sordaria* spp.	Turian, 1955
Ca^{2+}	*Achlya, Lagenidium, Phytophthora* and *Pythium* spp.	Elliott, 1986; Kerwin & Washino, 1986
	Chaetomium globosum	Turian, 1978
	Penicillium notatum	Pitt & Mosley, 1985
	Phytophthora parasitica, required with NO_3 but not with L-asparagine as N source	Elliott, 1989
	Rhodosporidium toruloides	Miyakawa *et al.*, 1986
Mn^{2+}	*Aspergillus nidulans*	Zonneveld, 1975
NH^{4+} or glycine	*Trichoderma viride*, essential to light-induced rhythmic sporulation	Ellison *et al.*, 1981
Zn^{2+}	*Neurospora* spp.	Turian, 1955

properties; in *Penicillium notatum* similarly induced sporulation is linked to disruption of glycolysis and the TCA cycle. In *Sclerotium rolfsii* sclerotium formation results from a shift in reducing potential arising from a truncation of the TCA cycle and stimulation of the pentose phosphate pathway by action of sulphydryl group antagonists (Willetts, 1978). Such changes in metabolic cycles can alter adenylate charge and availability of cyclic AMP. Increased cyclic AMP levels inhibit oosporangiogenesis in *Lagenidium* and *Phytophthora*, and conidiation in *Neurospora crassa*, but stimulate fruiting in *Coprinus macrorhizus*, *Schizophyllum commune* and other basidiomycetes (Gold & Cheng, 1979; Elliott, 1988).

Some nutrients are required only at specific reproductive stages, or under particular environmental conditions, or in order to respond to specific reproductive triggers (Table 5.2). For example, protoperithecial morphogenesis in *Sordaria macrospora* will not proceed past the ascogonium-core hypha stage in the absence of biotin (Hock *et al.*, 1978). This kind of requirement extends to the greatest degree in many symbionts. For instance, in the sheep rumen, haem-containing compounds such as haematin and haemin induce sporulation and zoospore formation in the gut-inhabitant *Neocallimastix frontalis* (Orpin, 1978). Here, haem-containing compounds derived from green plant material require prior degradation by other rumen microorganisms to become active.

Nutrient depletion, pH and secondary metabolites

Nutrient limitation may result from either substrate shortage or lack of enzymic competence. Inorganic nutrients may also be unavailable if substratum pH is too high, or become toxic if it is too low. Extremes of, or fluctuations in, pH may have additional direct effects on growth (see chapters 2 and 3). Nutrient-limited restriction of balanced vegetative growth, and its consequent effects on both primary and secondary metabolic pathways, are linked directly to reproduction. Indeed, several secondary metabolites might be considered as being hormones. Further, secondary metabolic pathways often result in pigment formation which, as in the case of melanin, may have a role in the protection of reproductive structures.

Commonly, the presence of a mechanical barrier to mycelial expansion stimulates sporulation or sclerotium production because, in the restricted growth zone, increased interhyphal competition for nutrients or the accumulation of secondary metabolites may occur. For instance, a transient decline in extension occurs during sclerotium formation in a rhythmic isolate of *Sclerotinia sclerotiorum*, and growth arrest is a prerequisite for sexual reproduction in *Phycomyces blakesleeanus* in response to hormones (Humpherson-Jones & Cooke, 1977b; Drinkard *et al.*, 1982). Similarly, pH-induced growth inhibition, mediated by changes in availability and flux of ions and alterations in endogenous electrical currents, stimulates zoospore formation in *Blastocladiella emersonii* (Stump *et al.*, 1980).

Another common reproductive trigger is nitrogen exhaustion, which stimulates sporulation in some *Penicillium* species, basidioma initiation in *Lentinus edodes*, and oospore formation in *Phytophthora*. Amino acid depletion is particularly important in induction of sporulation in *Blastocladiella emersonii* (Correa & Lochi, 1986). Carbon exhaustion can also have a major effect, for instance in stimulating basidioma formation in *Coprinus*, and sporangiogenesis in *Phytophthora* species (Rao & Niederpruem, 1969; Elliott, 1989).

Changes in nutrient quality, particularly in carbon : nitrogen ratios, occur naturally during resource degradation, or as migration to new habitats takes place. Different proportions of macroconidia, microconidia and chlamydospores are produced by: *Fusarium oxysporum* f.

sp. *elaedis* on media with different carbon : nitrogen ratios; sclerotium and cord formation by *Sclerotium delphinii* and *Sclerotium rolfsii*, and rhizomorph production by *Armillaria* are all stimulated by growth from high nitrogen to low nitrogen environments when constant amounts of carbon are available (Thompson, 1984; Oritsejafor, 1986; Punja, 1986). Cord and rhizomorph formation in response to nitrogen step-down favours translocation, migration to new areas and, for *Sclerotium rolfsii*, increases the area over which its sclerotia are produced (Fig. 5.4). Staled culture media, either lacking in nutrients or containing enzymes or products of secondary metabolism – for example the phenol derivative sclerin – induce sclerotium formation in *Sclerotinia sclerotiorum* and *Sclerotium rolfsii*. In *Sclerotinia sclerotiorum* unidentified organic acids may also be involved (Humpherson-Jones & Cooke, 1977c; Cooke, 1983). This kind of behaviour relates to the natural situations in which sclerotia are produced, when exploitation of host tissues is nearing completion, and where both secondary metabolite production and nutrient deprivation occur.

Metabolic switches related to reproduction are accompanied by changes in quality and quantity of RNA and protein, but details of these interrelationships are poorly understood. In *Saccharomyces cerevisiae*, following sporulation induction, total RNA levels increase but then decrease during sporulation. Turnover of RNA is rapid, and up to 50% of rRNA present on completion of sporulation is newly synthesized (Haber *et al.*, 1977). In *Blastocladiella emersonii* total RNA levels increase during zoosporangium formation to then decrease during differentiation, the mRNA synthesized early in sporulation being degraded, but that formed later becoming stored in the zoospores. Similarly, several proteins associated with ribosomes and free RNA particles are synthesized late in zoosporogenesis, and are then also stored within zoospores. This may reflect a general mechanism for preserving RNA in dormant spores (Johnson & Lovett, 1984; Jaworski & Harrison, 1986). In *Aspergillus nidulans*, about 1300 new mRNA sequences are synthesized during conidiation, considerably more than the 45–150 genes for this species suggested by genetic studies. However, this large number of sequences may result in subtle phenotypic changes not detectable visually, as well as producing some stored mRNA sequences which may be used during germination. In *Schizophyllum commune* only about 35 new specific mRNAs and 37 additional polypeptides are synthesized by dikaryotic fruiting mycelium (Timberlake, 1980; Wessels *et al.*, 1985; Ruiters & Wessels, 1989). Sporulation-specific proteins have been detected in *Aspergillus nidulans* as well as in a range of other fungi. Sclerotium-specific proteins occur in *Sclerotinia* species, but their function is not yet clear (Dahlberg & van Etten, 1982; Petersen *et al.*, 1982; Léjohn &

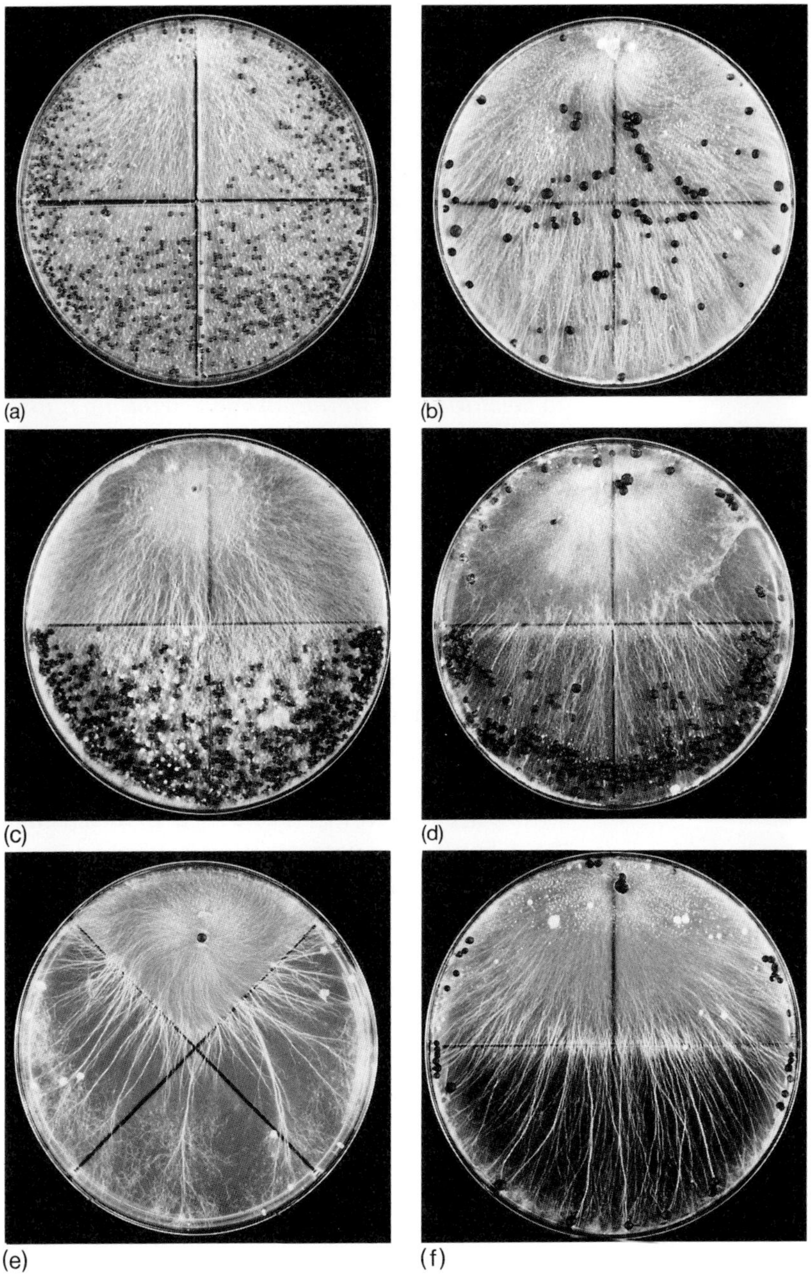

Fig. 5.4 Patterns of sclerotium and cord formation on agar. (a, c) *Sclerotium rolfsii*; (b, d–f) *Sclerotium delphinii*; (a) and (b) have uniform levels of carbon and nitrogen; (c) and (d) have reduced nitrogen levels in the lower half of each dish; (e) has no nitrogen in the lower region of the dish; (f) has soil extract in the lower region (from Punja, 1986, © British Mycological Society).

Braithwaite, 1984). There is also evidence that small intramycelial morphogenetic factors controlling fruitbody formation pass from mycelia to the sporophore in some fungi, but these have not yet been characterized (Breton, 1978).

Carbon dioxide and aeration

As fungi grow, they are likely to encounter regions of variable aeration. With poor air exchange, CO_2 will accumulate and O_2 levels decrease, and these changes may result in stress-induced alterations in growth and metabolism which either stimulate or inhibit reproduction, depending on the habitat requirements of the species involved. Sexual and asexual reproduction of septate fungi is O_2 dependent, but to varying degrees. For instance, at 0.5% O_2 ascogonia of *Neurospora sitophila* form but perithecia do not; protoperithecia require at least 1% O_2 and low CO_2 levels for development (Turian, 1978). In *Neurospora crassa* O_2 requirements for reproduction are in the order macroconidia > microconidia > ascogonia, this sequence being related to the activity of different metabolic pathways in mycelium of different ages, and a switch from apical to lateral development in older parts of the mycelium.

Carbon dioxide concentrations greater than those in the normal atmosphere stimulate ascoma development in *Aspergillus nidulans* and *Chaetomium globosum*, and conidiation in *Alternaria tagetica*, but inhibit differentiation and sporophore formation in *Agaricus bisporus*, *Penicillium* species and *Schizophyllum commune* (Long & Jacobs, 1974; Sietsma *et al.*, 1977; Cotty, 1987). In *Aspergillus nidulans* α-1,3-glucan is the reserve carbohydrate used as the energy source for cleistothecium formation, and low CO_2 levels inhibit its synthesis during vegetative growth, and also decrease activity of the α-1,3-glucanase required for its mobilization once competence is reached (Zonneveld, 1975, 1988). Reduced sclerotium production by *Sclerotium rolfsii* in the presence of elevated CO_2 levels, combined with increased production in response to light, could account for the more abundant occurrence of sclerotia near the soil surface in nature (Punja, 1986).

Controlled aeration is essential for commercial production of *Agaricus bisporus* basidiomata (Flegg *et al.*, 1985). Following the spawn run, restriction of ventilation in the growing house raises CO_2 levels, so inhibiting basidioma initiation and enhancing CO_2 fixation by the mycelium. It also raises the temperature to 20–25°C, which is near the optimum for mycelial extension. When colonization of the compost is nearly complete it is cased with a mixture of peat and chalk, which becomes colonized from below. Ventilation is then increased, the temperature falls to 16–20°C, and CO_2 concentration drops to between 0.04 and 0.08%, which favours basidioma initiation and subsequent maturation. Microorganisms in the casing layer may also aid basidioma initiation by lowering levels of volatile self-inhibitors produced by *Agaricus*. Provided that the compost remains moist, and the atmosphere humid, mushroom production then takes place in repeated flushes.

Temperature

Fungi are adapted to grow and reproduce within specific temperature regimes and, depending on other prevailing conditions, changing these regimes in the laboratory can often produce a switch in the type of reproduction taking place. This suggests that, in nature, fungi have an inherent flexibility to react to environmental changes via reproductive pleomorphism.

Stress, in the form of high temperature (37°C) and low O_2 concentration, induces chlamydospore formation by *Fusarium sulphureum* (Barran *et al.*, 1977). In other fungi, exposure to prolonged high temperature stress or large temperature shifts induces spore swelling, microcycle conidiation and morphic switches (see also Chapter 3). If these responses occurred in nature, they would maximize the chances of survival until conditions ameliorated. In *Ceratostomella fimbriata* and *Eurotium herbariorum* ascomata are produced at high temperatures and conidia at low temperatures, whereas in *Neurospora crassa* ascoma production occurs only at low temperatures (Turian, 1978). These temperature optima may reflect adaptations to different habitats but, like the varied temperature requirements for basidioma initiation, are not always easily related to the natural environment (Manachère, 1980).

However, differences in growth and reproduction in response to changes in temperature which can be associated with habitat occur between equine-pathogenic and non-pathogenic isolates of *Basidiobolus haptosporus* (Zahari & Shipton, 1988). Saprotrophic isolates grow poorly or fail to grow at 40°C (near the blood temperature of the horse), whereas a pathogenic isolate grows well. Optimum formation of conidia and zygospores in the pathogenic isolate occurs at higher temperatures (30–35°C) than in saprotrophic isolates (30°C), suggesting that distinct ecotypes exist in this species. Similarly, interactions between temperature and leaf age are important in controlling spore production by *Cronartium quercuum* f. sp. *fusiforme* on northern red oak (*Quercus rubra*) (Kuhlman, 1987). Aeciospore infection of young (50–75% expanded) leaves at 16°C leads to rapid urediniospore production, dissemination and disease spread early in the year. By contrast, aeciospore infection of fully mature (succulent) leaves at 21°C encourages teliospore formation later in the year.

Water availability Water availability is a major controlling factor during reproduction. With gradual desiccation, secondary metabolic pathways may be activated, so that until water stress fatally damages cellular processes, reproduction may be promoted. By contrast, long-term inundation may be lethal to many terrestrial fungi such as *Sclerotium cepivorum* (Leggett & Rahe, 1985). It is, however, essential for reproduction in many zoosporic species. Thus, levels of water availability required for fruiting and spore discharge are highly variable and related to habitat and life strategy.

Biomass production by *Paecilomyces farinosus*, a pathogen of the brown planthopper (*Nilaparvata lugens*), decreases linearly with decreasing water potential over the range −2.1 to −7.5 MPa, but blastospore formation is stimulated sixfold over the range −2.1 to −4.7 MPa before subsequently decreasing (Inch & Trinci, 1987). Blastospores are produced in the haemocoel of the insect, and their production may relate to the water potential of this environment. In *Basidiobolus haptosporus*, as water availability decreases, conidium production ceases at −1.1 to −1.6 MPa, but zygospore formation continues down to −3.9 MPa, and vegetative growth is maintained down to −4.8 to −5.6 MPa (Zahari & Shipton, 1988). Significant conidium formation can therefore only be expected in wet-season conditions, whereas vegetative growth and zygospore formation can continue in drier conditions, so enabling survival between wet seasons. This explains the common appearance of *Basidiobolus haptosporus* in amphibian and reptile dung during wet winter periods, when there has been consumption of arthropods which have grazed on mycelium and conidia. Similarly, the mucorine species *Syzygites megalocarpus*, which occurs on decaying basidiomata, produces sporangia early in growth, when water is freely available, but zygospores later when water availability decreases due to desiccation of the substratum (Kaplan & Goos, 1982).

In India in the rainy season *Taphrina maculans*, the leaf pathogen of turmeric, exhibits a biphasic diurnal cycle of ascus development and spore release (Upadhyay & Pavgi, 1979). Ascus elongation and ascospore discharge from its three to five layers of ascogenous cells begin early in the morning, being induced by leaf surface moisture and low temperature. Discharge then declines as temperature rises and humidity decreases, with a second peak occurring from the remaining ascogenous cells in the late evening when cooler, moist conditions return. By contrast, *Taphrina deformans*, which occurs in temperate regions, produces only a single layer of ascogenous cells each day, and these complete ascospore discharge in a single evening.

Irradiance Reproduction in the light is, for most fungi, a way of ensuring maximum chance of dispersal. Nevertheless, the importance of light for induction, developmental control and maturation of reproductive structures varies widely amongst fungi, and can do so even between closely related species or isolates of the same species. For example, *Ascochyta viciae*, *Phoma trifolii* and *Pilobolus crystallinus* require light for sporulation, whereas *Ascochyta gossypii*, *Pilobolus sphaerosporus* and some isolates of *Ascochyta pisi* do not. In the homothallic fungus *Phoma caricae-papayae*, light-dependent and light-independent sporulating strains are known, and similar properties are controlled by a single gene in *Cochliobolus miyabeanus* (Tan, 1978; Chang, 1980; Honda & Aragaki, 1983). Most Basidiomycotina require light for either basidioma induction or maturation, and in the microgravity conditions of the space laboratory, light is required for primordium formation in *Polyporus ciliatus* (Kasatkina *et al.*, 1980; Manachère, 1980).

Three distinct groups of wavelengths stimulate reproduction: ultraviolet (UV, 220–320 nm; peak 290 nm), near UV and blue (330–500 nm; peaks 370 and 450 nm) and yellow/red/far-red (550–675 nm; peak 600 nm). Near UV may extend from 300 to 390 nm, but its effects are generally linked to blue responses. Different wavelengths may be active in stimulating similar processes in fungi of the same family, and action spectra for the same wavelength groups can differ markedly (Table 5.3, Figs 5.5 & 5.6). In addition in some fungi, such as *Botrytis cinerea* and *Helminthosporium* species, sporulation may be reversibly promoted by near UV but inhibited by blue light, the degree of sporulation depending on the relative fluence rates of each wavelength.

The specific quality and intensity of irradiance, and the period of exposure required for induction, may change with time and stage of morphogenesis. For instance, fruiting in *Sphaerobolus stellatus* is promoted by blue light during the first 8 days of development, but between this time and glebal discharge at 14 days, blue light has no effect. By contrast, for 4–5 days preceding glebal discharge, yellow/red light can promote fruiting (Ingold & Peach, 1970). In *Sclerotinia sclerotiorum*, *Sclerotium delphinii* and *Sclerotium rolfsii*, sclerotium production increases with irradiance (Fig. 5.7). Primordia are induced on dark-grown colonies when these are subsequently exposed to white light, the degree of response decreasing with the duration of the previous dark incubation, and sensitivity being greatest during a period before primordia form naturally in either continuous light or continuous dark (Fig. 5.8). As the time for autonomous induction approaches, colonies become less sensitive, until they fail to respond (Humpherson-Jones & Cooke, 1977a). In the Chytridiomycete *Monoblepharis macandra*, dark-grown cultures develop gametangial thalli, whereas light-grown cultures develop only sporangial thalli (Marek,

Table 5.3 Examples of irradiance wavelengths stimulatory to reproduction (from Tan, 1978, © Edward Arnold).

Ultraviolet (200–320 nm)	Near ultraviolet and blue (330–500 nm)	Yellow/red/far-red (550–675 nm)
Conidiation	Sporangium initiation	Ascospore formation
Alternaria chrysanthemi	*Phycomyces blakesleeanus*	*Saccharomyces carlsbergensis*
Helminthosporium oryzae		*Leptosphaeria avenaria*
Stemphylium botryosum	Conidiation	
Pyricularia oryzae	*Aspergillus ornatus*	
Botrytis cinerea	*Penicillium isariiforme*	
	Trichoderma viride	
Pycnidium formation	Circadian rhythm of conidiation	
Ascochyta pisi	*Neurospora crassa*	
Septoria nodorum		
	Coremium formation	
Perithecium formation	*Penicillium claviforme*	
Pleospora herbarum		
Leptosphaerulina trifolii	Perithecium formation	
	Gaeumannomyces graminis	
Ascospore formation	*Nectria haematococca*	
Leptosphaerulina spp.		
	Ascospore formation	
	Saccharomyces carlsbergensis	
	Saccharomyces cerevisiae	
	Fruitbody initiation	
	Favolus arcularius	
	Schizophyllum commune	
	Sphaerobolus stellatus	
	Sclerotium initiation	
	Sclerotinia sclerotiorum	
	Sclerotium rolfsii	

1984). Light and dark cycles, rather than continuous light or dark, promote fruitbody formation in *Coprinus macrorhizus* and *Gelatinospora reticulospora* (Tan, 1978).

Light effects on reproduction may change with nutritional and other environmental conditions. For example, both sclerotia and apothecia of *Pyronema domesticum* form on a weak inorganic salts medium exposed to low light levels, but under relatively intense light apothecia only are formed. On rich media, sclerotia are formed in darkness, but apothecia fail to form even if mycelia are exposed to otherwise favourable light regimes (Moore-Landecker, 1987). Reduced aeration prevents light-induced basidiomata of *Lentinula edodes* from maturing, and in *Alternaria cichori* near UV light inhibits sporulation at 28°C

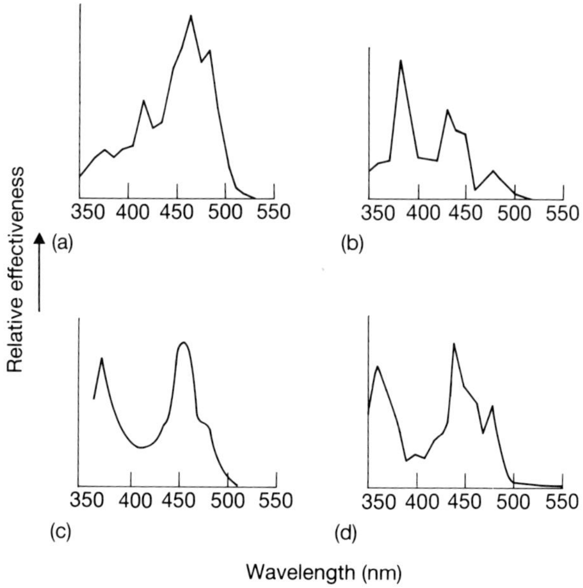

Fig. 5.5 Action spectra of some blue photoresponses. (a) Inhibition of the circadian rhythm of conidiation in *Neurospora crassa* (from Sargent & Briggs, 1967, © American Society of Plant Physiologists); (b) conidiation in *Trichoderma viride* (from Kumagai & Oda, 1969, © Japanese Society of Plant Physiologists); (c) coremium formation in *Penicillium claviforme* (from Faraj Salman, 1971, © Springer-Verlag); (d) perithecium formation in *Nectria haematococca* (from Curtis, 1972, © American Society of Plant Physiologists).

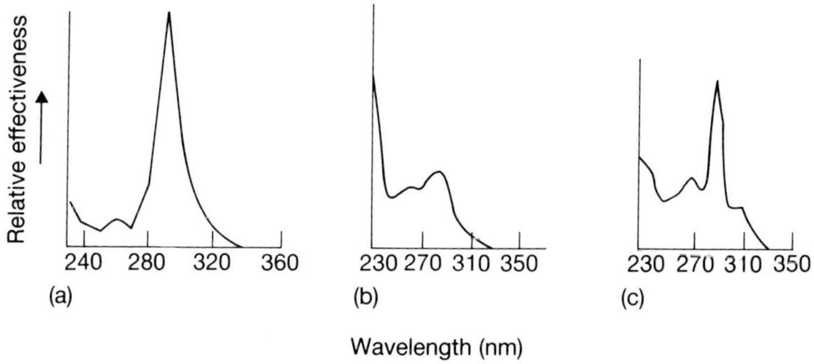

Fig. 5.6 Action spectra of some UV photoresponses. (a) Pycnidium formation in *Ascochyta pisi* (from Leach & Trione, 1965, © American Society of Plant Physiologists); (b) conidiation in *Alternaria dauci*; (c) perithecium formation in *Pleospora herbarum* (from Leach & Trione, 1966, © Pergamon Journals).

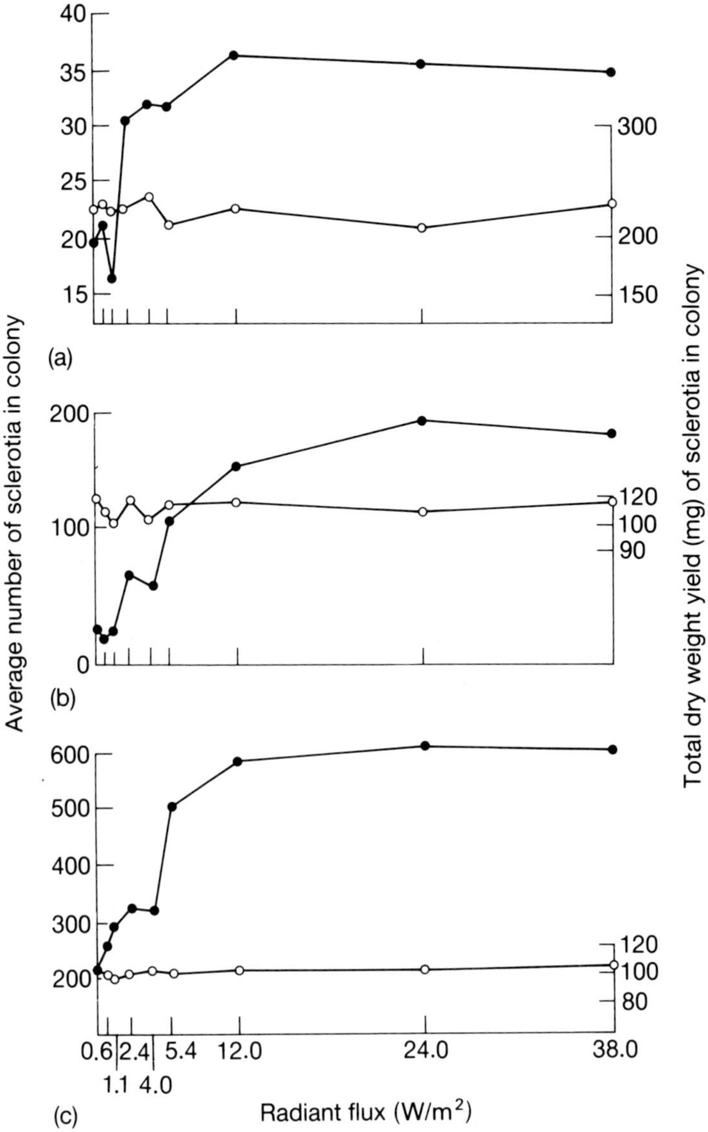

Fig. 5.7 Effects of continuous light of varying irradiance on sclerotium number (•) and total dry weight yield of sclerotia (o) in agar-grown colonies. (a) *Sclerotinia sclerotiorum*; (b) *Sclerotium delphinii*; (d) *Sclerotium rolfsii* (from Humpherson-Jones & Cooke, 1977a, © *New Phytologist*).

but has no effect at 17°C (Vakalounakis & Christias, 1986; Leatham & Stahmann, 1987).

Several types of photoreceptors have been identified. Near UV and blue photoreceptors are the most widespread, and consist of

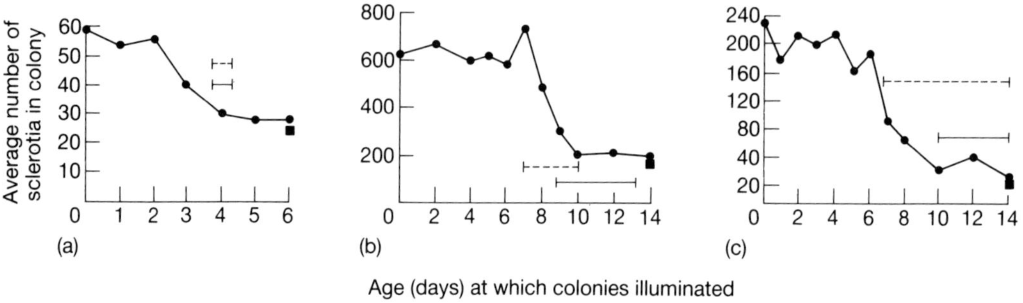

Fig. 5.8 Sensitivity of dark-grown colonies of different ages to subsequent exposure to continuous irradiance at 38 W/m^2. (a) *Sclerotinia sclerotiorum*; (b) *Sclerotium rolfsii*; (c) *Sclerotium delphinii*. Unbroken and broken horizontal bars indicate duration of sclerotium formation in continuous dark and light respectively. The solid square indicates the final number of sclerotia formed in continuous dark (from Humpherson-Jones & Cooke, 1977a, © *New Phytologist*).

membrane-bound flavoprotein moieties. Either a single near UV/blue photoreceptor (mycochrome), or possibly two different molecules, are involved in the reversible near UV/blue light control of conidiation in *Alternaria tomato*, *Botrytis cinerea* and *Helminthosporium oryzae*. It has been suggested that a far-red/red reversible photoreceptor might interact with mycochrome, or that a multiwavelength receptor, could be active in controlling sporulation in some fungi (Tan, 1978; Vakalounakis & Christias, 1985; Kumagai, 1988). UV photoreceptors consisting of a family of substituted cyclohexenone mycosporines have been found in over 70 species of Zygomycotina, Asco-Deuteromycotina and Basidiomycotina (Fig. 5.9). Light absorption by mycosporines is thought to lead to modifications in sterol metabolism (Arpin & Bouillant, 1981).

The primary events triggered by light absorption which then lead to induction of reproduction are unclear. In reactions involving blue light, flavoprotein–cytochrome b receptor complexes in the plasmalemma, endoplasmic reticulum and mitochondria may initiate a cascade of responses, including rapid changes in membrane potential, adenylate charge and, most notably, cyclic AMP levels (Table 5.4). Changes in cyclic AMP would have diverse effects on cellular metabolism, including a protein phosphorylation cascade. This would lead to different gene expression, protein synthesis, substrate utilization, secondary metabolite production and, consequently, reproduction (see Chapter 3). In this regard, changes in mRNA populations have been detected within 8 hours of light-induced conidiation in *Helminthosporium carbonum* (Thevelein *et al.*, 1987; Hannau *et al.*, 1989). Blue light triggers carotene synthesis in *Phycomyces blakesleeanus*, and retinol, which has a range of regulatory actions in animals – including induction of mRNAs and specific proteins – is

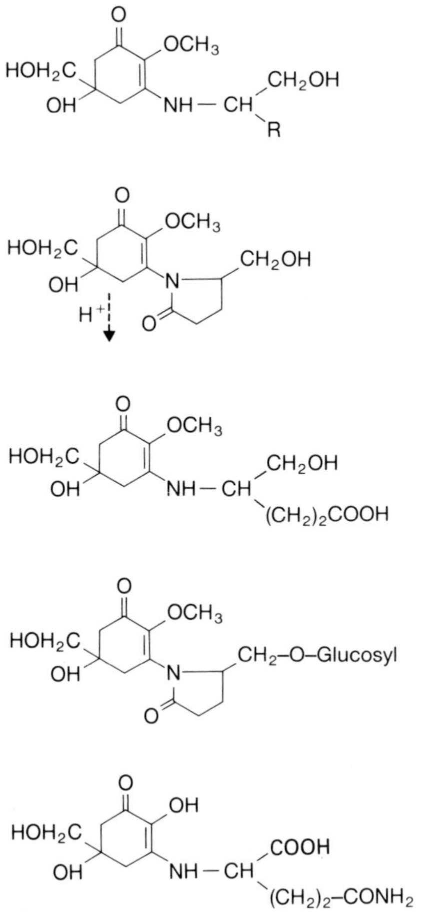

Mycosporine 1 (R = CH_2OH)
Mycosporine 3 (R = COOH)

Mycosporine 2 (ring form)

Mycosporine 2 (open form)

Mycosporine 2 glucoside

Normycosporine–glutamine

Fig. 5.9 Structures of some mycosporines (from Arpin & Bouillant, 1981, © Academic Press).

then formed. Retinol can also enhance carotene levels in *Phycomyces*, producing a self-stimulatory pathway. This may be part of the system responsible for the blue light stimulation of sporangiophore initiation and phototropism in this species (Galland & Lipson, 1987).

Rhythms and cycles

Rhythmic behaviour is most frequently seen in the form of regular periods of alternating growth and reproduction, and can be viewed as reflecting short-term adaptations to biotic or abiotic factors which themselves alter periodically. Most commonly, rhythms are controlled exogenously by the cyclic influence of light, temperature or humidity, and they consequently have a basic period of 24 hours (Fig. 5.10). However, longer cycles can occur depending on the interplay of environmental and nutritional influences.

More rarely, fungi exhibit endogenously controlled circadian rhythms

Table 5.4 Interrelationships between light, membrane changes, nucleotide levels and fruiting.

Species	Observation	References
Coprinus macrorhizus	Cyclic AMP increases fruiting	Uno & Ishikawa, 1976
Lagenidium giganteum	Cyclic AMP inhibits oosporogenesis	Kerwin & Washino, 1986
Neurospora crassa	Blue light causes photoreduction of cytochrome b in plasma membrane, ER and mitochondria.	Borgeson & Bowman, 1985
	Light induces changes in plasma membrane potential and decreases cyclic AMP.	Potapova *et al.*, 1984
	Light induces no changes in cyclic AMP levels.	Show & Harding, 1987
	Cyclic AMP inhibits conidiation	Harding, 1973
Phanerochaete chrysosporium	Cyclic AMP increases fruiting	Gold & Cheng, 1979
Phytophthora cactorum	Cyclic AMP inhibits oosporogenesis	Elliott, 1988
Phycomyces blakesleeanus	Blue light changes plasma membrane potential and decreases intracellular pH	Weiss & Weisenseel, 1990
Saccobolus platensis	Light induces fruiting and increases in cyclic AMP levels	Galvagno *et al.*, 1984
Schizophyllum commune	Cyclic AMP increases fruiting	Schwalb, 1978
Trichoderma viride	Light causes photoreduction of flavins and cytochromes.	Horwitz *et al.*, 1986
	Light induces increases in ATP levels and changes in plasma membrane potential	Gresik *et al.*, 1988
Several Deuteromycotina	Near UV light decreases membrane permeability	Mani & Swamy, 1981

with a periodicity of 24 hours. For example, rhythmic isolates of *Sclerotinia sclerotiorum* grown in darkness and at constant temperature produce sclerotium primordia in successive zones, with corresponding regular fluctuations in the rate of colony expansion (Figs 5.11 & 5.12). Locally, the greatest decrease in mycelial extension occurs in that area of the colony which is positioned immediately in front of any developing sclerotium (Fig. 5.13). This implies competition for some internal component needed for both sclerotium formation and hyphal growth. Localized reduction in the availability of such a component could also account for the stimulation of sclerotium formation at a mechanical barrier (Humpherson-Jones & Cooke,

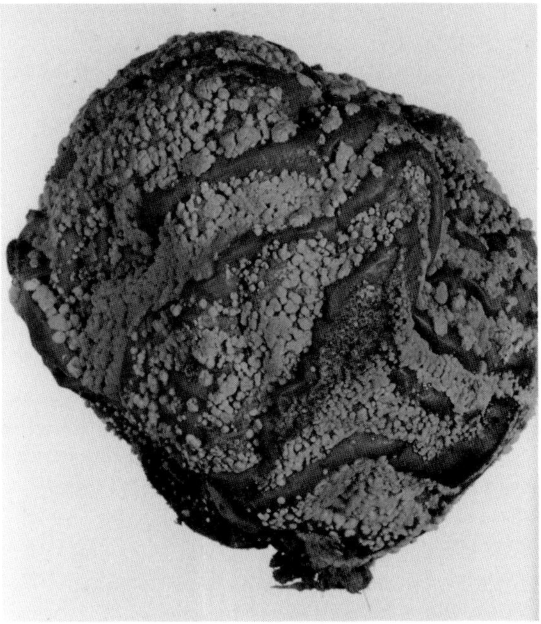

Fig. 5.10 Rhythmic sporulation of *Monilinia fructigena* on apple. Prominent raised zones of pale, fertile hyphae are separated by darker areas of sterile mycelium.

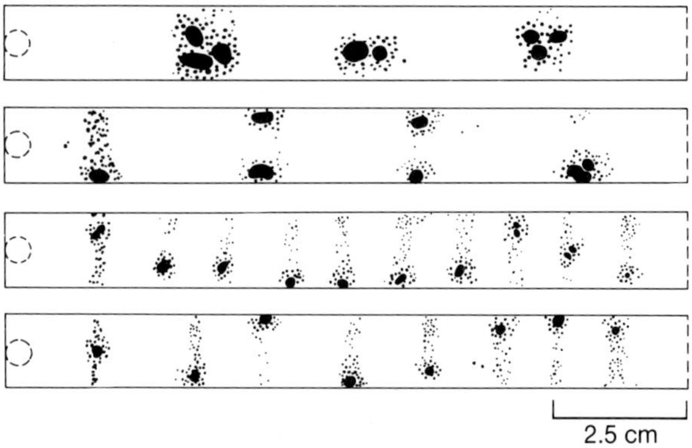

Fig. 5.11 Surface view diagrams of rhythmic sclerotium production by four isolates of *Sclerotinia sclerotiorum* growing in tubes in darkness. The inoculum disc is to the left, growth proceeding to the right. Sclerotia are black, and piled mycelium stippled (from Humpherson-Jones & Cooke, 1977b, © *New Phytologist*).

1977b). Internal competition for amino acids between coremial initials in *Penicillium claviforme* is thought to determine the regular spacing of fertile zones on agar (Watkinson, 1977).

Reproductive rhythms based on nutrient acquisition cycles can also be recognized, and may range in duration from days in laboratory

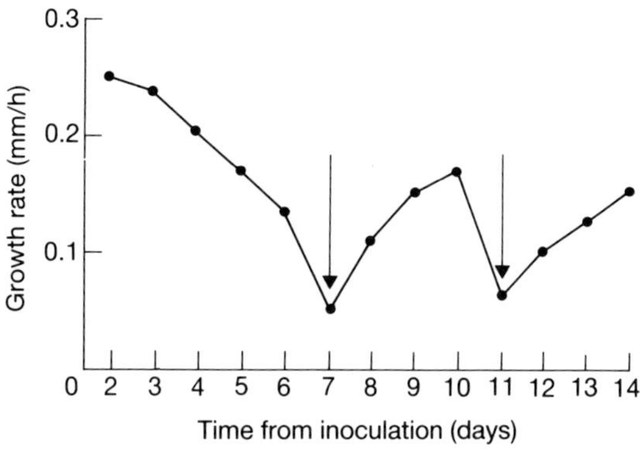

Fig. 5.12 Changes in rate of hyphal extension at the colony margin during sclerotium formation by a rhythmic isolate of *Sclerotinia sclerotiorum*. Arrows indicate the time of primordium appearance behind the margin (from Humpherson-Jones & Cooke, 1977b, © *New Phytologist*).

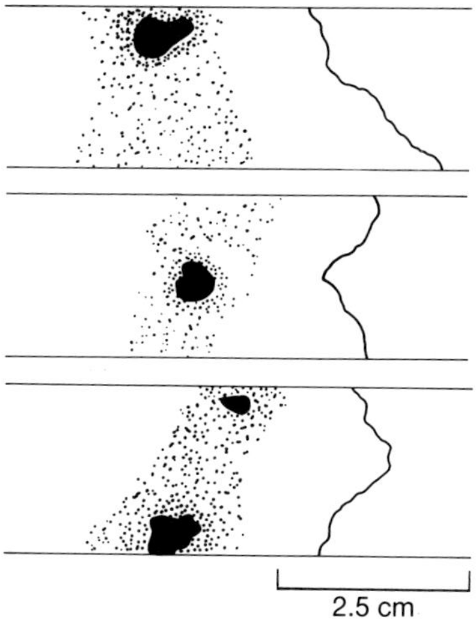

Fig. 5.13 Sclerotium formation by *Sclerotinia sclerotiorum* growing in tubes, showing the shape of the colony margin in relation to the position of the most recently formed sclerotium. Sclerotia are black, and piled mycelium stippled (from Humpherson-Jones & Cooke, 1977b, © *New Phytologist*).

culture to a yearly cycle in the case of some, often perennial, basidiomata on trees or in fairy rings in grassland. For example, *Fomitopsis pinicola* must utilize 0.2 m^3 of standing timber to provide enough nitrogen for annual basidioma production; the time taken to occupy and utilize this quantity of wood limits the rate at which the basidioma can be produced (Merrill & Cowling, 1966). Fairy rings formed by *Marasmius oreades* are characterized by two or three concentric annuli of abnormal grass growth, which spread out on a yearly basis with basidioma production occurring in regions of dead grass. Early in the year, nitrogen and phosphorus are mobilized from the substratum by the young expanding mycelium, leading to increased availability of these nutrients. The grass here consequently grows taller and greener, identifying the site of future basidiomata production. Subsequently, as differentiation and fruiting occur, the demand for nitrogen and phosphorus increases, resulting in the soil in this region becoming depleted. The vegetation, unable to satisfy its nitrogen and phosphorus requirements, weakens and dies, the effects of starvation being compounded by cyanide production by the fungus, and fungally induced water repellency of the soil. The following year the mycelium extends to unexploited regions, leaving dead or highly vacuolate hyphae behind (Fisher, 1977).

In nature, light–dark cycles are probably the most effective environmental factors that establish vegetative rhythms at or near the substratum surface which, in turn, determine rhythmic reproduction. After growth from an inoculum in the dark, light inhibits extension of surface hyphae, which causes increased O_2 consumption by the mycelium as a whole. Hyphae within the substratum, protected from the light but subjected to lack of O_2, continue to elongate at a constant rate. When darkness returns, hyphae within the substratum, which have by now grown under and beyond the inhibited surface hyphae, emerge and continue to extend at a constant rate until the onset of light conditions. The resultant bands of hyphae of differing structure and physiology have different propensities for reproduction (Fig. 5.14). A gradient of O_2 alone, as well as changes in temperature and humidity, can thus inhibit extension growth of surface hyphae and, if cyclic, may impose a reproductive rhythm.

Inhibition appears to involve changes in membrane permeability and ion fluxes. In septate fungi, hyphae lose polarity, the Spitzenkörper disappears, substrate uptake decreases leading to depletion of endogenous reserves, and staling begins. The consequence of these events is that inhibited hyphae become capable of reproductive differentiation. Here, formation of reproductive structures depends on rhythms induced in vegetative mycelia in situations where vegetative growth would not otherwise be inhibited by nutrient shortage alone. Such rhythms produce repeated hyphal zones capable of reproductive effort, and this increases the total number of reproductive structures that can be formed from a single colony. Further, reproductive potential is increased over the whole of the lifetime of the colony rather than merely at the end of vegetative growth (Schrüfer & Lysek, 1990).

The relationship of reproductive rhythms to endogenous and exogenous factors is at its most complex in some basidiomycetes (Manachère *et al.*, 1983; Ross, 1985). In *Coprinus congregatus* the

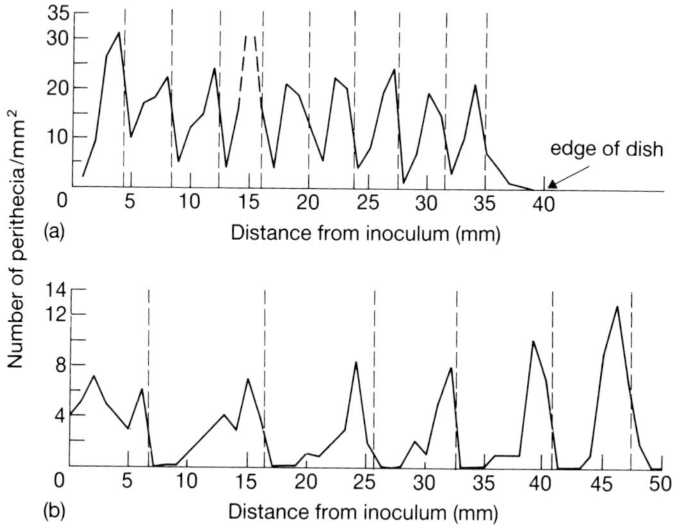

Fig. 5.14 Rhythmic production of mycelial bands and perithecia in: (a) *Chaetomium globosum*; and (b) *Pestalotia rhododendri*. Edges of mycelial bands are marked by vertical dashed lines (from Lysek, 1974, © *Naturwissenschaftliche Rundschau*).

timing of events differs with the strain studied but, when competence is gained, light and nutrient depletion are the two primary stimuli for rhythmic primordium induction. During a 12-hour light–12-hour dark cycle, primordia are formed every 4 days. The peripheral growth zone can be committed to primordium initiation by exposure to a single short light period, but another exposure to light is required 3 hours later to inhibit further primordium initiation. This inhibition lasts for 3 days, producing the rhythmic pattern of basidioma production. It has been suggested that hyphae within the zone of incipient primordium formation become deficient in soluble carbohydrate, such that cyclic AMP formation occurs, thus stimulating organogenesis. The appearance of cyclic AMP at the time of initiation has been observed in *Coprinus macrorhizus* (Uno & Ishikawa, 1976).

Other agarics also produce basidiomata in distinct flushes, and in *Agaricus bisporus* rhythmic changes in available carbohydrate in mycelia and compost are involved in the control of fruiting. Mathematical behavioural models indicate that endogenous substrates accumulate in the mycelium until a threshold level is reached which initiates basidioma formation. Basidioma growth then proceeds at the expense of these substrates until their levels fall below that required for initiation; fruiting then ceases (Chanter, 1979). In this regard, glycogen and trehalose are depleted during fruiting and recover between flushes, basidioma formation being accompanied by an increase in extracellular endocellulase activity, which falls only when the mushrooms are harvested (Hammond & Nichols, 1979; Wells *et al.*, 1987). This fall in enzyme activity may occur because, after picking, carbohydrate levels within the mycelium recover sufficiently to impose catabolite repression of endocellulase (Wood *et al.*, 1988). Consequently, it appears that flushing is controlled by regular cycling in the activity of pathways involved with the degradation of extracellular substrates, and with intracellular carbohydrate translocation, metabolism and storage.

Periodicity of spore release

By and large, spore release amongst fungi takes place at random, but annual, seasonal and daily cycles can occur. Seasonal cycles may correlate with fruitbody production under the influence of yearly changes in environmental conditions. For instance, basidiospores in the atmosphere are most numerous during the autumn with a smaller

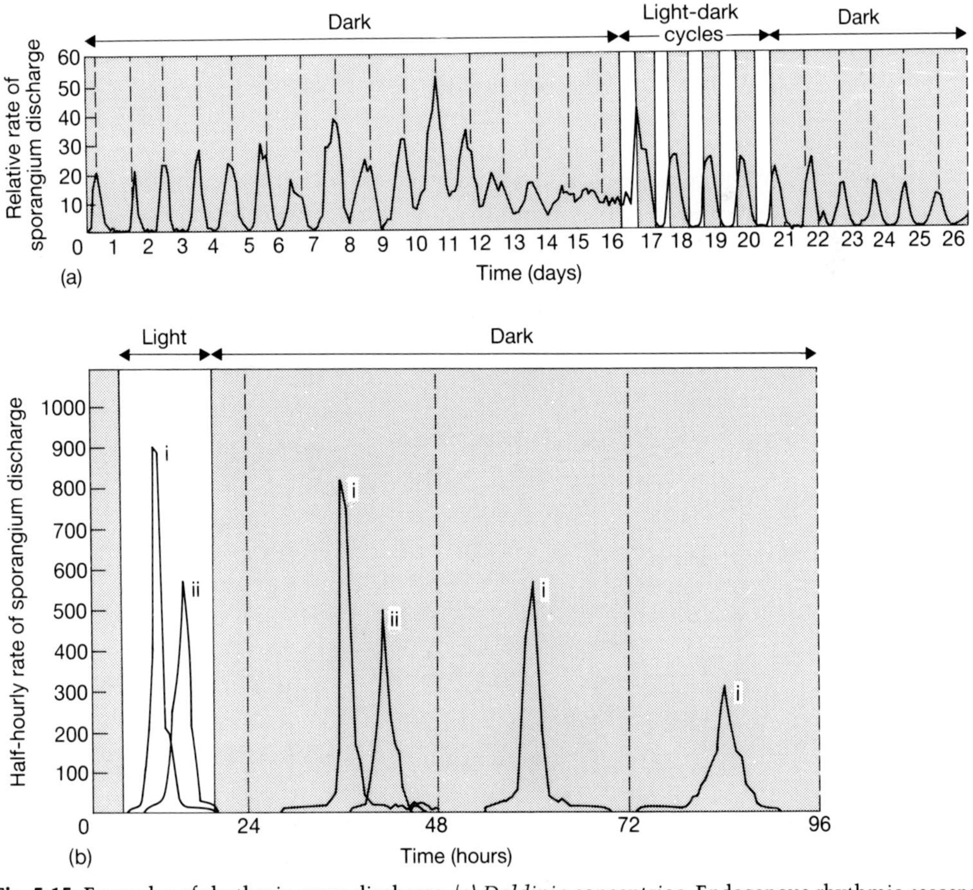

Fig. 5.15 Examples of rhythmic spore discharge. (a) *Daldinia concentrica*. Endogenous rhythmic ascospore discharge occurs in the dark following an initial 12-hour light–12-hour dark cycle. The rhythm decreases with time, losing the night-time maximum, but can be restored by renewing the light–dark cycle. Dashed lines indicate midnight (from Ingold, 1959, © *Nordic Journal of Botany*). (b) *Pilobolus sphaerosporus* (i) and *Pilobolus crystallinus* (ii) exhibiting a daytime maximum endogenous and light-dependent sporangium discharge respectively. Dashed lines indicate midnight (from Uebelmesser, 1954, © Springer-Verlag).

peak in the spring, which corresponds with the appearance of basidiomata in these seasons. The summer maximum for conidia of *Cladosporium* species is associated with higher summer temperatures favourable to mould growth, and spring and summer peaks for ascospore release from *Venturia inaequalis* and *Venturia pirina* with the first impact of rain on mature ascomata (Hyde & Williams, 1953; Latorre *et al.*, 1985).

However, most rhythmic systems have a period of 24 hours, and may be under either endogenous or exogenous control. In *Daldinia concentrica* and *Pilobolus sphaerosporus*, rhythmic spore release is maintained in darkness for several days at constant temperature and humidity following a previous 12-hour light–12-hour dark cycle (Fig. 5.15, p. 137). By contrast, *Pilobolus crystallinus* and the majority of other *Pilobolus* species have an absolute requirement for light to initiate sporangium development and to so facilitate spore release (Schmidle, 1951; Uebelmesser, 1954). Light does, however, stimulate sporulation in all these species and, in *Daldinia concentrica*, can be used to re-entrain the rhythm when it is lost after long periods in darkness.

Periodic spore release is often dependent on cyclic changes in light, temperature, humidity and surface moisture, and is frequently compounded by random variations in wind velocity. For example, light is the major factor controlling ascospore release in *Sordaria fimicola*, with almost no discharge occurring in short or long dark periods (Fig. 5.16). Daily temperature switches between 25 and 20°C can

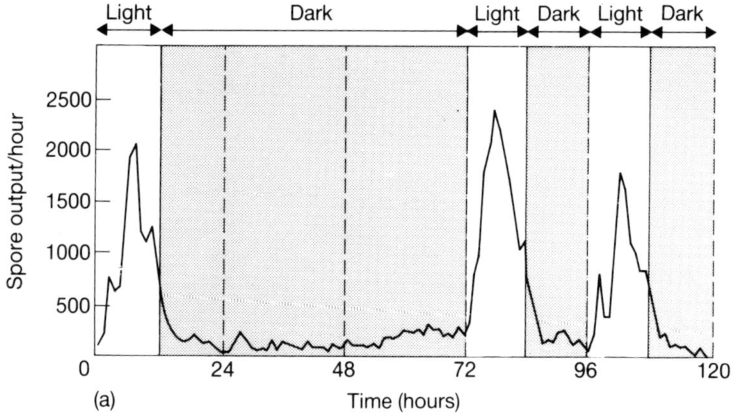

Fig. 5.16 Ascospore release by *Sordaria fimicola*. (a) Light-dependent daytime optimum for discharge (from Ingold & Dring, 1957, by permission of Oxford University Press); (b–d) interaction between temperature switches of 25 and 20°C and light, and their effects on discharge; (e, f) interaction between switches of 8 and 20°C and light; light is unable to induce discharge at 8°C (from Ingold, 1965, by permission of Oxford University Press).

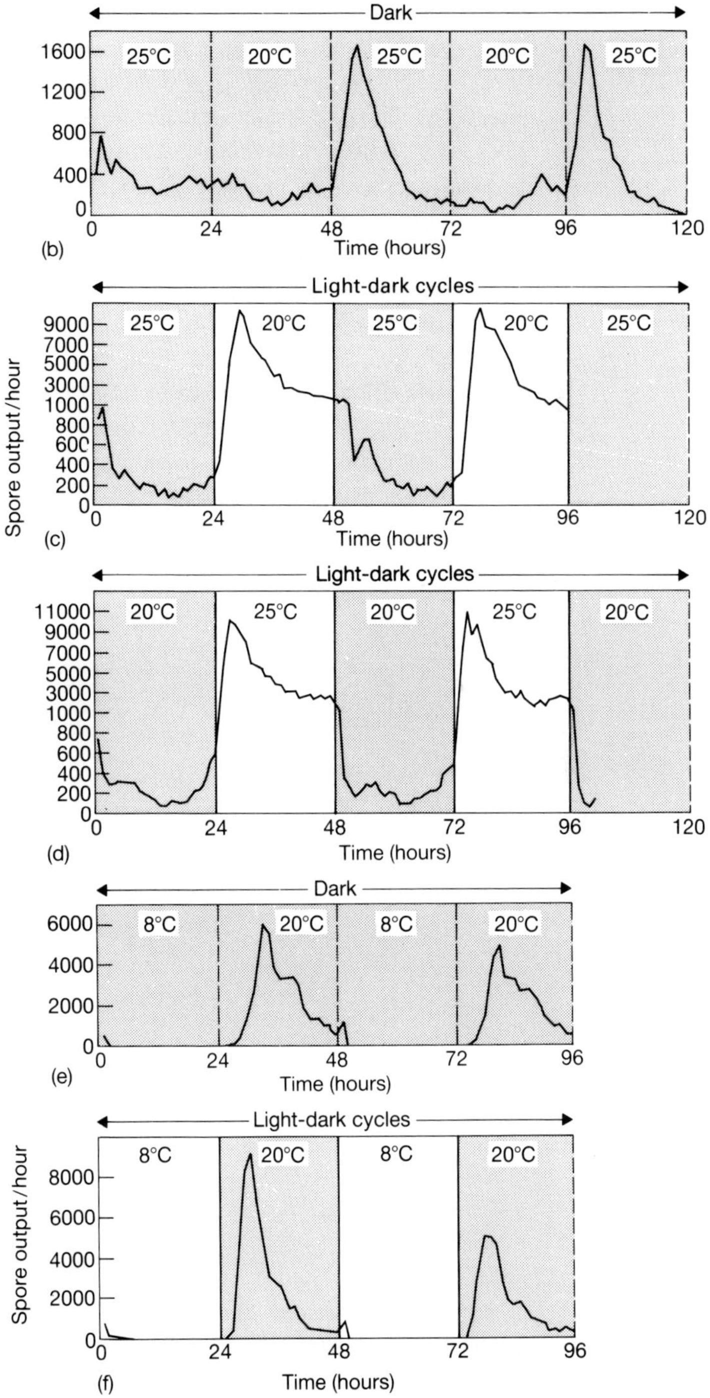
Dark
25°C
20°C
25°C
20°C
25°C
1600
1200
800
400
0
0
24
48
72
96
120
(b)
Time (hours)
Light-dark cycles
25°C
20°C
25°C
20°C
25°C
9000
7000
5000
3000
1000
800
600
400
200
0
Spore output/hour
(c)
Time (hours)
Light-dark cycles
20°C
25°C
20°C
25°C
20°C
11000
9000
7000
5000
3000
1000
800
600
400
200
0
(d)
Time (hours)
Dark
8°C
20°C
8°C
20°C
6000
4000
2000
0
(e)
Time (hours)
Light-dark cycles
8°C
20°C
8°C
20°C
8000
6000
4000
2000
0
Spore output/hour
(f)
Time (hours)

induce spore release in darkness, but exposure to light overrides this temperature-based rhythm. However, if the temperature falls to 8°C, spore release ceases even if normally suitable light conditions obtain (Fig. 5.16). Thus, in spring and summer, the daytime peak for ascospore release is light-, rather than temperature-controlled. By contrast, *Hypoxylon fuscum* exhibits a mainly nocturnal periodicity, with immediate inhibition of discharge by light of even low intensities (Fig. 5.17). More ascospores are also discharged by *Sordaria verruculosa* in darkness than in light, but here release is stimulated by a switch from darkness to light, and is inhibited by the reverse change (Fig. 5.17; Ingold & Marshall, 1963).

The role of daily changes in humidity and temperature in controlling rhythmic ascospore formation and release in *Taphrina* species has already been outlined, but there are numerous other examples of humidity-controlled spore discharge which differ in periodicity. For instance, *Peronospora tabacina* shows maximum spore release in

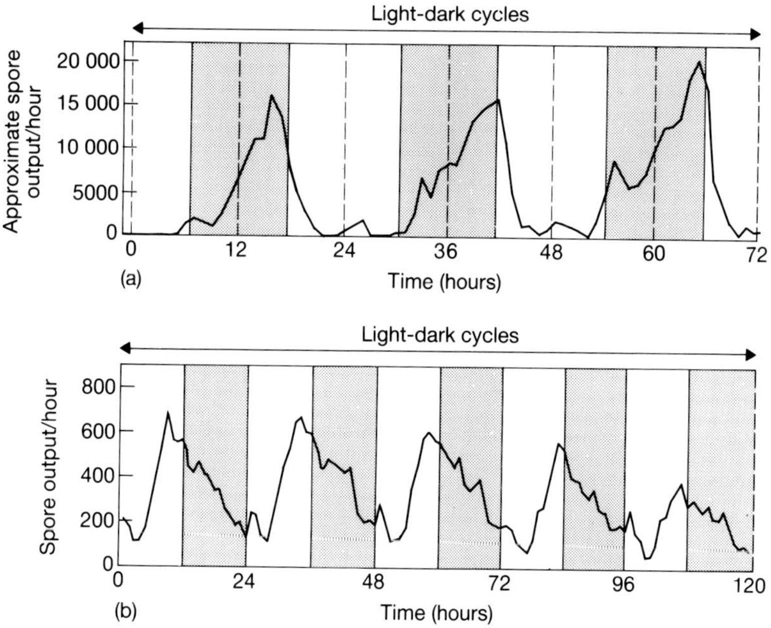

Fig. 5.17 Rhythmic ascospore discharge. (a) *Hypoxylon fuscum*, showing a nocturnal optimum and light-inhibited release (from Ingold, 1933, © *New Phytologist*); (b) *Sordaria verruculosa*, showing a daytime maximum of release, but with a greater total number of ascospores discharged in darkness (from Ingold, 1965, by permission of Oxford University Press).

the morning as the twisting of drying conidiophores leads to spore secession (Waggoner & Taylor, 1958). In temperate climates, in the absence of rain, dry-spored species such as the powdery mildews *Erysiphe cichoracearum*, *Erysiphe graminis* and *Podosphaera clandestina*, release the bulk of their conidia between 11.00 and 16.00 hours, usually a period of lowest RH, highest temperature, lowest leaf surface moisture and greatest wind velocity (Khairi & Preece, 1978). By contrast, the predawn maximum for ballistospore release in *Sporobolomyces* is related to high humidity (Zoberi, 1964).

In the rusts *Uromyces psoraleae* on the legume *Psoralea digitata*, and *Puccinia andropogonis* on *Zanthoxylum americanum* (prickly ash), aeciospores are released in darkness mainly between 20.00 and 24.00 hours. Before dusk they are dry and shrunken within the aecium, but as leaf surface moisture increases during darkness, they swell rapidly with a force sufficient to expel them from the constriciting peridium (Kramer *et al.*, 1968). Basidiospores of *Puccinia malvacearum* are released over a similar period, and this has survival value for these desiccation-susceptible xenospores (Carter & Banyer, 1964). By contrast, sporidia of *Chrysomyxa abietis* on Norway spruce (*Picea abies*), and of *Cronartium ribicola* on *Ribes* species, are released between 24.00 and 4.00 hours. This may both enhance their survival and facilitate maximum dispersal of them, either from vegetation around lakes and in swamps to upland areas, or from open to closed vegetation, on low-speed stable winds that occur at this time of night (Collins, 1976).

Periods of maximum wind velocity in the middle of the day are associated with maximum spore release from some fungi. For example, numbers of ustilospores of *Ustilago nuda* (loose smut of barley) reach a peak above heavily infected crops under these conditions (Sreeramula, 1962). Here, infected individuals are taller than surrounding healthy plants, and the spores, exposed as dry powdery masses above the general crop level, are easily dispersed by wind. Winds of low humidity, occurring characteristically around midday, effect greater release of dry-spored species such as *Trichothecium roseum* than those of high humidity (Zoberi, 1961).

Free water, in the form of rain rather than dew, is required for ascospore discharge in *Gaeumannomyces graminis*, *Venturia inaequalis* and *Venturia pirina*, with maximum release occurring during light periods for the latter two species (Washington, 1988). Splash dispersal by rain is important in slime-spored species, where spores enter and are carried by water droplets, although short-term peaks of spore release of dry-spored species, such as *Botrytis cinerea* and

Erysiphe graminis, can occur through the physical action of leaf disturbance by rain impact (Fitt *et al.*, 1989).

6 Propagules: factors affecting survival

Spores have attracted far more attention than any other fungal structure and this is reflected in the massive literature concerning their form and function (see Madelin, 1966; Weber & Hess, 1976; Turian & Hohl, 1981; van Etten *et al.*, 1983; Cole & Hoch, 1991). A great deal of this information is of relevance to ecophysiology, and it is in the area of spore biology, perhaps, that most common ground exists between the ecologist and physiologist. Accordingly, a generalized approach is adopted here and in the next chapter, the aim being to first categorize spores into functional groups and, second, to outline the means whereby their several functions are discharged. However, when considering the ways in which spores allow fungi to spread and survive, the contribution of alternative propagules should not be overlooked. Such units of dissemination and perennation, sometimes referred to confusingly as diaspores, commonly consist of individual or aggregated vegetative cells. They are not confined to asporogenic fungi, but are additionally produced by, and have ecological significance for, many typically sporogenic species.

Functional classification

The term spore is a broad one referring to a reproductive unit. What is or is not a spore is usually so self-evident that greater precision becomes unnecessary. Nevertheless, a general description, rather than a strict definition, has been attempted, a fungal spore being envisaged as a nucleate unit delimited from its parent thallus, lacking cytoplasmic streaming and vacuoles, having a low water content and metabolic rate, and being specialized for dispersal, reproduction or survival (Gregory, 1966). Putting aside the fact that some spores are vacuolate, and substituting the presence of enhanced levels of such energy-rich storage compounds as glycogen, lipids and trehalose, then equally well this description could apply to other propagules, for example sclerotia. These similarities at the cellular level indicate that all propagules, irrespective of gross structure, might be drawn into a single functional scheme, and that furthermore, given this, some important fundamental mechanisms are probably shared by them.

Spores can be divided into two functional groups: memnospores, which remain at or close to their place of origin and facilitate survival during periods of adversity, and xenospores, which become dispersed to more distant locations, there either initiating new mycelia or mating with existing individuals (Gregory, 1966). Each group can be further distinguished by means of a number of functionally correlated general properties (Table 6.1). Oospores, zygospores and chlamydospores are typical memnospores, whilst xenospores are represented by zoospores, sporangiospores, conidia, spermatia and basidiospores.

Table 6.1 General properties of spores assigned to two functional groups (after Gregory, 1966).

Characteristics	Memnospores	Xenospores
Physical	Various sizes but with a tendency to be large and thick-walled	Various sizes but with a tendency to be small, often minute, and thin-walled
	Some variation in shape but with a tendency to be spherical	Highly variable in shape, often with adaptations for flight, flotation and impaction
	May remain attached to parent mycelium, detachment often effected via lysis	Become quickly separated from parent mycelium, often by means of a specific launching mechanism
Physiological	Survival period relatively long, often considerably so	Survival period relatively short
	Often incapable of immediate germination even in favourable environmental conditions	Usually capable of immediate germination in favourable conditions
	Commonly dormant, requiring specific environmental stimuli to induce germination	Dormancy much less common

Obviously many fungi can produce spores of both functional types, and some may possess more than one kind of each type. In the latter category are macrocyclic rusts; some Ascomycotina, for instance *Nectria* species, the anamorphs of which may produce macroconidia, microconidia and chlamydospores; and some Hymenomycetes, for example *Coprinus* species, that form oidia and chlamydospores in addition to basidiospores. Spores can themselves give direct rise to further spores of either the same or different functional type. Thus, rust teliospores (memnospores) produce basidiospores; zygospores germinate via sporangium formation; conidia and zoospores can generate further xenospores by means of secondary spore formation and diplanetism respectively; and cells of some conidia, for instance of *Fusarium* species, may become transformed to chlamydospores.

Useful though this classification may be, it must be emphasized that the two functional groups are not mutually exclusive, that a degree of overlap exists between them, and that there are numerous examples of spores having a combination of memnospore and xenospore characteristics. Ascospores have an important dispersal function but are also durable and frequently dormant. Basidiospores of some Hymenomycetes and Gasteromycetes, and conidia of many mycoparasites, although having the physical characteristics of xenospores, will not germinate until they have received specific environmental stimuli, a physiological property characteristic of memnospores.

Further combinations of features are imparted to spores normally disseminated en masse with the whole or part of the fruit body. In the coprophiles *Pilaira* and *Pilobolus*, the protective sporangial wall confers memnospore characteristics on the enclosed spores which themselves, whilst being xenospores, are dormant until they have passed through the herbivore gut. Similar protection of xenospores is provided by the cleistothecia of some powdery mildews, the peridioles of *Cyathus* species, and by the fruit bodies of subterranean Gasteromycetes that are spread by mycophagous mammals.

Taking into account these kinds of variations, propagules other than spores can usually be identified as possessing either memnosporic or xenosporic properties. Xenopropagules are represented by the soredia, hormocysts and goniocysts of lichens, the glebal gemmae of *Sphaerobolus stellatus*, and the basidioma-like gemmae of *Omphalia flavida*. Most sclerotia, especially those of *Claviceps* and *Typhula* which exhibit dormancy, are clearly memnopropagules. Resting mycelia and quiescent hyphal cords and rhizomorphs can be viewed in the same light. As with memnospores, these survival stages may give rise to xenospores under favourable conditions.

Determinants of viability and germinability

A propagule's success depends on it remaining viable for a period appropriate to its function; this involves much more than merely staying alive. Viability should strictly refer to the potential of a propagule to give rise to one or more individuals, either directly or by production of additional propagules, regardless of the conditions that may be necessary for it to do so. Propagules can therefore be alive but non-viable, that is incapable of germination whatever the circumstances. By the same token viable propagules can be non-germinable because either internal or external conditions, or both, are unsuitable. Germinability is the capacity of a viable propagule to demonstrate its viability under a given set of conditions. There is thus a clear distinction between viability and germinability, which will now be examined further with regard to non-motile spores.

Despite their theoretical and practical importance these two properties of spore populations have commonly been either confused or erroneously equated. It is necessary to emphasize that viability has an absolute value which can only be expressed, in a population context, as the percentage of spores that are capable of germinating in conditions deemed to be favourable, and notwithstanding the time required for germination to take place (Louis & Cooke, 1985). Assessing it is frequently difficult, if not impossible. For example, incubation periods required may be unrealistically long, or mycelia arising from those spores germinating first in a population may prevent observation of subsequent events, especially if germination is spread through a protracted period. Thus, for largely practical reasons it is germinability which is measured, this being the percentage of spores that germinate within a specified time. By contrast with viability, this has an arbitrary value, since there is subjective choice of the incubation period, and its level is dependent on the rapidity with which viable spores within the population germinate under particular conditions. Hence, whilst it holds that loss of viability will result in a decrease in germinability, low germinability of a spore sample need not necessarily indicate low viability.

Behavioural models

Irrespective of their function, all non-motile spores, including sporangia of some zoosporic fungi, pass through the same fundamental physiological cycle. Viability first is low at spore initiation, rises as development and maturation proceed, and reaches an eventual maximum. A constant level is then maintained for a time but, as the spore senesces, viability gradually declines and is finally extinguished. This cycle is, of course, accompanied by a corresponding one for germinability, but the latter may depart significantly in detail from that for viability. Furthermore, viability–germinability patterns for xenospores differ markedly from those for memnospores. The principal variations can be illustrated using theoretical models of the behaviour of xenospore and memnospore populations (Fig. 6.1). In these, the spores are assumed to exhibit the 'ideal' physiological characteristics listed in Table 6.1, and to so exemplify the two functional groups. A second assumption is that they are behaving in an environment which, with one important exception, remains stable and optimally favourable to germination.

On the basis of these assumptions, the viability cycle of a typical xenospore population is relatively rapid. Maturation occurs quickly, the maximum viability phase is brief, and the decline to extinction is steep, although it still greatly exceeds the duration of the maturation phase. Initially, germinability follows an identical pattern, all viable spores being capable of germination during the maturation and plateau stages. However, the decline phase of germinability is shorter than that for viability because, as viable spores senesce, the environmental conditions necessary to facilitate germination become increasingly exacting, so that failure rate rises. Suitable alterations in the environment, which are beyond the assumptions in the model, could extend the germinability decline phase to equal that for viability.

The typical viability cycle of memnospores is much more extended, almost infinitely so in some cases. Maturation is slow, and the durability of these spores results in a maximum viability phase of long duration. Decline is long drawn out, with the curve for it becoming

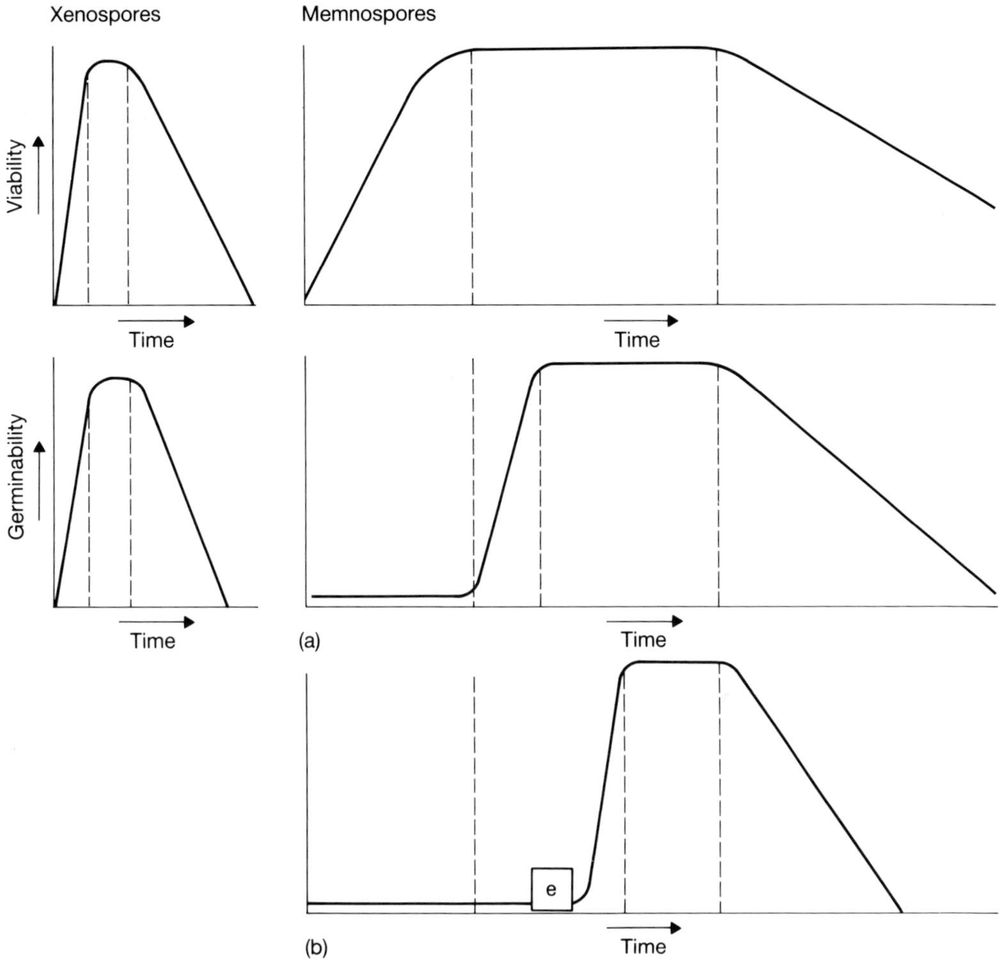

Fig. 6.1 Behavioural models for xenospores and memnospores illustrating changes in viability and germinability during development. For memnospores two examples are given for germinability. (a) Germination is achieved in the absence of a specific environmental stimulus; (b) germination requires occurrence of a specific environmental stimulus (e).

asymptotic as extinction is approached, a few individuals within the population retaining viability for a very lengthy period. By contrast with xenospores, germinability remains at zero during maturation and the first part of the maximum viability phase due to an inherent state of dormancy. Then, having achieved maximum viability, there are two possible germinability patterns. In the first, given the previously mentioned environmental stability, spores are slowly released from dormancy so that there is a gradual rise in germinability which is followed, after a period of maximum germinability, by a slow decline (Fig. 6.1). As with xenospores, and for the same reasons, this decline is much steeper than for viability, although some individuals will retain germinability for a considerable time. In the second situation, germinability remains at zero during the maximum viability stage until such time as dormancy becomes broken by a drastic, and often specific, alteration in prevailing environmental conditions (Fig. 6.1). On the return of conditions favourable to germination, germinability then rises virtually instantaneously, as all spores within the population have been simultaneously potentiated for germination. Due to the irreversible removal of their durability by the dormancy-breaking process, spores remain germinable for a relatively short period, and the decline phase is commensurately brief.

Longevity and survivability

Even typical xenospores can have a remarkably long life when stored under laboratory conditions where either availability of water is low or temperatures are reduced (see Sussman & Halvorson, 1966). For example, conidia of *Aspergillus* and *Penicillium* have been germinated after 10–12 years, basidiospores of *Tilletia foetida* after 35 years, and those of *Schizophyllum commune* after 52 years. Some of the longest-lived spores are those of Myxomycota, a span of 68 and 75 years being recorded for *Lycogala flavofuscum* and *Hemitrichia clavata* respectively. In the majority of cases such records probably have little relation to the ability to survive in nature. Data for survival periods in natural situations are scanty but, here too, longevity may be considerable. In soil, conidia of *Helminthosporium sativum* are still germinable after 20 months, and resting spores of the pathogenic chytrid *Synchytrium endobioticum* may persist for up to 30 years. This compares favourably with survival of soil-borne sclerotia, those of *Rhizoctonia tuliparum* for instance remaining germinable for 10 years or more. However, the ability of a spore which has so survived to then germinate in the laboratory may not be a true indication of its germinability in nature, where conditions are bound to be much more demanding. Estimates of longevity should therefore be approached with some caution unless germinability tests have been carried out under ecologically meaningful conditions. Similar reservations must be extended to those survivability studies which assess the resilience of spores subjected to environmental extremes, often far beyond those likely to be encountered normally, by means of post-treatment germination assays that employ regimes lacking in ecological relevance.

Longevity is governed by the group interaction of intrinsic and extrinsic factors. Intrinsic factors include spore size, shape, internal composition and organization, and wall structure, which for the most part can be regarded as being undynamic features. By contrast, extrinsic factors, such as temperature, irradiance, availability of water and the influence of microorganisms, are dynamic and hold the possibility for both wide and rapid fluctuation. The attributes of the ideal spore should therefore include an ability to survive potentially lethal extrinsic factors and a capacity to resist the deleterious effects of non-lethal, but generally unfavourable, conditions, together with the capability for rapid response to those combinations of factors that can support the germination process. However, the interactions involved are by no means fully understood. Moreover, it is often uncertain as to which intrinsic factors are truly adaptive with regard to longevity and which, whilst contributing to it, are fortuitous by-products of primarily dispersal-related evolutionary events.

Architecture, composition and metabolism

Spores have generally low metabolic rates. Dormant memnospores and dormant xenospores have been found to contain irregularly lobed mitochondria with low multiplication rates, which is possibly symptomatic of reduced endogenous oxygen tensions imposed by the impermeability of the spore wall (Hawker & Madelin, 1976; Furch, 1981). Other features that may contribute to low levels of metabolic activity are reduced internal hydration, and the spatial separation of nutrient reserves from the enzymes essential to their mobilization. As an additional generalization, it might be supposed that survival by spores of damaging environmental perturbations, or continuously imposed adverse conditions, would be contingent upon the maintenance of such a cryptobiotic state. Despite the wealth of available information on wall structure and composition, and to a lesser degree on cellular constituents, it is difficult to generate further broad statements on the interrelationships between intrinsic characters and survival. This is because, taken at large, spores are not necessarily either analagous or homologous structures, having not only widely differing functions but also an extremely varied genesis. The problem is exacerbated by a dearth of physiological data that can be related to specific architectural or organizational characteristics directly (see Hawker & Madelin, 1976).

In the absence of supporting physiological evidence, it is neither fruitful to directly presume purposes for the physical characteristics of a particular spore species, nor to assign them functions via analogy with the known roles of similar characteristics in other spores. This is especially true of cell wall features. The thick, and often multilayered, walls of many spores are undoubtedly biochemically inert, mechanically strong and impermeable, their main functions being to protect the cytoplasm from harmful extrinsic factors and to maintain the intrinsic cryptobiotic state. Equally, depending on their chemical composition, they can be highly permeable, thickness and structural complexity reflecting rather their roles as depots for either storage polymers or secondary metabolites. By contrast, as in powdery mildew conidia, the apparent delicacy of some thin cell walls belies their impermeable and highly protective character (McKeen *et al.*, 1967). With respect to secondary metabolites, it is frequently assumed that dark spore walls confer protection against short wave radiation. This may commonly be the case, but incorporation of pigments – particularly melanins and sporopollenins – into walls can more importantly impart both mechanical strength and antimicrobial properties to them.

Water relations and temperature

Water availability, temperature and the interplay of these, are of overriding importance for survival. The capacity of many spores to tolerate desiccation and subzero temperatures is widely exploited for maintenance of cultures and bulk inoculum production in various areas of fungal biotechnology. Fungal spores are not, however, exceptional in this respect so that, despite the long-standing fascination with the survival mechanisms involved, attention might instead be centred on the impact of more meaningful conditions. Spores that can be prepared successfully for long-term laboratory storage may frequently have a brief natural life span. Unfortunately much research on the effects of water and temperature on survival has been, and remains, concerned with environmental extremes and the influence of constant rather than fluctuating conditions. There is also considerable variation in methodology and terminology, especially regarding water relations, such that detailed interpretations and comparisons are made difficult.

Spore water and water availability. Ability to survive changes in external water availability will depend on the hydration characteristics of the spore and its equilibration behaviour with respect to its environment. It is accepted generally that spore water contents are below those of their parent mycelia, and that reduced hydration has potential survival value through mitigation of environmental stresses, particularly the effect of otherwise lethal temperatures. Numerous determinations of spore water contents have been made, mainly on air-borne xenospores of mesophilic fungi, and it has been shown that these vary within very wide limits (see Hawker & Madelin, 1976). However, such data are largely of only historical interest, since much of the observed variation can be attributed to differences in methods of culture and mensuration. Furthermore, water content estimations *per se* have little meaning unless accompanied by information on the degree to which the spores can equilibrate with external water conditions, the speed with which they do so and, most importantly, the solute potential of their aqueous component. The latter has obvious significance for the capacity of a spore either to restrain water loss to, or to take up water from, its environment. Determinations of these factors are rare, and for technical reasons are likely to remain so (Ayres & Paul, 1986). At the present time any discussion of spore water relations of necessity must be brief and tentative.

Most survival studies have revolved around the effects of atmospheric moisture on germinability of air-borne xenospores. The utility of relative humidity (RH) as an expression of water activity, and the problems involved in equating the latter with corresponding water potentials, have been pointed out previously; no such conversions have been made here (see Chapter 4). Bearing in mind the many differences in methodology, the non-uniformity of the RH regimes employed, and the limited number of species examined in detail, fungi seem to fall into four groups depending on the favourable effects on survival of either high RH, low RH or an intermediate RH; or on the unfavourable influence of an intermediate RH (Table 6.2). If these divisions can be accepted, then immediately it is apparent that: first, they correlate neither with spore type nor spore architecture; secondly, that fungi of contrasting life styles exhibit common behaviour; and thirdly, that fungi of similar life styles behave differently with respect to RH. These manifestly differential effects of RH defy detailed analysis, due to the aforementioned paucity of fundamental data on water relations, and ignorance as to spore metabolism during the survival phase. However, some broad conjectures can be made.

Dependence on high RH for survival probably reflects the damaging effects on spore membranes of low RH. For example, conidia of *Botrytis fabae* incubated at 42% RH suffer a rapid loss of infectivity compared with that at 92% RH. This is associated with a decrease in dehydrogenase activity and ATP levels, and an increase in loss of UV-absorbing compounds when conidia are subjected to leaching. These changes are indicative of disruption of membrane integrity, leading to a depletion of endogenous respiratory substrates. In *Botrytis cinerea*, leaching of conidia has been shown not only to remove carbon compounds but also to increase the respiration of those substrates that remain (Brodie & Blakeman, 1977; Harrison, 1983). These effects become more intense as temperature rises.

The conidia of many acervular, pycnidial and sporodochial fungi are produced and liberated within a mucilaginous matrix of exopolysaccharides and glycoproteins, which can influence spore behaviour in a number of ways (Ramadoss *et al.*, 1985). Amongst them is the conferment of protection against the damaging effects of low RH, even when the matrix is present as a highly attenuated film on individual dispersed spores (Nicholson & Moraes, 1980; Louis & Cooke, 1983, 1985). This beneficial effect on survival also extends to favourable RH

Table 6.2 Fungi grouped according to the relative humidity (RH) most favouring spore survival.

Grouping	Spore type	RH (%) for maximum survival	Temperature (°C) obtaining	References
High RH favourable				
Botrytis fabae	Conidia	92	22	Harrison, 1983
Leptosphaeria maculans	Conidia	100	25	Louis & Cooke, 1985
Some rust fungi	Urediniospores	40–100	Various	Leathers, 1961; Sussman & Halvorson, 1966
Some powdery mildew fungi	Conidia	40–100	Various	
Downy mildew fungi	Sporangia	53–76	10–25	Bashi & Aylor, 1983
Low RH favourable				
Alternaria porri	Conidia	14–38	Various	Rotem, 1968
Beauveria bassiana	Conidia	0 and 33	25	Clerk & Madelin, 1965
Helminthosporium oryzae	Conidia	20	31	Page, Sherf & Morgan, 1947
Microcyclus ulei	Conidia	0	24	Chee, 1976
	Ascospores	<80	24	Chee, 1976
Paecilomyces farinosus	Conidia	0 and 33	25	Clerk & Madelin, 1965
Sclerotinia sclerotiorum	Ascospores	33	25	Grogan & Abawi, 1975; Caesar & Pearson, 1983
Septoria nodorum	Conidia	<60	20	Griffiths & Peverett, 1980
Septoria tritici	Conidia	<65	25	Gough & Lee, 1985
Uromyces phaseoli	Urediniospores	31–43	Various	Schein & Rotem, 1965
Intermediate RH favourable				
Melampsora lini	Urediniospores	40–60	—	Cited by Teitell, 1958
Mycosphaerella pinodes	Conidia	62	25	Louis & Cooke, 1985
Intermediate RH unfavourable				
Aspergillus flavus	Conidia	32 and 85	29	Teitell, 1958
Aspergillus terreus	Conidia	32 and 85	29	Teitell, 1958
Endoconidiophora fagacearum	Ascospores	<50 and 95	24	Merek & Fergus, 1954
Fusarium moniliforme	Microconidia	33 and 100	25	Liddell & Burgess, 1985
Metarhizium anisopliae	Conidia	0–12 and 76–92	25	Clerk & Madelin, 1965
Rhizopus oryzae	Sporangiospores	0, 60 and 100	25	Akushie & Clerk, 1981

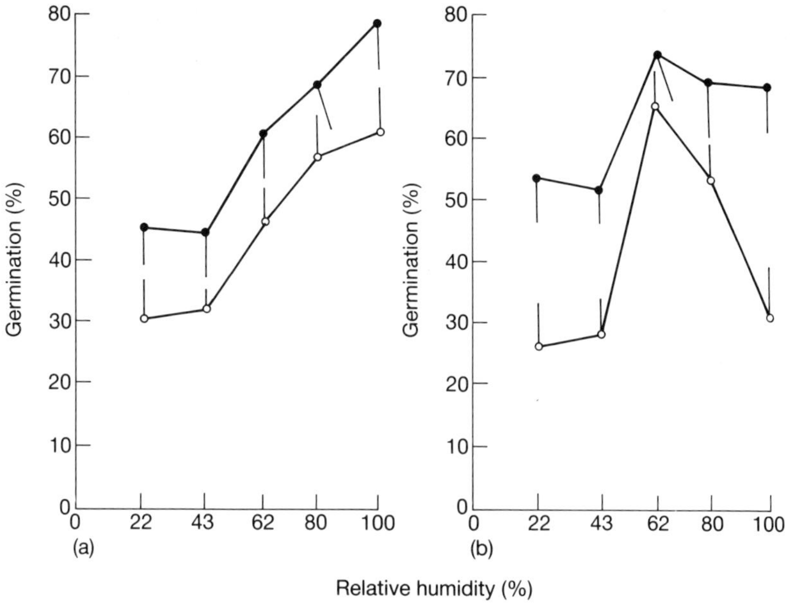

Fig. 6.2 Effect of conidial matrix on survival over a range of RH. Conidia with matrix (●) or with matrix removed by washing (○) were incubated at 25°C for 21 days at each RH and were then germinated on agar. (a) *Leptosphaeria maculans*; (b) *Mycosphaerella pinodes* (from Louis & Cooke, 1985, © British Mycological Society).

ranges (Fig. 6.2). The mechanisms involved are unknown, but the matrix may either protect the cell surface in some way so as to restrict water loss, or it might act as a metabolic inhibitor, so reducing the severity of stress-induced damage. Other kinds of spores, for instance ascospores, are also dispersed with either a coating or attachment of mucilage (Ingold, 1978). It is possible that this may have a similar protective role.

Where high RH has an adverse effect on longevity this is usually attributed to enhanced respiration rates due to increased endogenous hydration. There would then be a rapid depletion of respiratory substrates, losses accelerating with increasing temperature. If this is the case, then it points to the existence of a fundamental metabolic difference, rather than a structural one, between these kinds of spore and those that survive best at high RH. Such a view is supported by the observation that survival of *Septoria nodorum* conidia at high RH, which is normally poor, is improved by the presence of conidial matrix, possibly because of its inhibitory effect on respiration (Griffiths & Peverett, 1980).

Speculation becomes even more difficult when examples are considered of the beneficial or harmful results of incubation at intermediate RH ranges. Here can be found bimodal, and in the instance of

Rhizopus oryzae, trimodal RH-survival responses (Table 6.2). Especially perplexing are cases in which a very narrow median RH range is unfavourable to survival. For example, at 29°C, conidia of *Aspergillus flavus* and *Aspergillus terreus* survive poorly at 75% RH, but are affected only slightly at 73% and 77% RH. A temperature increase to 45°C shifts the adverse RH to 81% (Teitell, 1958). Survival of *Metarhizium anisopliae* conidia is shortest at around 45% RH but leaching, even for 1 minute at 0°C, raises the least favourable RH to 65%. These deleterious RH effects are reduced if incubation is carried out in an O_2-free or CO_2-enriched atmosphere. It is thus possible that an unfavourable RH releases those constraints on respiration that impose quiescence on the spore, but why this should occur at a median RH is not known. The additional effect of leaching might depend on the removal of osmotically active compounds, which would then, in turn, affect the hydration characteristics of the spore (Clerk & Madelin, 1965).

By contrast with the varied behaviour of air-borne xenospores, survival of memnospores generally seems to be favoured by low water potential. For example, chlamydospores and oospores of soil-borne plant pathogens have a greater longevity in dry than in wet soils, although in part this may be due to reduced microbial antagonisms at low potentials (Cook & Duniway, 1981). As with xenospores, memnospore structure may have little bearing on survival. This is the case with chlamydospores of some *Fusarium* species, between which there are no obvious differences, but which show contrasting degrees of resistance to desiccation (Sitton & Cook, 1981). With the notable exception of *Phymatotrichum omnivorum*, the sclerotia of which are killed by air drying, sclerotia resemble memnospores in that they are durable when dry but lose viability in moist soil (Abawi *et al.*, 1985). Some sclerotia become inviable most rapidly in conditions of fluctuating water availability. Possibly, a state of high and steady hydration diminishes nutrient reserves, whilst fluctuations induce highly wasteful metabolic shifts (Cooke & Al-Hamdani, 1986). Perhaps more seriously, wetting–drying cycles cause severe metabolite leakage, which not only exhausts reserves but also stimulates microbial antagonists within the surrounding soil (Smith, 1972; Coley-Smith *et al.*, 1974).

Adverse temperatures. Almost by definition, the majority of spores can be expected to retain viability at temperatures that would be inimical to vegetative development. For instance, the ubiquity of thermophiles indicates survival of their spores at median temperatures, and spores of many mesophilic species are capable of surviving both subzero temperatures and those above the normally lethal point of 60°C. General spore properties conferring durability have been listed already and, where adaptations to temperature extremes are concerned, to these may be added membrane characteristics and other features of the kind found in similarly adapted vegetative cells (see Chapter 4). What is perhaps surprising is the sensitivity of spores of some mesophiles to even mild chilling. For example, young conidia of *Botryodiplodia ricinicola* can germinate in distilled water, but if they are first chilled at 5–10°C for up to 9 hours in aqueous suspension, then germination becomes dependent on exogenous nutrients. Similar chilling of previously air-dried conidia eliminates germinability even in the presence of exogenous nutrients. Germinability can be restored by slow rehydration in moist air, but rewetting is fatal (Ogunsanya & Madelin, 1977). This behaviour indicates an extreme sensitivity of spore membranes to moderately depressed temperatures, which presumably results in catastrophic metabolite loss via leaching. *Rhizopus sexualis* sporangiospores similarly lose germinability if chilled at 3°C for 3 weeks, the effect increasing at 0 and −1°C. Again, this may be due to severe leaching, since the decline is retarded by provision of exogenous nutrients (Dennis & Blijtham, 1980).

The damaging effect of reduced temperatures is alleviated by the presence of conidial matrix. Removal of matrix from spores of *Leptosphaeria maculans*, *Septoria apiicola* and *Sphaerellopsis filum* renders them more susceptible to freezing at −5°C. As spores age, the protective role of matrix becomes more crucial to their survival; matrix-free, 10-day conidia of *Mycosphaerella pinodes* retain germinability at −5°C equally as well as those with matrix, but matrix-free, 17-day conidia fail to do so (Louis & Cooke, 1983, 1985).

Irradiance Several kinds of observation suggest that heavily pigmented spores are more resistant to damaging radiation than are hyaline spores. Floristic studies on phylloplane communities, desert soils receiving either high insolation or a spectrum of radiation from nuclear weapons, and γ-irradiated soils, have revealed a consistent preponderance of dark-spored species. Also, it has been demonstrated that, in *Cochliobolus sativus*, UV radiation is less damaging to a dark-spored than to a white-spored strain (Durrell & Shields, 1960; Tinline *et al.*, 1960; Johnson & Osborne, 1964; Pugh & Buckley, 1971). Melanins absorb various kinds of radiation and dissipate energy primarily by undergoing reversible increases in free radicals. This may spare the spore membrane from the damaging effects of irradiation-induced free radicals (Bell & Wheeler, 1986). Whilst it is obvious that wall pigments can shield the protoplast, it should not be assumed that pigmentation is an adaptation for that specific purpose, nor that it is the sole available defence. First, as has been mentioned already and will be discussed later, wall pigments possibly have a number of more important survival roles, and radiation absorption may be only incidental to these. Second, possession of physiological systems enabling rapid damage repair may be of greater value than a physical capacity for damage limitation. There is abundant evidence that repair of UV-irradiated DNA occurs via photoreactivation, a process which is dependent upon light-activated enzymes (Jagger, 1958). This applies not only to pigmented spores but also to lightly pigmented or colourless conidia, sporangia and ascospores (Buxton *et al.*, 1957; Sussman & Halvorson, 1966; Puhalla, 1973; Caesar & Pearson, 1983; Rotem *et al.*, 1985). Finally, an extracellular matrix, rather than the spore wall itself, can provide protection against UV radiation, possibly via absorption, but probably indirectly through some undetermined influence on spore metabolism (Louis & Cooke, 1983, 1985).

The fungicidal action of shorter-wave UV radiation (250–270 nm) together with the weaker but none the less damaging effects of UV at longer wavelengths (265–340 nm) have been demonstrated repeatedly (Owens & Krizek, 1980). Very few attempts have been made to bring these observations within the realm of ecological reality, and there seems to be little general awareness of either what kinds and amounts of short-wave radiation impinge on fungi in nature, or of how damaging they might be. These questions mainly relate to air-borne xenospores of all kinds, but possibly most importantly to phylloplane saprotrophs or leaf parasites, the spores of which may experience radiation at all stages of development from initiation to germination. Despite the huge volume of information on take-off, flight, arrival and establishment of leaf inhabitants, supporting data on survival are rare (Waggoner, 1983). However, it is apparent that, depending on circumstance, daytime exposure can reduce viability, and that sunlight has a direct damaging effect on spores quite distinct from its action via alteration of temperature and RH (De Weille, 1960; Bashi *et al.*, 1982).

Germicidal UV wavelengths (UV-C, 250–280 nm) are mostly filtered out by the atmosphere, so that solar UV radiation reaching the Earth's surface largely consists of a mixture of UV-B (280–320 nm) and UV-A (320–400 nm). The quantity and quality of UV dosage at any point depends on latitude, altitude, aspect and season, and is further locally modified by a multitude of factors, including meteorological conditions and vegetation characteristics (Johnson *et al.*, 1976). Although UV-B and UV-A radiations are much less damaging than UV-C, they still exert a strong, cumulative effect on spore survival. This is evidenced by the observation that summer-dispersed ascospores of *Sclerotinia sclerotiorum* distributed within a bean crop (*Phaseolus vulgaris*) lost viability much more rapidly on topmost leaves than on leaves within the canopy, this difference being independent of temperature. Sheltering topmost leaves with plastic films that reduced UV radiation within the range 300–400 nm significantly increased ascospore survival on them. In the laboratory, ascospores irradiated at 280–269 nm for just over 7 days with simultaneous photoreactivating light were severely affected, cumulative UV dosage being $1.76 \times 10^6\,J\,m^{-2}$. This was close to cumulative dosage rates of $2.42 \times 10^6\,J\,m^{-2}$ estimated for topmost bean leaves over a 6-day period in the field, the time at which ascospore survival in the field declined most rapidly (Caesar & Pearson, 1983).

Similar results have been recorded for conidia of *Alternaria solani* and urediniospores of *Uromyces phaseoli* exposed to solar radiation (Rotem *et al.*, 1985). Conidia were killed by a cumulative natural UV (>290 nm) dosage of $3.7–5.5 \times 10^6\,J\,m^{-2}$ received over 4–5 days, and urediniospores by $3.4–5.0 \times 10^6\,J\,m^{-2}$ over 2–3 days. Sporangia of the downy mildews *Peronospora destructor* and *Peronospora tabacina* have been found to be highly sensitive to solar radiation, remaining viable for only 6 hours in even moderate sunlight (280–630 $W\,m^{-2}$). Survival in strong sunlight (630–900 $W\,m^{-2}$) was not significantly shorter, which has been taken to indicate that the killing effect is not gradual, but that it acts abruptly after reception of a critical dose (Bashi & Aylor, 1983).

If, as seems likely, these examples of photosensitivity are typical of air-borne spores at large, then survival after a protracted flight in all but the most favourable insolation conditions may be much lower than might be supposed. Furthermore, although a spore may arrive successfully, the time-window for initiation of germination may be drastically narrowed on insolated substrata. Some of the ecological consequences of this kind of behaviour are illustrated by downy mildews of irrigated desert crops. Here, disease spread is favoured by cloudy weather, but is restricted by sunlight, since most sporangia are released early in the day and are liable to be killed by solar radiation. Sporangia remaining attached to their sporangiophores are less photosensitive, so that those liberated late in the day, whilst fewer in number than those detaching earlier, are more effective in inciting infection (Bashi *et al.*, 1982; Rotem *et al.*, 1985).

Microbial effects

As well as being influenced by abiotic factors, spore viability may be adversely affected by microbial activity. The degree to which this occurs will depend on physical and chemical habitat features, and the resultant variety and density of the general microbiota. Effects are most severe where spores remain for considerable periods in litter or soil horizons in which microbial activity is intense. However, spores spending even a brief sojourn in more sparsely inhabited locations, for example the phylloplane, may also suffer loss of viability (Omar & Heather, 1979). It is important here to distinguish between mycostatic factors and those that kill spores. Where it occurs, mycostasis is a more or less widely and continuously imposed influence generated by microbial activity, which prevents germination but does not necessarily, at least in the shorter term, result in loss of viability. It has been viewed historically as an aspect of microbial antagonism, but is perhaps better considered as a form of stress, the alleviation of which allows germination to proceed. Moreover, it might also be looked upon as being beneficial in situations where it prolongs longevity of spore populations without reducing viability by a significant degree. Given these features, mycostasis becomes an aspect of externally imposed constraints on germination, and will be dealt with as such in Chapter 8. By contrast, killing factors are the bases of intense, direct or indirect antagonisms that lead to spatially restricted destructive disturbance, resulting in loss of spore biomass.

A very broad range of microorganisms is capable of destroying fungal spores in a variety of habitats. It is, however, the soil which has commanded most attention, principally because spore antagonists may prove useful for biological control of soil-borne pathogens. For example, oospore populations of economically important species of *Phytophthora*, *Pythium* and other Peronosporales, are commonly extensively parasitized by Chytridiomycetes, Oomycetes, Deuteromycotina, Actinomycetes and bacteria, zoosporic antagonists being favoured by high soil water potentials (Sneh *et al.*, 1977; Wynn & Epton, 1979; Daft & Tsao, 1984; Sutherland & Lockwood, 1984). High water potentials also encourage the activities of spore-consuming Myxobacteria, for instance *Polyangium* species, and giant soil amoebae of the family Vampyrellidae (Homma & Cook, 1985). Variously, these can attack conidia of *Alternaria*, *Cochliobolus* and *Thielaviopsis*, and *Fusarium* chlamydospores. A plasmodium or pseudopodium contacts the spore and quickly perforates it by means of highly localized enzyme action, which frees a minute disc of wall material (Fig. 6.3). Spore contents, including septa, may then lyse, or fungal cytoplasm may be withdrawn to be absorbed by that of the antagonist. Plasmodia of Myxobacteria may flow into spores via the perforations (Old, 1977; Anderson & Patrick, 1978; Homma, 1984). In some situations, for example woodland habitats, large numbers of spores may be ingested and digested by myxomycete swarm cells, although the ecological impact of this activity has not been assessed (Madelin, 1984).

The impressive longevity of many soil-borne sclerotia should not be taken as meaning that they are any less susceptible to microbial antagonisms than spores. Indeed, their size and nutrient richness

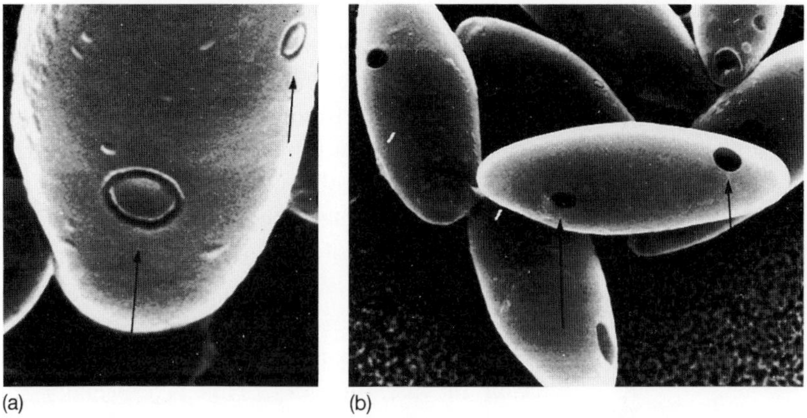

Fig. 6.3 Destruction of spores by soil amoebae. (a) Conidium of *Cochliobolus sativus* with annulations caused by contact with mycophagous amoebae; (b) conidia with large perforations resulting from removal of discs produced via annulation (from Anderson & Patrick, 1978, © American Phytopathological Society).

makes them potential habitats for other microorganisms, including fungi. Mention has been made already of the effect of wetting–drying cycles on nutrient leakage from sclerotia, and its stimulation of the local microflora, many members of which will be antagonists. There is evidence that similar naturally occurring fluctuations promote rapid sclerotial decay, especially in organic soils where a large, varied microbial population is active (Leggett *et al.*, 1983; Leggett & Rahe, 1985). More specifically, there are numerous reports of destruction of sclerotia by bacteria, antibiotic-producing fungi and mycoparasites (Aluko & Hering, 1970; Naiki & Ui, 1975; Baard *et al.*, 1981; Whipps *et al.*, 1988). Mycophagous amoebae can also perforate and destroy sclerotial cells, and larger sclerotia, for instance of *Sclerotinia* species, may become infested with fungus gnat larvae (Mycetophilidae) if already partially decomposed by microorganisms (Turner & Tribe, 1976; Homma & Ishii, 1984).

Soil-borne propagules are not without defence against microbial attack; sclerotia of *Rhizoctonia* species for instance produce a number of antibacterial and antifungal metabolites (Gladders & Coley-Smith, 1980; Burton & Coley-Smith, 1985). Furthermore, sclerotia and many conidia are melanized, and this probably contributes to their resistance. Melanins have been shown to physically strengthen fungal cell walls, and to protect them from lysis by inhibiting the action of chitinases and glucanases, enzymes which must have a key role in microbial destruction of propagules (Bloomfield & Alexander, 1967; Bull, 1970). Although sites of melanin synthesis and deposition, and its bonding characteristics with cell wall components, are not always clear, it does seem to have a wide and important ecological role (Baard *et al.*, 1981; Rast *et al.*, 1981; Griffiths, 1982). However, two points might be made here. First, other pigments, especially sporopollenins, may have similar functions, although little is known about their distribution amongst fungi (Gooday, 1981). Second, a notable feature of mycophagy in Myxobacteria and vampyrellid amoebae is that there appears to be a preference for melanized cells. An effective defence against one group of antagonists may, therefore, be useless against another.

Spores do survive in spite of the odds against them. Their success depends, at least in part, on high wastage rates being offset by production in vast numbers. This is especially so for xenospores, for which longevity largely reflects survival by lottery. For memnospores the situation is different, and many display insensitivity to environmental stresses or perturbations. Here, longevity reflects survival through dormancy, an important feature of propagule biology which will now be examined.

7 Dormancy and activation

Propagules have three interlinked roles: dissemination through space, survival in time, and transmission of genetic information to vegetative or sexual receptors. In nature it is commonly the memnopropagule which acts as a temporal bridge, this function often being associated with dormancy. The latter is a widespread biological phenomenon, in which there is preservation of the soma under conditions unfavourable for, or inimical to, its normal metabolism. However, although a scheme for a unified concept of dormancy amongst animals, plants and microorganisms has been promoted, its precepts with respect to fungi are questionable, and bear re-examination here.

Concepts Dormancy has been defined as any rest period or reversible interruption of phenotypic development of an organism; within it two states have been distinguished (Sussman, 1965a). First, there is constitutive dormancy, a condition in which development is prevented by some innate property of the dormant stage, for example a diffusion barrier, a metabolic block or a self-inhibitor. Second, there is exogenous dormancy, a condition in which unfavourable physicochemical factors in the environment of the dormant stage prevent development. Despite the sanctity which these terms have acquired, there are compelling arguments for abandoning them, at least with regard to fungal propagules, on the grounds that they stretch the notion of dormancy much too far (see also Cochrane, 1974; Tommerup, 1983).

Dormancy is here considered to be a special state, in which viability is coupled to zero germinability. Thus, although a propagule may have the potential to give rise to new individuals, it is dormant if it is unable to do so even under environmental conditions that normally favour vegetative development, including germination, of the fungus in question. Germination is prevented by one or more physical or biochemical factors inherent to the propagule. These may become modified gradually during the life of the propagule, so permitting germination eventually to take place, or they may be drastically, and more rapidly, altered by activation, that is reception of a specific environmental stimulus. Sometimes the severity of the stimulus required is such that it amounts to disturbance, since it would be destructive not only to the vegetative stages of most other fungi but also to any but the dormant phase of the fungus concerned. By contrast, dormancy may also be alleviated by metabolites originating from exogenous sources, these functioning strictly as activators rather than providing nutrients; germination is then rapidly initiated and effected. These three very different processes of slow achievement of germinability, activation via disturbance, or activation by exogenous non-nutrient metabolites, indicate that – at its extremes – dormancy may be a state of either profound rest, or of readiness to take quick advantage of favourable changes in environmental conditions. The latter should not be confused with exogenous dormancy which will now be considered.

Dormancy, as more strictly delineated above, embodies the idea of constitutive dormancy but excludes that of exogenous dormancy. The latter, as it has been applied to fungi, is largely a misnomer for a variety of situations in which germination is prevented other than through dormancy *sensu stricto*. The distinction is quite clear; on the one hand germination is constrained by some characteristic of the propagule itself, on the other simply delayed by unfavourable external physicochemical factors – in other words environmental stress. It is patently confusing collectively to describe inhibition of germination by extrinsic, physically, chemically or microbially induced factors as forms of dormancy. Under such conditions germinable propagules are merely quiescent, as indeed would be vegetative hyphae in the same situation (Tommerup, 1983). However, of greater importance than semantics is the fact that misuse of the term dormancy obscures important differences that distinguish those propagules surviving through time mainly by chance from those doing so by design.

Dormancy mechanisms

Dormancy occurs throughout all major fungal groups. It is not confined to memnopropagules, for instance oospores, zygospores and sclerotia, but also is found amongst xenospores – most commonly in ascospores and urediniospores – and in some sporangiospores, conidia and basidiospores. In most instances, possession of dormancy has an obvious ecological significance and is usually related to a need, at some stage during normal development, to bridge relatively extended periods during which conditions are inimical to vegetative growth, and are unfavourable to survival of non-dormant propagules. Such conditions may occur sporadically and be of potentially indefinite duration, for example when, as a random event, a substratum becomes exhausted of nutrients, or an essential symbiont dies. Alternatively, they may obtain regularly in both time and space, as with seasonal fluctuations in temperature, water potential and availability of colonizable substrata or compatible hosts.

Permeability and compartition

The distribution of dormancy fits well with life style, but no firm generalizations are possible regarding its relationship either to propagule structure or biochemical constitution. For instance, there seem to be no major physical differences between oospores of different *Pythium* species, nor between zygospores (or azygospores) within Entomophthorales; yet in some species these mennospores are germinable shortly after formation, whilst in others they are dormant (Ayers & Lumsden, 1975; Perry & Latge, 1982). Furthermore, in *Phytophthora fragariae*, there are similarly wide behavioural variations between oospores of different isolates (Duncan, 1985). All this suggests that genetic control of dormancy is exerted through systems of much finer resolution than those represented by gross anatomical or physiological features.

The spore wall and its associated membrane could act as an impermeable barrier isolating the protoplast physically from the external environment, and so maintain cryptobiosis, perhaps via anhydrobiosis, but evidence for this is equivocal (van Etten *et al.*, 1983; Ulanowski & Ludlow, 1989). Dormant spores are not necessarily impermeable to oxygen, water or nutrients, and although activated spores may become increasingly permeable, this change appears to be a fairly distant result of activation rather than a primary event. Very little is known of the ability of spores to resist equilibration with exogenous conditions, but there is no reason to suppose that, as a general feature, spores with dormancy mechanisms are more buffered from external influences than those without them. Reasons for the inactivity of processes crucial to germination must therefore be sought elsewhere, one possibility being the physical separation of essential physiological elements. However, few examples are known of spores in which dormancy is possibly maintained in this way, and even here this may be only one of several co-existing mechanisms. Germination of heat-activated ascospores of *Neurospora tetrasperma* depends on utilization of endogenous trehalose reserves held within the cytoplasm. In the dormant spore, trehalase is associated with the protein–carbohydrate matrix of the wall, where it is protected from the effects of heat and is prevented from contact with its substrate. Activation increases plasmalemma permeability, and so allows trehalose to diffuse to sites of hydrolase activity (Hecker & Sussman, 1973). However, compartition is not a feature unique to dormant spores. In quiescent conidia of *Botryodiplodia theobromae*, the encoded enzyme subunits for cytochrome oxidase are held in precursor form outside the mitochondria; transport of them into mitochondria takes place only when conditions favour germination (Brambl, 1980).

Inhibitors

On a more general basis, dormancy is frequently attributable to the action of autoinhibitors of germination, which may be located either inside or outside the spore. In instances of the latter, the inhibitor commonly is held within, or forms a large part of, a mucilaginous extracellular matrix secreted either by the spore itself or originating from sporogenous hyphae (Louis & Cooke, 1985). Where formed, matrix material is a constant feature of spore morphology, so that the non-germinable state imposed by its presence falls by definition within the realm of dormancy rather than quiescence. The distinction is none the less a fine one and, while determining where and when a spore can germinate, the function of such inhibitors may span the hinterland between imposition of dormancy and control of germination (see Chapter 8).

Autoinhibition is most evident within very dense spore populations and may be confined largely to these situations. The ecological implications of population dormancy, as opposed to that of individuals, are reasonably clear. Inoculum wastage is eliminated when spores are prevented from germinating in or on the fructification; similarly, intraspecific competition is reduced where large numbers of spores arrive simultaneously within a confined area of substratum, since few individuals are in a position which permits escape from inhibition.

Autoinhibition in urediniospores. Autoinhibition as a consequence of crowding is exemplified by the urediniospores of rust fungi, although the mechanisms through which their dormancy is imposed are probably unique to them. Here, the degree of inhibition is directly related to population density, and dormancy may be alleviated by subjecting spores to leaching (Macko, 1981). This, together with the fact that isolated, non-leached spores germinate, whilst those that are crowded do not, suggests that autoinhibition is contingent on the continual production – within dense populations – of water-soluble, diffusible factors at a rate sufficient to maintain a general level of dormancy. The corollary of this is that individual urediniospores cannot by themselves maintain the internal concentration of inhibitors necessary to prevent germination. Synergism is involved here, although its dynamics have not been established. In this regard, the effect of autoinhibition on infection may not be as straightforward as might be supposed. Comparisons have been made between germination of clusters and separated urediniospores of *Puccinia recondita* and *Uromyces appendiculatus* (Aylor & Ferrandino, 1986). Whilst percentage germination in clusters of two to six spores was less than that for separated spores, the percentage of clusters giving rise to at least one germination was greater than percentage germination of separated spores. Autoinhibition, therefore, may not always be strong enough to render clusters epidemiologically unimportant.

Inhibition is due to synthesis of cinnamic acid derivatives (Fig. 7.1). In *Puccinia graminis tritici*, the active compound is methyl *cis*-4-hydroxy-methoxycinnamate (methyl *cis*-ferulate: MF), whilst in other species, for instance *Uromyces phaseoli*, it is methyl *cis*-3,4-dimethoxycinnamate (MDC). Although their exact mode of action is not clear, it is evident that they influence the coincident germination

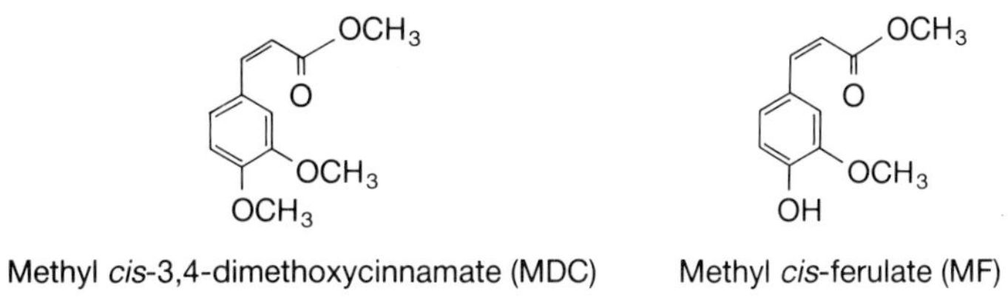

Fig. 7.1 Autoinhibitors occurring in dormant urediniospores.

events of germ pore plug dissolution and early germ tube growth (Hess *et al.*, 1975). Urediniospore germ pores are occluded by a deposit of mannoprotein, and the enzymes postulated as being necessary for its erosion may be reversibly inhibited by MF and MDC. However, no dissolution of plug material occurs when isolated spore walls are incubated in the presence of a homogenate prepared from spores in which plug erosion has begun (Hess *et al.*, cited in Macko, 1981). Breakdown of the plug seems to be effected by enzymes released from the tip of the newly formed germ tube (Fig. 7.2). Autoinhibition could, therefore, act via retardation of germ tube development, prevention of

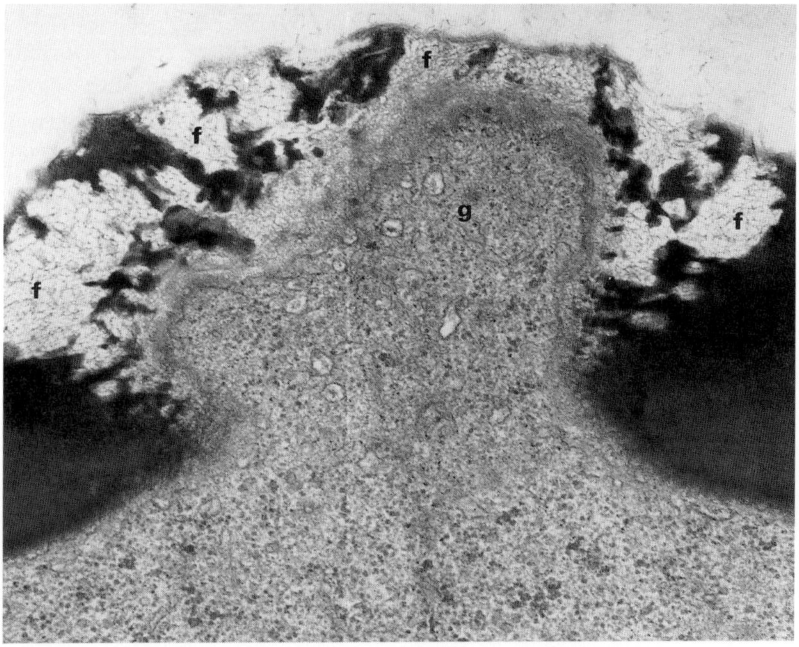

Fig. 7.2 Germ pore plug dissolution in the urediniospore of the bean rust fungus *Uromyces phaseoli*. Fissures (f) develop within the germ pore plug material, allowing the naked germ tube (g) to emerge (from Macko, 1981, © Academic Press).

enzyme release, inhibition of released enzyme, or a combination of all these.

It might be noted that where autoinhibition of other spore stages of rust fungi occurs, this has not been attributed to cinnamic acid derivatives. For example, they are absent from dormant aeciospores of *Cronartium comandrae* (Eppstein & Tainter, 1976).

Autoinhibition in other spores. Population dormancy of non-dispersed spores is widespread amongst fungi, the degree to which autoinhibition persists after dissemination being determined by the rate at which inhibitors are lost by spores through diffusion or become inactivated during metabolism. On a general level, failure of spores to germinate whilst in contact with their parent mycelia has been attributed, in a broad range of fungi, to nonanoic acid or its salts which, after spore dispersal, are removed by leaching (Garrett & Robinson, 1969; Hobot & Gull, 1980). Similarly, aldehydes, alcohols, amines and other volatile metabolites produced by sporulating mycelia have also been implicated in several fungi (McKee & Robinson, 1988; Robinson *et al.*, 1989). Clearly autoinhibitors probably abound, although few have so far been identified and, in addition, autoinhibition mechanisms may vary widely even within a single genus. For example, ustilospores of the bunt fungi *Tilletia caries* and *Tilletia controversa* exhibit population dormancy but, on dispersal, those of the former germinate rapidly whilst those of the latter remain dormant for several months due to their retention of autoinhibitors. These are probably spore surface lipids, especially linoleic acid (Trione, 1977; Trione & Ross, 1988). Differences also occur between species of *Colletotrichum*. In some, autoinhibition is attributable to the influence of an extra-conidial matrix which becomes attenuated during rain-splash dispersal so that germination can then take place (Louis & Cooke, 1985). By contrast, in others, dormancy may be imposed by an internal inhibitor which is removed by leaching (Lax *et al.*, 1985). The nature of matrix-determined inhibition is not known, but it has been suggested that the presence of high levels of elemental sulphur within conidial mucilage may be responsible (Pezet & Pont, 1975).

Activation For the most part, autoinhibition is alleviated as a matter of course when spores become scattered or leached during the dispersal process. However, there are numerous instances in which, in order for dormancy to be removed, more particular environmental stimuli are necessary, ranging from the reception of specific metabolites to exposure to violent physicochemical perturbation. The essence of the term activation is that it refers only to exogenous dormancy-breaking factors. Sometimes, propagules that can pass from dormancy to a germinable condition without the necessity for activation have been described as undergoing autoactivation due to substances associated with, or produced by, the propagules themselves (van Etten *et al.*, 1983). Although a number of endogenous germination stimulators are known, as will be seen later in this book, no similar substances with a strictly dormancy-breaking role in nature have as yet been identified.

Activation mechanisms

Commonly, at a gross level, there is an obvious relationship between any activation stimulus required and the ecology of the fungus concerned. Usually it is not only possible to predict, from some knowledge of life style, the kind of activation treatment likely to be effective, but also to perceive the selective advantages accruing from possession of specific dormancy–activation systems. The latter are, as a general rule, devices which ensure that release from dormancy is synchronous with the occurrence of biotic or abiotic environmental conditions favourable to vegetative development. Whilst it is possible to reach a broad appreciation of the significance of many kinds of activation, and although the major physical and physiological processes underlying germination and ensuing from activation have been mapped, there is almost total ignorance as to the target sites and fundamental events involved in activation. This situation matches, and arises from, doubts surrounding the exact nature of dormancy-imposing mechanisms.

Ecological certainty has often been undermined, rather than reinforced, by laboratory studies. The same spore can be activated by very dissimilar treatments. For example, urediniospores and ustilospores of rusts and smuts respectively can be activated by a wide and heterogeneous range of chemicals. Other kinds of spores can be activated variously by furans, detergents, organic solvents, lipids, heat, light and γ-irradiation (see Sussman, 1965b, 1976; Sussman & Halvorson, 1966). Many studies in this area are paradigms of the schism between ecological and physiological investigations, the latter commonly employing treatments far removed from conditions which any fungus could be expected to experience in nature, to ends which throw little light on ecological behaviour. For this reason, and before considering activation within an ecological context, some points should be made immediately concerning what is thought might happen when propagules receive an activation stimulus. It has been emphasized already that dormant propagules are structurally and functionally diverse, and therefore whether there is a universal mechanism through which activation operates seems arguable. In addition, the numerous instances of activation of a single spore type by several disparate methods indicate that treatments with no apparent ecological relevance can mimic those which might have relevance. Finally, for any spore type, there are probably few dormancy sites, and perhaps only one, but several possibilities for affecting them.

Despite these difficulties, various theories have been advanced as to the primary events of activation (see van Etten *et al.*, 1983). The foremost of these is based on observations made during heat activation of spores of *Dictyostelium discoideum* (Myxomycota: Acrasiomycetes) and sporangiospores of *Phycomyces blakesleeanus* (Cotter, 1981; Furch, 1981). It suggests that the activation stimulus acts upon hydrophilic regions of membrane proteins located on the inner face of the plasmalemma, the signal perhaps being transmitted to them via transient changes in the physical state of membrane phospholipids. Then, consequent alterations in membrane protein conformation patterns initiate a chain of events leading to germination, during which mitochondria may have an important early role. It is thought that in dormant spores a regulatory protein in the mitochondrial inner membrane has a 'restricted' conformation, which blocks electron transport and so prevents oxidative phosphorylation. Heat activation brings about conformational changes such that this protein acquires a 'relaxed' configuration, permitting oxidative phosphorylation and, ultimately, facilitating germination.

Temperature effects

Activation of propagules by shifts to high or low temperatures is a widespread phenomenon and, in nature, temperature is probably the most common dormancy-breaking factor. In the laboratory, minimum activation periods have been found to range from only a few minutes, in cases where high temperatures are a requirement, to several weeks or months where low temperatures are necessary. Effective temperatures are usually well beyond either the maxima or minima for vegetative growth and, in many instances, active mycelia would die if exposed to them.

Frequently, firm ecological conclusions can be drawn from temperature activation studies. For example, long-term, low-temperature requirements are typical of, but are perhaps not confined to, fungi of north-temperate regions. Here, autumn-produced, and initially dormant, propagules are activated by winter cold, and then remain quiescent until higher vernal temperatures permit germination to take place. It should, however, be noted that concomitant seasonal cycles in, for instance, water availability and radiant flux may also act as activators, either together with or independent of temperature. By contrast, it is often more difficult to account for the necessity for heat shock, particularly when this needs to be of only brief duration. Simply, this may reflect ignorance of the natural history of the fungi in question. *Phycomyces blakesleeanus*, for instance, is encountered uncommonly during floristic surveys, but has been recorded from lipid-rich substrata, bread and dung. Such fragmentary knowledge of its habitat relationships does not help to explain why its sporangiospores can be activated by incubation at 50°C for 3–4 minutes. However, as was pointed out above, this treatment may have no ecological significance, but could be substituting for an entirely different and more relevant factor.

Overwintering and aestivation. For many fungi, overwintering involves much more than mere survival through prolonged periods of depressed temperatures. Possession of a dormancy mechanism which can be broken only by fairly lengthy exposure to cold reduces inoculum wastage by preventing germination during temporary unseasonably warm periods, and ensures that it occurs, for the most part, when seasonal warming trends have become established. This has particular importance for symbiotic fungi because, in this manner, the existence of germinable propagules becomes coincident with renewed host activity or development, thus maximizing opportunities for infection. Such dormancy–activation systems have been found in the oospores, ascospores, ustilospores and sclerotia of plant pathogens and in the basidiospores of agarics. In the case of the latter this may be related to the establishment of mycorrhizas during root formation in spring (Kneebone, 1950, cited in Sussman & Halvorson, 1966). Although data are generally lacking on the time required for activation of specific propagules over a range of temperatures, it seems that a distinction can be made between those requiring freezing temperatures and those which respond to chilling, that is temperatures from just above 0 to around 10°C. Activation of some dormant ascospores and basidiospores is brought about by cooling at 0 to −17°C whilst, generally, chilling suffices for oospores, zygospores, ustilospores and sclerotia (Coley-Smith & Cooke, 1971; Perry & Latge, 1982; Perry & Fleming, 1989).

Details of requirements for, and subsequent effects of, low-temperature activation are few, but general behaviour patterns for propagules may be exemplified by the ergots of *Claviceps purpurea.* In north-temperate regions these sclerotia require chilling for activation, temperatures below zero and above 10°C being ineffective. Within the activating range, the longer the chilling period the higher is subsequent final germination, and the higher the activating temperature, the longer the period required to induce maximum germination (Fig. 7.3). However, too long a chilling period may result in a reduction in germination levels. In the laboratory, 4–6 weeks' incubation at 0 to 5°C is sufficient to activate 80–100% of an ergot population, carpogenic germination via stroma production, followed by ascospore discharge, then being most free at 10–25°C (Mitchell & Cooke, 1968).

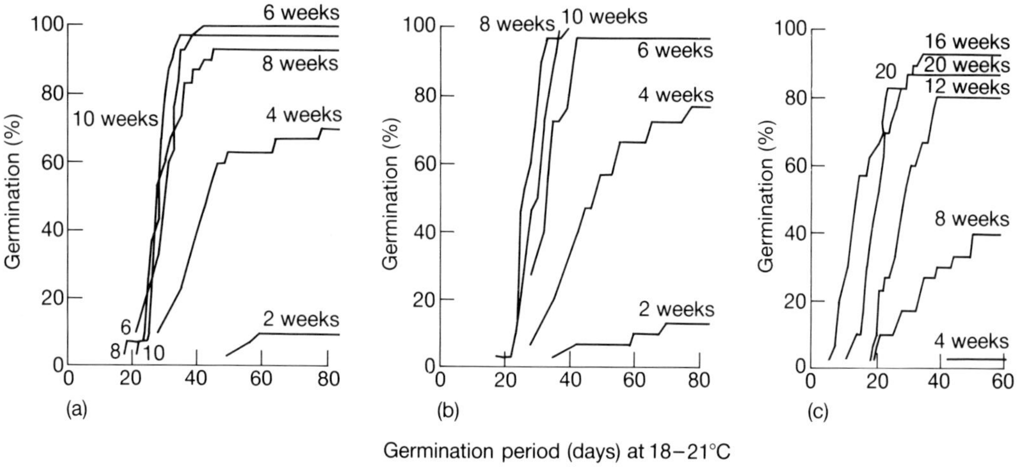

Fig. 7.3 Effect of chilling at (a) 0°C; (b) 5°C; and (c) 10°C for various periods on subsequent germination of sclerotia of *Claviceps purpurea* (from Mitchell & Cooke, 1968, © British Mycological Society).

This suggests that, in nature, sclerotia are probably fully activated by late autumn or early winter, but remain quiescent while it is too cold for germination to take place. Stroma formation during seasonal warming then synchronizes ascospore release with anthesis of host Gramineae, the exposed stigma being the sole entry portal for this pathogen. Beyond north-temperate climates, winter dormancy is not necessarily linked to cold activation. For example, in California citrus soils, chlamydospores and oospores of *Phytophthora parasitica* exhibit winter dormancy but are presumed to be activated when soil temperatures rise above the 9–12°C winter levels (Lutz & Menge, 1986).

Within north-temperate zones, cold-adapted fungi remain active throughout the winter period and become quiescent during the warmer months. It is to be expected that at least some of these species possess appropriate temperature activation–dormancy mechanisms. This is so for *Typhula incarnata*, a psychrophilic basidiomycetous pathogen causing snow rot of winter barley, which is favoured by cool, moist autumn weather followed by a period of snow cover in winter. Infection is brought about during autumn by the myceliogenic germination of soil-borne sclerotia formed as disease spreads during the winter and early spring. These do not germinate in spring, even when conditions obtain which are similar to those that permit autumn germination. Summer dormancy persists for 3–6 months but is broken by exposure to temperatures of 15–30°C for a week or more. Then, summer warmth imposes quiescence on activated sclerotia until the return of autumn weather favourable to germination (Woodbridge, 1988).

Perturbation and disturbance. Mention has been made of activation via heat shock which, in numerous scattered instances, appears to have no obvious relevance to the natural history of the fungi involved. It does, however, have an important bearing on the ecology of some well-defined communities, the best known being the characteristic inhabitants of decomposing herbivore faeces, and the colonizers of newly burnt ground.

Coprophilous fungi may be loosely defined as being those species that are confined to, or are typical of, the dung habitat, their establishment within faeces usually requiring, or resulting from, prior passage of their spores through an animal. These spores, especially ascospores, are commonly dormant, but are activated – either thermally, chemically, enzymically, or by a combination of all these processes – as they pass through the gut (Webster, 1970). The degree of thermal shock to which spores are exposed is relatively low and, in small animals, is comparatively brief. The amount of perturbation experienced during passage would seem to fall far short of the violence of disturbance; this is evidenced by the survival of ingested non-dormant spores of a variety of non-coprophiles (Johnson & Preece, 1979).

By contrast, thermal disturbance is an essential prerequisite for the establishment of typically postfire Ascomycotina. These pyrophilous, or carbonicolous, fungi are ruderals which are confined to burnt sites because of their inability to compete within a complete soil microflora. The latter is destroyed by fire, which also renders the soil more favourable to colonization by postfire species by raising pH, releasing nutrients, and alleviating mycostasis (El-Abyad and Webster, 1986a,b; Wicklow, 1988a). At particular distances below fires, and at their edges, suitable conditions obtain for heat activation of dormant ascospores. For *Trichophaea abundans*, for instance, this involves exposure to 50–70°C for several minutes. Germination is thus cued to occur as soon as soil temperatures ameliorate. This necessity for heat shock during the spore stage contrasts markedly with the mesophilic nature of subsequent mycelial growth. It has been noted that the ascomycete flora fruiting on burnt prairie soil can include typical dung inhabitants, which presumably have arisen from soil-borne ascospores. This suggests that the activating effects of grassland fires can provide an alternative to passage (Wicklow, 1975). Spore discharge by these fungi coincides with the emergence of new prairie vegetation; ascospores attaching to this are then ingested by herbivores and become reintroduced to faeces.

Diffusates and contacts

There are numerous examples where germination has been found to be stimulated by contact with living or dead microbial, plant and animal cells, or by compounds diffusing from them. For the most part, these effects have been attributed simply to nutrient enhancement of the surroundings but also, in many cases, have been explained in terms of dormancy breaking. However, investigators have not always exercised sufficient rigour to enable dormant populations *sensu stricto*, as defined above, to be distinguished from the merely quiescent. Such failures have promoted confusion between activation-dependent germination and that resulting from the removal of exogenous restraints via alleviation of environmental stress. It is of fundamental importance to recognize that the former is linked to a specific response to one or more triggers, whilst the latter arises from more generalized reactions to the presence of germination-promoting metabolites and nutrients. This does not preclude nutrients from functioning as activators but, on both evolutionary and ecological grounds, it is more likely that non-nutrient chemicals will be involved. Any nutrient-based activator would be required to have high specificity for its target propagule, otherwise it would be competed for, and probably sequestered by, non-dormant microbial cells sharing the same habitat. On the other hand, a selective advantage would be conferred on propagules with adaptations for the reception of activator molecules that were either not metabolizable, or were only slowly metabolized, by microorganisms at large. This would additionally facilitate the certain identification of activating signals against a background of intense biochemical noise. Similar fine tuning to specific germination cues is also shown by some quiescent propagules when under the influence of mycostasis (see Chapter 8).

Soil-borne symbionts. By far the largest volume of literature on the stimulatory effects of diffusates relates to the soil environment and, in particular, to those of root emanations. However, critical examination of data concerning both antagonistic and mutualistic fungi indicates that, in most instances, roots influence propagules by releasing them from quiescence rather than by removing dormancy. There is a paradox here since, on the grounds outlined above, specific chemical activation mechanisms might be expected to be widespread in soil, especially so since many root-associated symbionts have a restricted host range. Yet the evidence for such mechanisms, for example in the resting spores of *Plasmodiophora brassicae*, and in oospores, is frequently equivocal (Macfarlane, 1970; Dernoeden & Jackson, 1981; Morgan, 1983).

In the light of this, the long-cherished notion that precise dormancy–activation systems are involved in ensuring root infections needs to be re-examined. It is, in fact, unnecessary to invoke activation by root diffusates even if the fungus in question is highly host specific.

Activation of soil-borne propagules may be brought about in a variety of other ways, some of which have already been described, the only requirement then being the rapid response of quiescent propagules to particular germination-enhancing host metabolites. Furthermore, there is evidence that dormant propagules produced in soil, or liberated to it, can become activated rapidly in the absence of both roots and wide environmental fluctuations. For example, when oospores of *Pythium ultimum* are incubated at constant temperature with non-sterile soil extract, 20% are activated within 10 days and 70% by 30 days (Hemmes & Stasz, 1984). This is possibly due to the action of microbial enzymes on the thin, hydrophobic outer layer of the complex oospore wall, the dissolution or modification of which is followed by thinning or disappearance of the thick, inner wall layer. This change signals the achievement of a germinable condition (Fig. 7.4). Germination is then stimulated by exogenous nutrients which, in nature, would be provided by host roots or hypocotyls.

Interfungal communication. Host activation is to be expected in those interfungal symbioses in which one fungus is obligately nutritionally dependent on another. Similar systems might also be anticipated where the fusion of two mating types, one being spore-borne, is necessary for sexual reproduction. However, as is too often the case within the field of dormancy, available evidence is both scanty and difficult to interpret.

Turning first to ecologically obligate mycoparasites, spores of many – if not most – species of even highly specialized haustorial and contact biotrophs germinate freely in the absence of a host, sometimes doing so in distilled water. Generally, in those cases where host extracts or diffusates stimulate germination, nutrient supplementation appears to be the cause. For instance, in *Laterispora brevirama*, although culture filtrates of its host *Sporidesmium sclerotivorum* induce over 80% conidial germination, levels of 25% are achieved on yeast extract agar alone (Ayers & Adams, 1985). There are, however, a very few examples of the apparent involvement of specific non-nutrient triggers. A water-soluble factor, mycotrophein, is required by many contact biotrophs for axenic development and, by implication, by some for conidial germination. Neither its identity nor mode of action are known, and although these mycoparasites generally have narrow natural host ranges, mycotrophein is present in a wide spectrum of non-host species, and in a variety of organic substrata, for instance tree barks (Gain & Barnett, 1970; Hwang *et al.*, 1985). It is therefore difficult to relate dependence on mycotrophein to either mycoparasitism in general or host preference in particular.

Activating factors have a less elusive role in *Mycogone perniciosa*,

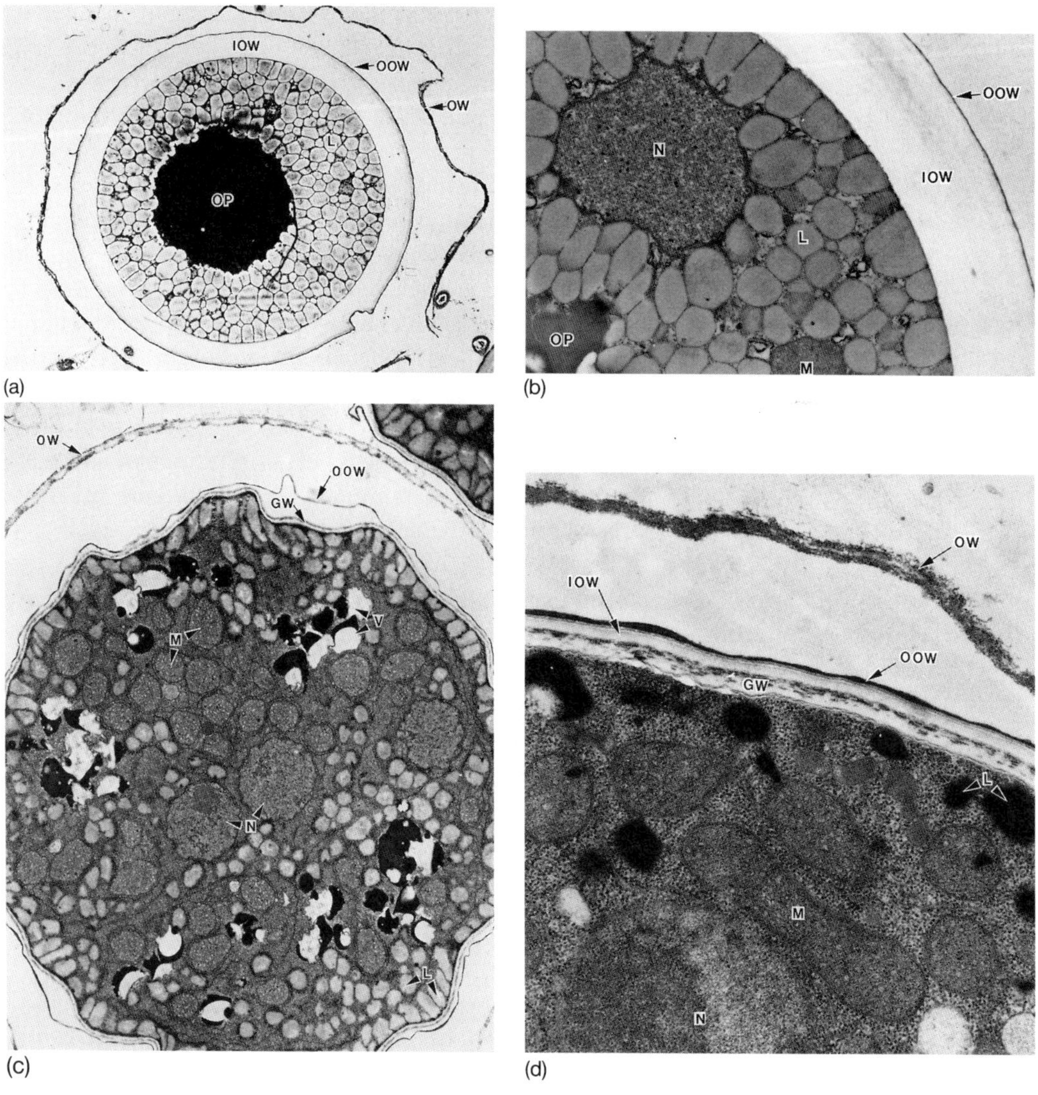

Fig. 7.4 Structure of the dormant and activated oospore of *Pythium ultimum*. (a) Dormant spore, ooplast (OP) with a thin outer wall (OOW) and a thick inner wall (IOW), the whole surrounded by the oogonium (OW); (b) detail of the two walls and nucleus (N), ooplast (OP), mitochondrion (M) and lipid bodies (L); (c) activated spore after disappearance of the inner wall showing the persistent outer wall (OOW), the newly formed germination wall (GW) and vesicles (V); (d) detail of the remains of the inner wall (IOW) and the germination wall (GW) which will give rise to the germ tube (from Hemmes & Stasz, 1984, *Mycologia*, **76**, © 1984 New York Botanical Garden).

the wet-bubble pathogen of cultivated *Agaricus* species. This mycoparasite produces a number of spore types, a major survival propagule being a bicellular conidium consisting of a thick-walled, dormant upper cell and a smaller, thin-walled lower cell. Soon after conidiogenesis, the latter dies, but the remaining cell is capable of surviving for a further 2 years. Activation occurs if it comes into contact with either a host basidioma or vegetative mushroom mycelium (Fig. 7.5). In the laboratory, activation can also be brought about by basidioma extracts, and filtrates from cultured mycelia (Holland *et al.*, 1985; Holland, 1988). Water-soluble organic factors are involved and, though these have not been identified, it seems that activation occurs in two stages, the first dependent on an uncharged compound or compounds, the second on charged compounds. Spores are not activated when treated solely with either uncharged or charged fractions from host tissues, but they become so if they are treated with them either simultaneously or sequentially (Rawlins, 1989). Similar activation of conidia of the mycoparasite *Spinellus macrocarpus* during contact with agaric fruitbodies has been attributed to the presence of ascorbic acid, but its mode of action is unknown (Watson, 1964).

Activation as a means of mediating interfungal communication is central to mating processes in some agarics, where it facilitates

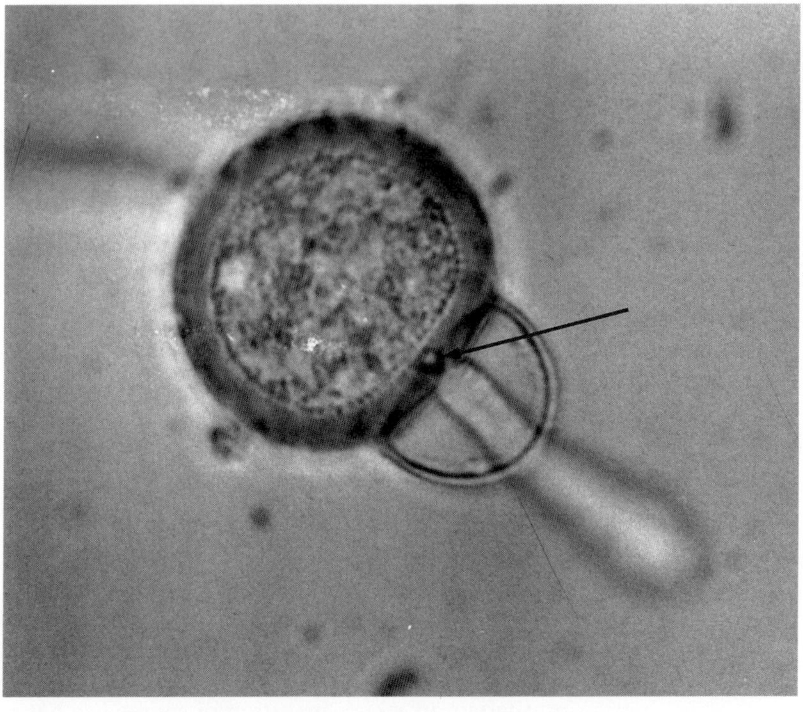

Fig. 7.5 Conidium of *Mycogone perniciosa* germinating in response to a mycelial extract from *Agaricus bisporus*. The lower conidal cell is empty. The germ tube arises from the thick-walled dormant cell which is separated from the dead lower cell by a plugged pore (arrowed) which remains closed, the germ tube in this instance emerging near to it (photograph by A. Rawlins).

recognition between basidiospores and mycelia. Dikaryotic hyphae of *Leccinum scabrum* show directed growth towards germ vesicles that have been produced by dormant basidiospores in response to activators diffusing from the dikaryon. Identical responses occur in *Paxillus involutus*, in which the activators are volatile, and in *Laccaria laccata* (Fries, 1978, 1979, 1983). At some levels taxon specificity operates so that *Leccinum* species, for example, can be divided into reaction groups according to the response of basidiospores to the presence of conspecific and congeneric dikaryons (Table 7.1).

Miscellaneous factors

On the evidence of an albeit imperfect literature, the most common and widespread dormancy–activation systems seem to be those that depend upon either temperature changes or substratum-derived chemicals. Possibly, to these may be added frequent examples of activation through wetting and drying cycles, although this treatment may simply mimic the impact of activating temperature shifts on the water relations of protoplasts. Perhaps more worthy of consideration are the effects of light and of passage through the digestive tracts of animals.

There are numerous instances of the apparently activating effects of light, mainly involving ustilospores and urediniospores (see Sussman & Halvorson, 1966). However, as with other putative activators, it is seldom clear whether dormancy or quiescence is being affected. Furthermore, data on the necessary irradiance thresholds and action spectra are either absent or, if available, are subject to question. Despite this, it might be expected that fungi with particular life styles would have evolved light-activated dormancy mechanisms. For example, in temperate climates, *Entomophthora*-induced epizootics in insect populations are seasonal, the main overwintering stage being the resting spore. In *Entomophthora aphidis* resting spores have basal germination levels of 1–3%, but up to 50% become activated by light periods of 14 hours or more within a 24-hour day (Wallace *et al.*, 1976). This probably has a direct bearing on the occurrence of disease outbreaks during summer months.

The spores of many fungi can pass unharmed through the alimentary canals of endothermic and poikilothermic vertebrates, and a wide range of invertebrates. Some ecological groups, for instance coprophiles and hypogeous Ascomycotina and Basidiomycotina, have co-evolved with grazers and predators to reach a state of dispersal mutualism in which ingestion of spores has become essential for their wide dissemination (Pirozynski & Malloch, 1988). Within these groups it is evident that, in some species, dormancy becomes broken during passage. The most quoted examples are coprophiles, which not only experience elevated temperatures during their journey through the gut,

Table 7.1 Activation (+) and non-activation (−) of basidiospores in *Leccinum* species in the presence of dikaryons (after Fries, 1981).

	Mycelia					
Basidiospores	*L. scabrum*	*L. holopus*	*L. aurantiacum*	*L. versipelle*	*L. variicolor*	*L. vulpinum*
L. scabrum	+	+	−	−	−	−
L. holopus	+	+	−	−	−	−
L. aurantiacum	−	−	+	+	+	+
L. versipelle	−	−	+	+	+	+
L. variicolor	−	−	+	+	+	+
L. vulpinum	−	−	+	+	−	+

especially in ruminants, but are also exposed to digestive enzymes (Webster, 1970). The relative contribution of temperature and digestion to activation is not clear, although it has been demonstrated that, in non-coprophiles, enzymes alone can break dormancy. For example, oospores of *Phytophthora megasperma* become activated during passage through the land snail *Helix aspersa*, an effect directly attributable to the action of β-glucuronidase (Salvatore *et al.*, 1973).

The discussion here of dormancy and activation has been deliberately restricted to general phenomena possessing broad ecological significance. However, since fungi have a wide ecological amplitude, within which there is great habitat diversity and complexity, it is reasonable to suppose that dormancy–activation systems exist which are unique to small groups of species or even to individual species. There are strong hints in the literature that this is so, but it would seem to be inutile to discuss particular examples in the absence of evidence that dormancy *sensu stricto* is involved in the first place.

8 Germination

Germination is the process by means of which a propagule discharges its ultimate role, that is to give rise directly to a new individual or individuals. For most spores its end point is the emergence of a germ tube which then branches to form a germling mycelium, but previous to this final expression of germination, intermediate developmental phases may intercede. Before considering germination physiology, some of these require brief description.

Intermediate germination modes

Within particular fungal groups, certain types of spore germinate by producing further spores, familiar examples are: sporangium production by oospores and zygospores, zoospore release from sporangia in Peronosporales and from cysts in diplanetic and polyplanetic Saprolegniaceae, repetitive conidium formation in Entomophthorales, and generation of basidiospores by teliospores and ustilospores of rusts and smuts respectively. Sometimes, as in some entomogenous and phylloplane fungi, whether a germinating spore will give rise to further spores can be environmentally determined, this mode being adopted under conditions of stress (Dickinson & Bottomley, 1980; Van Roermund *et al.*, 1984; Horn, 1989).

Such behaviour is termed iteration, and during it a sporophore is differentiated directly without the interpolation of a phase of significant germ tube growth. It is important to distinguish between iteration and microcyclic sporulation, which it superficially resembles. The latter term should be used strictly to describe the situation where a spore first gives rise to a minute mycelium which then rapidly sporulates. Additionally, microcyclic sporulation appears to be a response to disturbance, usually heat shock, to the spore prior to germination, rather than to stress during germination.

However, in either case there is obvious survival value in spores being able to produce a further generation of propagules quickly, especially when exogenous resources are in short supply, through early and total commitment of endogenous reserves, although these may be barely sufficient for that purpose. For example, conidia of some *Penicillium* species germinate in oligotrophic soil conditions by each giving rise to a single, dwarf phialide bearing further conidia of much reduced volume (Sheehan & Gochenaur, 1984). A germination mode akin to iteration occurs in some Hymenomycetes when basidiospores bud to form yeast-like cell groups, which presumably can act as xenopropagules, or produce blastoconidia which themselves bud (Ingold, 1985, 1988). Iteration also has parallels in the sporogenic and carpogenic germination of sclerotia, where sporophores and carpophores respectively emerge directly from sclerotial tissues to disseminate their spores. Although iterative germination usually results in xenospore production this is not always so, and memnospores may be formed. For instance, basidiospores of the fairy-ring agaric *Marasmius oreades* can germinate through either internal differentiation of a chlamydospore or by chlamydospore formation within a truncated germ tube. Whilst it would seem that, in general, iteration results in depauperate sporulation – increasingly so if there is repetition, with the daughter spore giving rise to a further spore – in some fungi it can, on the contrary, be surprisingly intense. The filiform ascospores of *Epichloë typhina*, which causes 'choke' of pasture grasses, can germinate only by iteration (Bacon & Hinton, 1988). Closely spaced conidiophores arise along the whole length of the ascospore and, since each conidiophore can produce several conidia, considerable numbers of the latter are generated.

Basic processes Despite the morphological variety of intermediate germination modes, it is possible that, whatever the outcome, all propagules hold a number of basic processes in common, if only during initiation of the intermediate or ultimate germination event. There is, however, little evidence to support this view, due largely to the fact that research has been concentrated upon elucidation of the steps leading to germ tube emergence. Such an emphasis is quite proper since, at least for mycelial fungi, this marks the arrival of the new individual.

It is not easy to set confines to the germination process, but it begins at some unidentifiable point where transition from dormancy or quiescence becomes irreversible. At the opposite temporal extreme it can be said to be ended by germ tube emergence which, at the same time, signals a germination success irrespective of its subsequent fate. However, it might also be considered that germ tube growth to the point of transformation to vegetative hypha, as indicated by the onset of branching, is also part of the process. There would seem to be little virtue in this view, because there is no evidence for major differences between the mechanism of germ tube elongation and that of other somatic hyphae, so that developmental distinctions are difficult to make at the cellular level (see Chapter 3).

The germ tube A common, but by no means universal, prelude to germ tube formation is swelling of the spore (Fig. 8.1). Where this does happen it has often been attributed to imbibition causing expansion of the elastic spore wall. Imbibition must certainly occur, but the consequent increase in cell volume is probably accommodated by isodiametric wall growth rather than simply by stretching (Beakes, 1980; Bartnicki-Garcia, 1981).

Even in nutrient-rich conditions, germ tube formation normally

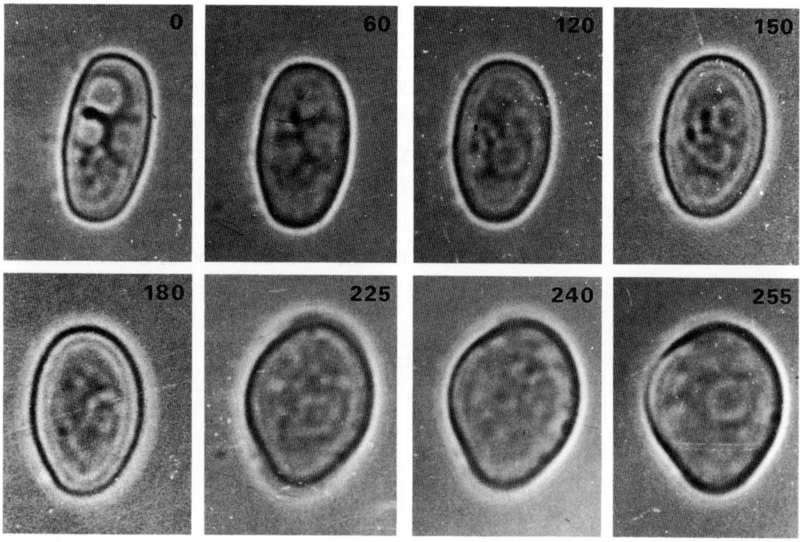

Fig. 8.1 Time-lapse sequence of early stages of germination in a conidium of *Ceratocystis adiposa* showing increase in volume prior to germ tube emergence. Time in minutes after imbibition is shown at the top right-hand corner (from Hawes, 1980, © British Mycological Society).

involves massive commitment of endogenous resources. Germ tubes are usually delicate structures which are highly vulnerable to disturbance or to the sudden onset of stress, and their demise almost invariably results in death of the parent spore or spore compartment, its reserves having become too depleted to support further synthetic processes. This general lack of a capacity for repeated germination may seem perplexing in view of the ecological advantages that would arise from it. However, any disadvantages are more than offset by either intense production and wide dissemination of randomly germinating spores, or through production of those in which germination is attuned to particular environmental cues. Although repeated germination appears to be rare, it has been observed in some fungi, and may be more widespread than might be supposed, especially perhaps, if the spore is large. The massive unicellular spores of the vesicular–arbuscular mycorrhizal species *Gigaspora gigantea* each give rise to several germ tubes. If these are excised, a further crop is produced; this process can be repeated several times over a period of a month or more, and may be an adaptive response to the potentially lethal effects of grazing by mycophagous invertebrates (Koske, 1981).

Germ tube initiation and development have received a great deal of attention at both structural and metabolic levels, but the disparate nature of much of the available data makes generalizations difficult (see van Etten *et al.*, 1983). For instance, although it has been suggested that fungi can be categorized into distinct groups according to the way in which the germ tube originates, sufficient variations exist to make this view untenable (Bartnicki-Garcia, 1968). There is a further complication in that interpretations of ultrastructural events during germination can differ widely, depending on which fixation and staining techniques have been selected (Beakes, 1980). Nevertheless, it is possible to describe some broad features of germ tube genesis with the caveat that, in respect of fine detail, significant departures from these can occur.

Spore walls frequently have more than one layer, often several, and various germ tube ontogenies are possible, each being determined by the developmental relationship of the young germ tube wall to existing wall architecture. Basically there are two possibilities. First, the germ tube may arise from the pre-existing spore wall. This takes place in such widely different spores as the azygospores of *Gigaspora margarita*, which have a four-layered wall, chlamydospores of *Phytophthora palmivora*, with a two-layered wall, and encysted zoospores of *Phytophthora* and *Pythium*, which have a wall with a single layer (Bartnicki-Garcia, 1981; Hemmes & Lerma, 1985; Siqueira *et al.*, 1985; Cho & Fuller, 1989). In all cases, there is first a clustering of cytoplasmic vesicles at the site or sites from which a germ tube will eventually appear. In cysts, the localized synthesis of new wall material then generates the germ tube, but in spores with multilayered walls the process is more complex. Vesicle accumulation is followed by the laying down of fresh material over a restricted area of the innermost wall layer, this sometimes being preceded by erosion of that layer. The new material constitutes the germ tube initial and, via apical expansion and extension, this develops into the tubular germ tube, which then pushes through the spore wall using a combination of physical force and enzyme action. Where germ tubes emerge via germ pores, for instance in rust urediniospores, enzymic digestion of the pore plug begins adjacent to the plasmalemma and progresses rapidly outwards. Removal of material causes large radial fissures to appear within the plug which eventually disappears almost entirely (see Fig. 7.2).

The second possibility is that the germ tube may originate from a new, entire wall – the germination wall – formed interior to the existing spore wall. This appears to occur much more frequently than does localized synthesis of germ tube initials and has been described for oospores, sporangia, sporangiospores and conidia (Hemmes & Hohl, 1969; Beakes, 1980; Hawes, 1980; Hemmes & Stasz, 1984; Monte &

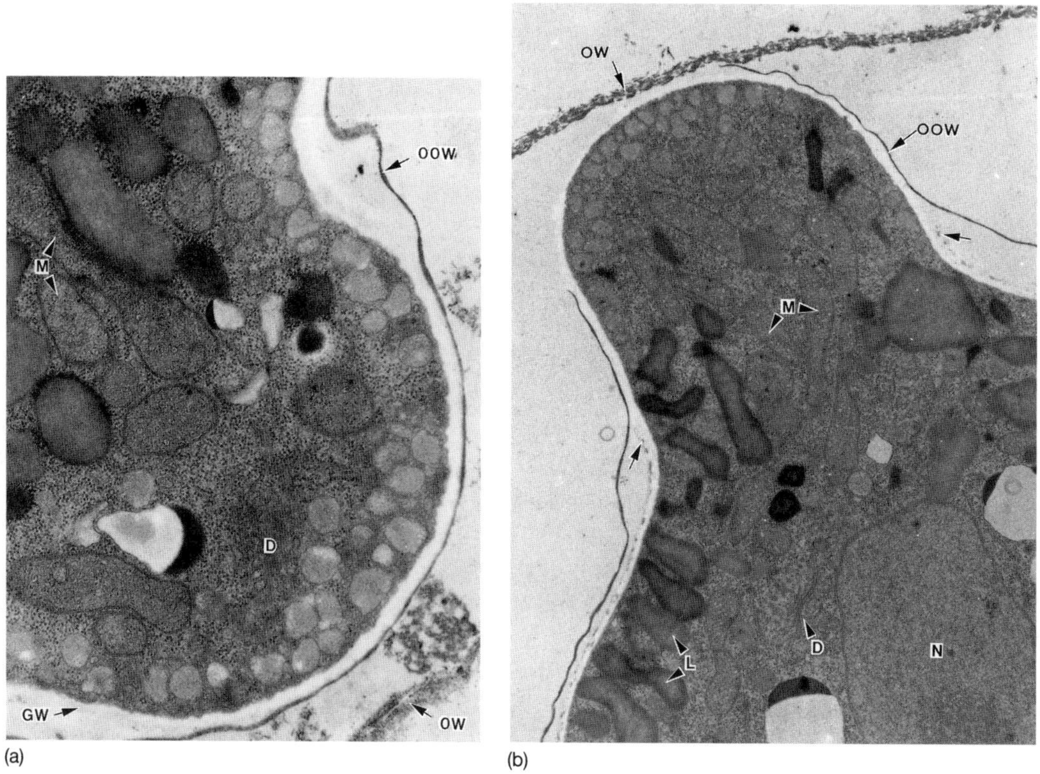

Fig. 8.2 Germ tube formation by the oospore of *Pythium ultimum*. (a) Localized bulging of the cytoplasm with a leading edge of vesicles, dictyosomes (D) and mitochondria (M). There is a well-developed germination wall (GW) and remains of the oogonial wall (OW). The outer oospore wall (OOW) is still intact. (b) Later stage, showing lipid bodies (L) and nucleus (N), in which the germ tube has ruptured the outer oospore wall and the outer part of the germination wall (arrowed). For details of intact oospore structure see Fig. 7.4 (from Hemmes & Stasz, 1984, *Mycologia*, **76**, © 1984 New York Botanical Garden).

Garcia-Acha, 1988; Dute *et al.*, 1989). Germination wall formation is accompanied by vesicle accumulation in the peripheral cytoplasm and, after its completion, localization of the same or similar vesicles generates the germ tube as has been described above (Figs 8.2 & 8.3). During oospore germination in *Phytophthora megasperma*, the inner oospore wall is first extensively eroded, its β-1,3-glucan components being mobilized to supply carbon and energy for the germination process, the germination wall being formed only after this digestion has begun (Beakes & Bartnicki-Garcia, 1989).

Irrespective of the details of germ tube initiation, a key event is polarization of cell wall synthesis, which is manifested in the gathering of vesicles at specific sites within the spore, and which occurs before there is any physical evidence of the germ tube. Polar-

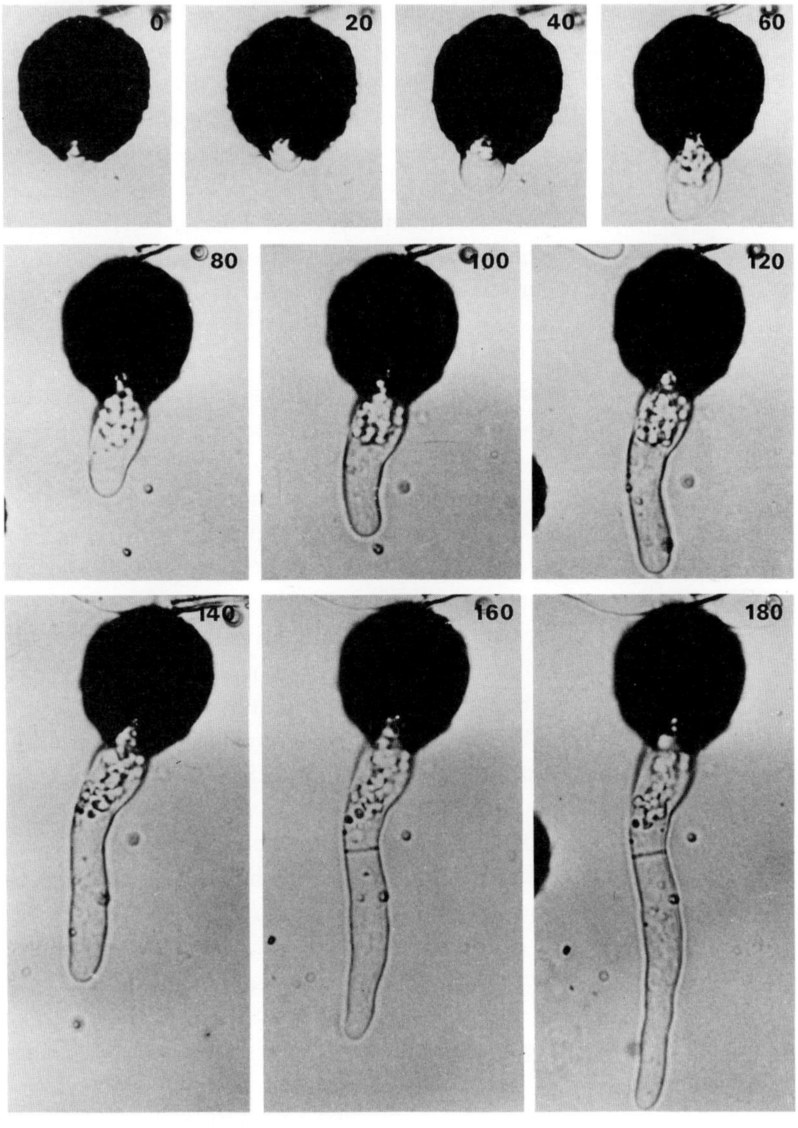

Fig. 8.3 Time-lapse sequence of germ tube emergence from a conidium of *Ceratocystis adiposa*. Time in minutes is shown at the top right-hand corner (from Hawes, 1980, © British Mycological Society).

ization must be brought about by a directional change in the migratory patterns of chitosomes and other wall-associated organelles (see also Chapter 3) and this, in turn, is linked to the metabolic and ultrastructural changes that occur when a spore passes from quiescence to the germination phase.

Changes in physiology and metabolism

The physiological and metabolic trends that accompany germination are, in general, similar to those that have long been known to attend the transition of any cell from quiescence to activity and growth (Fig. 8.4). Thus there is increased respiration, utilization of storage compounds, generation of reducing power, enhanced protein synthesis, and biogenesis of essential cell components, including new membrane materials (Brambl, 1981; Furch, 1981). A great deal of information exists relating to such changes, but largely it is based on the study of very few fungi, in particular, species of *Neurospora*, *Phycomyces blakesleeanus* and dimorphic *Mucor* species, and it is difficult to assess what proportion of it is unique to spores. Putting such general considerations aside, recent work on the metabolic attributes of germinating spores has focused on two areas: the controlling role of cyclic AMP and patterns of protein synthesis.

Germinating sporangiospores of *Phycomyces blakesleeanus* rapidly take up water and concomitantly synthesize large amounts of glycerol, which then leaks from the cell, so lowering its internal osmotic potential. Glycerol synthesis is at the expense of trehalose breakdown this, in turn, being under the control of cyclic AMP, such that increases in the latter cause an enhancement of trehalase activity (Verbeke & van Laere, 1986; van Laere & Hulsmans, 1987; van Laere *et al.*, 1987). By contrast, in sporangiospores of *Pilobus longipes* trehalose is mobilized via decompartition (see Chapter 7) and not by cyclic AMP-dependent phosphorylation, although cyclic AMP is involved in glucose transport into the cell. Exogenous glucose can act as a germination activator and, shortly after its application, there is a sharp rise in cyclic AMP (Bourret, 1986; Bourret & Smith, 1987; Bourret *et al.*, 1989). In *Mucor genevensis* and *Mucor mucedo*, cyclic AMP prevents germ tube development. Spherical growth of sporangiospores of these fungi is associated with large pools of cyclic AMP with high binding activity by unidentified proteins which are specific for particular stages in morphogenetic development. Germ tube emergence is marked by small pools of cyclic AMP with low binding activity of these proteins (Orlowski, 1988).

During imbibition there is also rapid protein and RNA synthesis associated with an increase in nuclear volume, and an increase in the size of mitochondria or the fragmentation of large mitochondria into numerous smaller ones. There may be depletion of amino acids as rates of protein and RNA synthesis accelerate during germ tube emergence (Orlowski & Sypher, 1978; Beilby & Kidby, 1982). In encysted zoospores of *Phytophthora palmivora* the amount and translatability of mRNA increases before germination occurs, this being attributable to transcription of new mRNA rather than to modification of stored RNA (Pennington *et al.*, 1989).

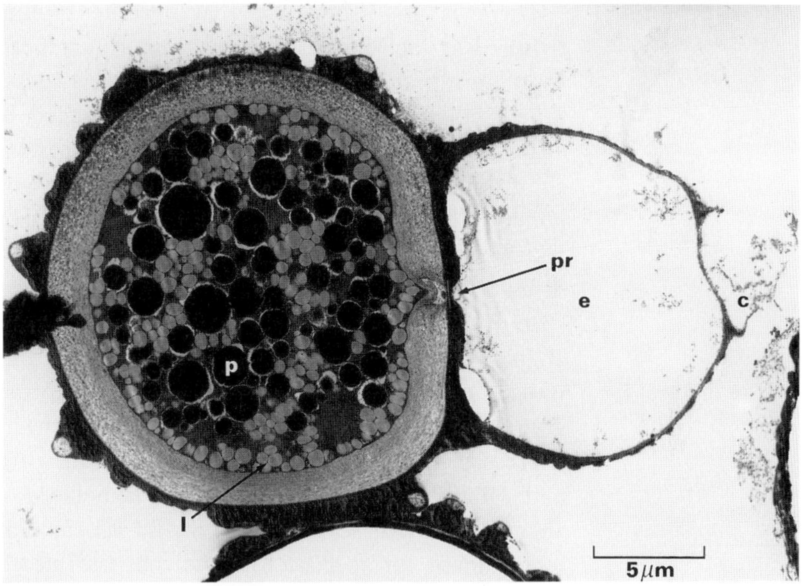

(a)

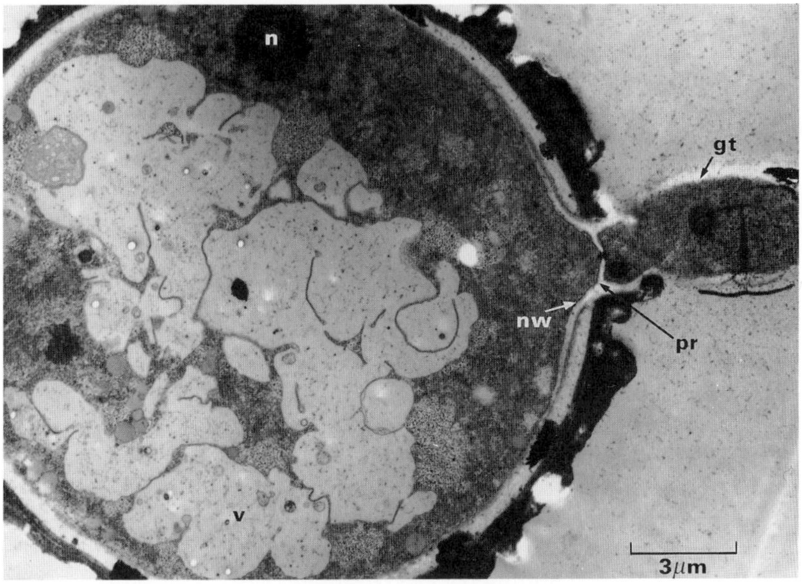

(b)

Fig. 8.4 Structure of dormant and germinating conidia of *Mycogone perniciosa*. (a) Dormant conidium with the viable, thick-walled apical cell to the left, and the dead, thin-walled basal cell which is devoid of contents to the right (e). The apical cell contains a dense, peripheral layer of lipid bodies (l) and protein-rich inclusions (p). The two cells are connected by a plugged pore (pr) and the remains of the conidiophore (c) are visible. (b) Germinating conidium. The emergent germ tube (gt) is formed from new wall material (nw) and is arising close to and below the plugged pore (pr). Reserve lipids and proteins have largely disappeared to be replaced by prominent vacuoles (v). One nucleus (n) is evident (photographs by A. Rawlins).

There is good evidence that specific proteins and mRNAs are required at each phase of the germination process. During sporangiospore germination in the dimorphic species *Mucor bacilliformis* and *Mucor rouxii*, spherical growth is succeeded by the emergence of either a germ tube or a yeast-like bud, depending on environmental conditions. Whichever is the case, proteins and RNA increase exponentially through all stages of germination, there being neither a lag phase nor a change in rate. Suitable inhibitors can halt germination at discrete steps, pointing to the operation of morphogenetic–state-specific proteins and mRNAs (Cano & Ruiz-Herrera, 1988). In sporangiospores of *Mucor racemosus* there is storage of the mRNA molecules necessary for both germ tube formation and yeast cell growth, environmental conditions then determining which of these is expressed (Linz & Orlowski, 1987). Returning to *Mucor bacilliformis* and *Mucor rouxii*, DNA synthesis is delayed until the end of the spherical growth stage, and it would appear that this is related to a requirement for specific mRNAs that initiate the onset of polarized growth.

Autostimulation

In the previous chapter autoinhibition was discussed in relation to the imposition of quiescence and dormancy. This aspect of germination control has an opposite in the phenomenon of autostimulation, in which spores produce specific germination-promoting metabolites (Macko, 1981). As with autoinhibition, the expression of autostimulation is associated with crowding, or at least the close proximity of one spore to another. For instance in *Geotrichum candidum*, *Mucor plumbeus* and *Rhizopus stolonifer*, paired spores germinate earlier than those not associated in pairs, there being some evidence that respiratory CO_2 is the stimulatory factor (Robinson *et al.*, 1968; Robinson, 1973a; Robinson & Thompson, 1982). Crowding sometimes also results in the simultaneous exhibition of autostimulation and autoinhibition, as for most urediniospores, where percentage germination is reduced but germ tube growth is accelerated (Yarwood, 1956). The reverse is true for ascospores of *Sclerotinia curreyana*, germination showing stimulation and germ tube growth being inhibited (Aggab & Cooke, 1981).

Several endogenous autostimulants have been identified (Fig. 8.5). Nonanal appears to occur commonly in rust urediniospores, but can also stimulate germination of smut ustilospores and conidia of *Penicillium* species. One of these compounds, cinnamaldehyde, is a partial structural analogue of urediniospore autoinhibitors (see Fig. 7.1). Where autoinhibitors are also present, autostimulators facilitate germination without, apparently, reacting with them (Macko, 1981).

$CH_3(CH_2)_7CHO$
Nonanal

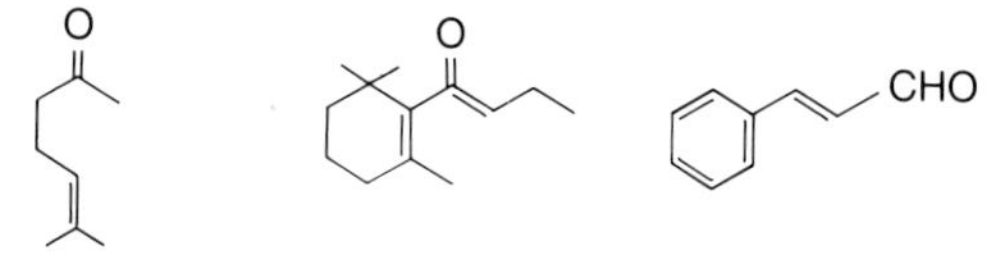

6-methyl-5-hepten-2-one   β-ionone   *trans*-cinnamaldehyde

Fig. 8.5 Autostimulators occurring in rust urediniospores and spores of other fungi.

The mechanisms of autoactivation are unknown, and there exists the suspicion that it might be largely a laboratory artefact. As will be amplified below, the environment of the spore has an overwhelming influence on its germination behaviour, and in most natural habitats this will be swamped by numerous biochemicals and microbial metabolites, both stimulatory and inhibitory, against which background autostimulants must operate. It is difficult to envisage, for example, the additional value of autostimulation via self-produced respiratory CO_2 in situations where there are other active microorganisms. In this case, and perhaps in many others, 'autostimulation' by a specific compound is a simple reflection of the general stimulatory effect of that compound (see, with reference to 'non-microbial inhibitors and stimulators', below).

Environmental factors

Germination behaviour can be strongly influenced by prehistory. Not only does spore age have a direct bearing on germinability, as was outlined in Chapter 6, but the latter is also affected by environmental conditions that obtain during propagule formation and maturation. It might be expected that experience of stress over this period would be adverse but, whilst this may well be generally true, the outcome is not always predictable. For example, when the powdery mildew *Leveillula taurica* and the rust *Puccinia striiformis* sporulate on drought-stressed hosts, germinability of conidia and urediniospores respectively is significantly reduced (Gopalan & Manners, 1984; Caesar & Clerk, 1985). By contrast, in similar circumstances, conidia of *Erysiphe pisi* and *Erysiphe graminis hordei* show increased germinability, those of the latter species being better able to infect a new generation of drought-stressed plants. Such adaptive changes, although small, are ecologically important in that they may have a considerable cumulative effect if maintained or repeated over several generations (Wyness & Ayres, 1985; Ayres & Paul, 1986). Pregermination stress can also affect germination mode. For instance chilled sclerotia of *Sclerotinia minor* subsequently germinate by giving rise to a diffuse mycelium, whilst those subjected to subzero temperatures eventually germinate eruptively by means of a plug of aggregated hyphae (Wymore & Lorbeer, 1987). Again, these differences have ecological significance, the first mode reflecting a partial allocation of resources to wide-ranging habitat exploration, the second involving total commitment of reserves to the formation of a discrete infection structure.

Irrespective of the degree to which germination behaviour is environmentally preconditioned, the ensuing germination process is impacted upon by a wide range of biotic and abiotic factors that determine both its rate and final outcome. The effects of many of these can be assessed accurately in the laboratory, but even under strictly controlled conditions the interactive effects of two or more factors can be difficult to interpret. Even more difficult is the extrapolation of laboratory findings to natural situations, in which multiple factors can fluctuate simultaneously if not entirely independently. Indeed, as with mycelial development, the ecological relevance of much work on the environmental control of germination, especially with regard to temperature and pH, should be vigorously questioned. There has been over-emphasis on the establishment of cardinal points for germination, for these to then be used as indicators of natural behaviour. Largely, this approach is useful only in revealing extreme cases, for instance psychrophilic and thermophilic tendencies, in which there are special requirements. The fact is, that when given laboratory conditions, the vast majority of fungi exhibit mesophily, developing optimally at around 25–30°C and at an approximately neutral pH. Determining performance ranges is not as important as gaining information on how wide and frequent natural departures from the optimum can be, and appreciating that, nevertheless, fungi can successfully confront conditions that are far from ideal.

Possibly germination has more exacting environmental requirements than any other developmental phase, and can reach completion only if suitable conditions obtain for an appropriate period. In many habitats, this means that germination is possible only when transient favourable windows occur within fluctuating, and generally constraining, circumstances. However, having achieved this, subsequent vegetative development – though it might be slow or aberrant – is often possible in conditions that are too severe to support germination.

Water relations and temperature

Availability of water to the protoplast is essential for germination and, given sufficiency of water as an absolute requirement, all other environmental factors, however important, can be conceived as having a mainly regulatory role. The corollary of this is that the amount of available water will influence strongly the response of germination to these factors. Additionally, at least in terrestrial habitats, temperature will govern the amount of water that can exist both within the spore and in its immediate environment.

In view of this, studies on germination – temperature relations that employ fully hydrated propagules are frequently of dubious ecological validity, with only a minority significantly illuminating natural behaviour. For example, sclerotia of *Colletotrichum coccodes,* the tomato anthracnose pathogen, germinate myceliogenically at 28°C, but at 22°C do so sporogenically via conidium formation. Thus, spores are mainly produced in cooler periods, when it is more likely that they will achieve dispersal through rain splash, and subsequently meet moisture conditions suitable for their germination (Dillard, 1988). In a similar vein, on the basis of temperature requirements it is possible to distinguish between two closely related species, *Phytophthora cactorum* and *Phytophthora syringae,* both of which occur in apple orchard soils and cause collar rots. Optimum temperatures for oospore germination are 20°C in *Phytophthora cactorum* and 12–15°C in *Phytophthora syringae,* which correlates well with peaks in activity of the former species during summer months and of the latter in the autumn and spring (Harris & Cole, 1982).

Water requirements for germination range from complete dependence upon free water, as in zoosporic fungi and some Entomophthorales, to independence of an exogenous supply, as in many Erysiphales. Conidia of the latter can germinate at zero RH, this being facilitated by a combination of a high endogenous water content and very impermeable cell walls (Yarwood, 1950). In addition, they contain large lipid bodies, the respiratory oxidation of which may generate additional endogenous water during germination (McKeen, 1970). Within these extremes, spores of most fungi will germinate in the absence of free water, provided that RH is high. With regard to water potential tolerance, in terrestrial fungi the minimum water potential for mycelial expansion is an approximate guide to the minimum for germination (Hocking & Pitt, 1979; and see Table 4.1). However, in many instances, the minimum potential for germination has been found to be significantly below that for hyphal growth. The reasons for this cannot be explained fully, but when such a situation obtains there is usually an extended lag phase, and development often fails to proceed further than the germ tube stage (Magan, 1988; Wheeler *et al.*, 1988a).

Largely for reasons of convenience, germination has been studied mostly in relation to osmotic potential rather than matric potential, with comparisons between the two being made rarely. Where this has been done, for instance for some species of *Fusarium*, *Penicillium* and *Trichoderma*, germination and germ tube extension have been found to occur over a wider range of osmotic than matric potential. By contrast, *Gliocladium* spores behave identically in both systems (Magan, 1988). However, spores of some fungi will germinate at very low matric potentials; those of the vesicular–arbuscular species *Acaulospora laevis*, *Gigaspora calospora* and *Glomus caledonium* are able to do so in soil at <−2.5 MPa, due to their ability to take up water from the vapour phase. Since soil matric potentials of <−1.0 MPa will not support plant growth, germination in dry soils may have an adverse effect on spore populations of these mycorrhizal fungi (Tommerup, 1984). Establishing the limits of water potential tolerance using osmotica is not a straightforward process since, for any fungus, the range is affected by the solute employed. For example, controlling solute potential with sugars usually gives a lower limit for germination by comparison with salts of monovalent ions (Dupler *et al.*, 1987). This is probably due to some solutes intervening directly in spore metabolism rather than acting indirectly via the exertion of osmotic effects. This would seem to be the case with NaCl. In studies on spoilage fungi of dried, salted fish, the minimum osmotic potential for germination in *Wallemia sebi* and several *Aspergillus* species was lower using glucose and fructose as osmotica than was that using NaCl. Over and above this osmotic effect, germination was also modified by salt tolerance which varied from species to species (Wheeler *et al.*, 1988a,b,c).

At non-optimum temperatures and pH there is a marked decrease in water potential range due to an increase in the minimum water potential that will support germination. Very low water potentials may also induce aberrant behaviour of germ tubes, so that these may be abnormally short, broad and vacuolated, or thickened and septate, or even globose. Promycelia arising from germinating teliospores of rusts may elongate but fail to produce basidiospores, or produce basidiospores which are not discharged due to an inability to form Buller's drop (Anikster, 1988).

Fluctuations in water availability either before or during germination can be severely damaging. If sporangiospores of *Rhizopus* species are exposed to an RH below 66% they become susceptible to chilling injury during subsequent rehydration (Dennis & Höcker, 1981). There is also some evidence that if air-dried conidia experience chilling, then this renders them fatally sensitive to rewetting (Ogunsanya & Madelin, 1977). Once germination has begun, even a brief interruption of water supply can be fatal. For instance, when pycnidiospores of *Leptosphaeria maculans* are in the lag phase, a drying period as short as 20 minutes is lethal (Vanniasingham & Gilligan, 1988).

An excess of water can have physiological effects as large as those arising from lack of it, and leaching is an important aspect of water relations. Leaching removes endogenous carbon compounds from propagules and stimulates respiration, which leads to increasing dependence on exogenous nutrients (Brodie & Blakeman, 1977; Filonow *et al.*, 1983; Hyakumachi & Lockwood, 1989; Jasalavich *et al.*, 1990). In soils, the effect is greatest at saturation and reduces with decreasing matric potential. Saturation not only imposes increasing nutrient stress but also allows the development of competing microorganisms that are better adapted to such conditions. These can utilize spore leachates, provided that sufficient oxygen is available (Filonow & Arora, 1987; O'Leary & Lockwood, 1988). A moderating factor will therefore be soil texture. It has been found that leaching from conidia of *Cochliobolus victoriae*, chlamydospores of *Thielaviopsis basicola*, and sclerotia of *Macrophomina phaseolina* is greater in coarse-textured soils than in clay soils. However, subsequent germination is also higher on coarse soils than on fine-textured ones. This would seem to indicate that germination depends on a complicated balance between the amount of endogenous material that is lost and that remaining in the vicinity of the spore, but which is unadsorbed by clay particles and so is available to the spore (Filonow & Lockwood, 1983b).

Mycostasis

In oligotrophic soils, leaching of metabolites coupled with their rapid uptake by the soil microbiota may induce a condition of mycostasis, in which spores and other propagules fail to germinate and, unless there is enrichment disturbance, become increasingly incapable of doing so; small, slowly germinating spores are more susceptible than larger, more rapidly germinating ones (Lockwood, 1988). Mycostasis can be annulled by: addition of carbon-rich nutrients; or by reducing the microbial population by autoclaving, centrifugation, ultrafiltration or treatment with antibiotics; or through diluting soil with an inert, sterile material. Addition of microbial populations to sterile soil confers mycostasis, as does the mixing of non-sterile soil particles with inert substrata, such as washed sand (Ho & Ko, 1982, 1985, 1986; Filonow & Lockwood, 1983a; Epstein & Lockwood, 1984).

When introduced to mineral soil, propagules quickly release carbon-rich diffusates (Epstein & Lockwood, 1983). Over a short period this may not affect germinability but prolonged leakage, which is exacerbated by leaching, together with additional and greater respiratory carbon losses, causes increasing debility; decreasing germinability is in direct proportion to carbon depletion (Arora *et al.*, 1985; Arora, 1988; Lockwood, 1988; Hyakumachi & Lockwood, 1989). The presence of soil bacteria and Actinomycetes, some of which are attracted to spores and colonize their exteriors, creates the steep diffusion gradient that is responsible for rapid carbon loss (Fradkin & Patrick, 1982; Arora *et al.*, 1983; Ho & Ko, 1986; Lim & Lockwood, 1988).

Mycostasis has been envisaged as a mechanism by which propagules are protected from spontaneous germination in the absence of potentially colonizable substrata, such protection being at the expense of debilitation (Lockwood, 1988). This raises the question of why mycostasis-susceptible fungi have adopted such a high-risk, energetically profligate means of synchronizing germination with enrichment, while others have evolved dormancy-activation systems – which have the overwhelming benefit of energy conservation – to achieve the same end. The answer would seem to be that, far from being a survival mechanism, mycostasis is simply a phenomenon, synonymous with starvation, that is an inevitable consequence when species with a low degree of *S*-selection encounter the high stress conditions obtaining in unenriched mineral soils.

Microbial inhibitors and stimulators

In some situations propagules in soil fail to germinate even where enrichment satisfies their energy requirements. For instance, in alkaline soils the degradation of nitrogen-rich residues, especially chitin, may result in the release of inhibitory levels of ammonia (Schippers & Palm, 1973). However, responses vary widely, conidial germination being sensitive in *Botrytis cinerea*, *Penicillium nigricans*, *Trichoderma hamatum* and *Trichoderma koningii*, but insensitive in *Gliocladium roseum*, *Penicillium chrysogenum* and *Trichoderma harzianum* (Schippers *et al.*, 1982). The mode of action of ammonia in this respect is unclear; although it induces metabolite leakage from *Phymatotrichum omnivorum* sclerotia, it does not do so for conidia of *Cochliobolus victoriae* and *Fusarium solani*, even though their germination is inhibited (Rush & Lyda, 1982; Loffler & Schippers, 1984).

Germinating spores require not only energy sources but also a supply of essential cations, including iron. Availability of the latter is strongly influenced by bacterial siderophores, the effects of which have been most studied in relation to root colonization. Here, specific strains within the *Pseudomonas fluorescens–Pseudomonas putida* group possess high affinity siderophores which sequester iron, so making it unavailable to other rhizosphere–rhizoplane microorganisms (Leong, 1986). In the rhizosphere of cucumber (*Cucumis sativus*) there is a direct correlation between inhibition of *Fusarium oxysporum* chlamydospores and siderophore production by fluorescent pseudomonads, the effect being counteracted by addition of excess iron to the soil (Sneh *et al.*, 1984; Elad & Baker, 1985a,b). In the *Phaseolus vulgaris* rhizosphere, the larger chlamydospores of *Fusarium solani* are not inhibited, which may indicate their independence of exogenous iron. The phylloplane too can be an arena for intense competition for iron. Sporangiospore germination in *Peronospora hyoscyami* (tobacco blue mould) is reduced by cation deprivation, the effects being alleviated if cations are applied directly, or if spores are provided with either chelators or microbial siderophores. This may indicate an ability to increase cation supply by making use of facilitating compounds of host or microbial origin (Johnson, 1988). A contrasting situation is found in *Colletotrichum musae*, where germination of conidia on banana fruit is stimulated by fluorescent pseudomonads and their purified siderophores but is suppressed by addition of iron (McCracken & Swinburne, 1979, 1980). It has been proposed that, within the conidia, iron is held at specific binding sites where it has an inhibitory role. Siderophores act by removing it from these sites, as do a variety of chelating agents, whilst permitting it to remain within the spore (Harper *et al.*, 1980).

The stimulatory effects of microbial activity seem to be as widespread as the inhibitory influences. For example, spore germination of *Glomus* species *in vitro* is stimulated by rhizosphere bacteria and Actinomycetes (Mayo *et al.*, 1986; Azcón, 1987; Mugnier & Mosse, 1987). In nature, it is possible that these vesicular–arbuscular mycorrhizal fungi are susceptible to heat-labile inhibitors released from senescing roots, and that spore surface microorganisms are able to alleviate their effects (Tommerup, 1985). Basidiospores of the ectomycorrhizal species *Hebeloma crustuliniforme* are stimulated by *Pseudomonas stutzeri* isolated from its own basidiomata, and by *Corynebacterium* from both basidiomata and mycorrhizal *Salix* roots. Basidiospore germination in *Laccaria laccata*, *Hebeloma crustuliniforme* and *Paxillus involutus* has been found to be similarly stimulated by an unidentified soil bacterium (Ali & Jackson, 1989).

The above examples of microbial effects suggest that there are clear-cut positive and negative influences on germination which, depending on circumstance, can be related directly to the ecological behaviour of the fungi concerned. This may well be so, but it should be borne in mind that, in general, studies have been based on observations of fungus–microorganism pairings *in vitro*. In many natural habitats, the composition of the microbial flora is probably such that whether or not germination can take place will depend on where the balance lies between stimulatory and inhibitory factors. This, in turn, will be affected by non-microbial compounds in the environment, for instance diffusates from plant tissues, which may either modify microfloral composition and activity or directly intervene in the germination process.

Non-microbial inhibitors and stimulators

Taken in the broadest sense, the number of non-microbial compounds that are capable of influencing germination is virtually limitless, and encompasses nutrients, fungicides and industrial pollutants. Even when these are excluded, the scope remains very wide, there being, for example, a large literature on inhibitory and stimulatory factors in leaves and leaf litter, wood, seeds, roots, and enriched and unenriched mineral soils (Schroth & Hildebrand, 1964; Mitchell, 1976; Irvine *et al.*, 1978; Hardie, 1979; Lockwood & Filonow, 1981; Nelson, 1991). However, in the vast majority of such studies it has not been possible to assign roles to specific compounds, so that it is often difficult to distinguish between factors – usually nutritional – that affect fungi generally, and those that are ecologically discriminatory in that they determine which assemblages or individual species will establish themselves in particular habitats. Consideration here is restricted to examples of the latter; but it needs to be emphasized that, although some substances of non-microbial origin can impact directly on spore physiology, others must first be converted to different active compounds by a microbial flora. The distinction between microbial and non-microbial factors then becomes blurred.

The greater part of research in this area has been concerned with behaviour in soil, this bias being clearly related to the agricultural importance of soil-borne plant pathogens. Within this complex environment it is to be expected that many fungi, particularly species with specialized host or substrate requirements, possess recognition mechanisms (different from activation systems described in the previous chapter) which ensure that germination takes place largely, or exclusively, in circumstances that are favourable to further development. In numerous situations this might simply involve a response to the proximity of a suitable nutrient source, but (putting aside those gross nutrients which would affect all fungi within their zone of influence) this being mediated by either non-nutrient or minor nutrient metabolites arising from that source. For instance, eruptive myceliogenic germination of *Sclerotium rolfsii* sclerotia in soil is triggered by volatile compounds emanating from dead plant tissues that have been rewetted after a drying period. These can act at a considerable distance, and are also responsible for directing growth of hyphal aggregates towards the tissues, both effects being attributable to a range of alcohols and aldehydes with, perhaps, isopropyl and butyl alcohol as major constituents (Punja *et al.*, 1984). Similarly, conidia of *Alternaria alternata* and *Fusarium solani* exhibiting mycostasis in mineral soil germinate in response to volatile peroxidation products of fatty acids. In the same circumstances, chlamydospores of *Fusarium solani* germinate if the soil atmosphere is enriched with octonal, nonanal, decanal or undecanal at concentrations far below those necessary for the exertion of a nutritional effect (Harman *et al.*, 1980).

Many such volatile germination stimulators of great structural diversity have been identified (Table 8.1). Whilst a number of them are

Table 8.1 Examples of volatile germination stimulators (after French, 1985).

Species	Propagule	Stimulator
Agaricus bisporus	Basidiospore	Isovaleric acid; isoamyl alcohol
Alternaria alternata	Conidium	2,4-hexadienal; 2,4-nonadienal; octanal; inonanal
Colletotrichum musae	Conidium	2,3-dihydroxybenzoic acid; anthranilic acid; catechol
Diaporthe perniciosa	Pycnidiospore	*p*-coumarylquinic acid; chlorogenic acid; caffeic acid
Entomophthora culicis	Conidium	Oleic acid (with chitin)
Fusarium solani	Conidium Chlamydospore	2,4-hexadienal; 2-hexene-1-ol; 2,4-nonadienal; nonanal
Penicillium sp.	Conidium	Nonanal

released from either intact or degraded plant tissues, for the most part their possible specific ecological roles within soil or litter remain conjectural, although it is probable that several facilitate substratum recognition. Of special interest in this regard is nonanal, previously mentioned in relation to autostimulation (see Fig. 8.5) which, together with some of its derivatives, appears to be both widespread and highly effective.

In some habitats, perhaps especially those characterized by high levels of phenolic compounds, inhibitors rather than stimulators may determine community structure. For instance, with regard to colonization of fallen conifer needles, ferulic acid stimulates germination in *Dothichiza pityophila* and *Thysanophora penicilloides* (both needle colonizers) but inhibits it in fungi that are typical saprotrophs within angiosperm leaf litter (Black & Dix, 1976).

As might be anticipated, specificity of response within soil is most highly developed amongst specialized symbionts. For example, sporangia of pathogenic *Pythium* species germinate rapidly in response to volatiles, principally ethanol, released by germinating seeds, whilst germination of encysted *Phytophthora* zoospores is induced or accelerated by the pectic constituents of root mucigel (Grant *et al.*, 1985; Nelson, 1987). Narrower host-specific effects are also known. Because of mycostasis, sclerotia of *Sclerotium cepivorum*, the onion white rot pathogen, will not germinate in soil unless they are in the presence of *Allium* species. Root exudates of the latter contain non-volatile alkyl- and alkenyl-cysteine sulphoxides which are metabolized by soil microorganisms to liberate volatile sulphur compounds, for instance 1-propyl and 2-propenyl compounds, which are effective germination stimulants (King & Coley-Smith, 1969; Esler & Coley-Smith, 1983). Sclerotia of *Stromatinia gladioli* behave similarly with respect to species of Iridaceae, although the metabolites involved have not been identified (Jeves & Coley-Smith, 1980). These and similar fungi have adopted highly effective methods to conserve inoculum in the absence of hosts. Such behaviour is not confined to pathogens, and occurs amongst mutualistic fungi. On synthetic media, the basidiospores of many ectomycorrhizal species either fail to germinate or do so only poorly. Root exudates from appropriate host plants may promote germination, the older parts of roots being more effective in this regard and with activity residing, apparently, in ninhydrin-positive compounds (Birraux & Fries, 1981; Theodorou & Bowen, 1987; Ali & Jackson, 1988).

Aerial plant parts may also contain germination stimulators which have specific roles. For example, teliospores of *Puccinia carthami* respond to volatile polyacetylenic hydrocarbons from *Carthamus tinctorius* (safflower) seedlings (Binder *et al.*, 1977, 1978). In laboratory conditions, self-inhibition of rust urediniospores within pustules is banished by many flavour compounds, including β-ionone (see Fig. 8.5), nonanol and nonanal, but their function in the natural infection process, if any, is yet to be determined (French, 1985).

Photocontrol In many fungi, germination is to some extent light-sensitive. However, in most instances, the ecological significance of photocontrol mechanisms is far from clear. Generally, light appears to be stimulatory, but there is a dearth of information on the spectra required and their energy contents. What evidence there is suggests that wavelengths within the blue portion of the visible spectrum are involved. For instance, conidial germination in *Erysiphe pisi* is enhanced *in vitro* by irradiance of 430 and 496 nm and is inhibited at 551 nm (Ayres, 1983). Stimulation also occurs in spores which would, in nature, germinate in the dark: oospores of *Phytophthora citricola* responding to blue light in the 400–420 nm range, but also to near UV light at 350–400 nm (Pluorde & Green, 1982). Similarly, oospores of *Phytophthora megasperma* are stimulated by wavelengths of 450–475 nm (Förster *et al.*, 1983). There is some evidence to suggest the presence of a flavin-type photoreceptor within light-stimulated oospores (Banihashemi & Mitchell, 1989). In oospores, light conditions during maturation may determine subsequent response so that, in *Phytophthora megasperma*, spores matured in the light do not germinate as well as those matured in darkness, whilst in *Phytophthora parasitica* the reverse is the case (Schechter & Gray, 1987; Ann & Ko, 1988).

Photoinhibition has also been observed, for example, in *Verticillium agaricinum* conidia where the active wavelengths are in the near UV at 320–430 nm (Osman & Valadon, 1981). This contrasts with the situation for some rust urediniospores which are inhibited by far red at 720 nm. Simultaneous irradiation with red light at 653 nm and inhibitory far red nullifies the effects of the latter, which indicates the operation of a photoreversible pigment as a receptor (Lucas *et al.*, 1975). Washing urediniospores relieves inhibition, which apparently acts at some point prior to germ tube emergence, as does autoinhibition (Knights & Lucas, 1980). For urediniospores of *Puccinia graminis*, and perhaps also for those of many other rusts, the usual germination-limiting factor is availability of water on the leaf surface. Rusts have a diurnal pattern of spore release, so that inoculum deposition is most likely to occur in daylight. Here, photoinhibition prevents the initiation of germination during transient periods of leaf wetness. In the field, urediniospores begin to germinate at dusk, night, with its normally lower temperatures and the possibility of dewfall, providing more reliable germination conditions (Knights & Lucas, 1981; Subrahmanyam *et al.*, 1988). A further facet of photoinhibition is exhibited by urediniospores of *Phakospora pachyrhizi* (soybean rust) which when illuminated unilaterally produce germ tubes only from the shadowed side of the spore, germ tube growth then being directed away from the light (Koch & Hoppe, 1987).

9 Orientation

The new individual arising from germination enters a vegetative phase which culminates in propagule production. During this, the major part of its cycle, it must overcome the sequential problems inherent in establishment, habitat exploration and exploitation and, finally, the positioning of sporophores so as to ensure escape via dissemination. Successful completion of each stage will depend on a large element of chance, but will also be favoured by an ability to react to appropriate stimuli that can determine non-random behaviour with respect to salient habitat features. Orientation systems essential for bringing gametes together were examined in Chapter 5, but other kinds of directed movement and growth in response to environmental cues are widespread, the latter embracing inorganic compounds, metabolites from animals, plants and microorganisms, light, gravity, and the structural features of substrata. These various aspects of sentience have been well reviewed in great detail, some a number of times, so that not only is there a wealth of familiar information on phenomenology, but also its ecological significance is, in many instances, fully understood (see Carlile, 1970, 1975; Hickman, 1970; Gooday, 1975; Wynn, 1981).

Whilst this is to be welcomed, there has been a tendency to focus on cellular mechanisms, with all responses to a specific stimulus, for example light, being considered within the same physiological context. However, the ecological reality is that the influence of any single factor will be qualified by several other interacting, and often conflicting, physicochemical stimuli, the nature of which will differ at various stages of development. Accordingly, emphasis here will be shifted towards a more general account of orientation as a requisite for success during habitat selection, colonization and sporulation.

Directed arrival

As a general rule, and with the possible exception of highly *C*- and *S*-selected species, colonization of a virgin substratum ahead of competitors first requires the establishment of positional primacy. In this regard, a major advantage would accrue from an ability for directional response to metabolites diffusing from such a living or dead substratum; selection of suitable colonization sites would be facilitated and also, perhaps, avoidance of unfavourable ones. Since optimum conditions for establishment may not obtain for very long, speed is of the essence here in order to avoid the possible onset of stress, disturbance or the arrival of other potential colonists. An additional problem encountered by many symbiotic fungi is the necessity to locate a specific entry route into the host; here physical, rather than chemical, guidance may often be involved.

Zoospore taxes Zoospores are unique amongst propagules in having a capability for autonomous site selection, although the extent to which this can be expressed depends on much more than simply the availability of an adequate water film; their performance under idealized conditions may depart markedly from that in nature. This is particularly so with respect to the distances that they are able to travel under their own volition.

Following discharge, maximum motility periods vary widely but, in optimum laboratory systems, have been estimated to be about 24 hours for uniflagellate and 20–30 hours for biflagellate spores (Lange & Olson, 1983). However, it seems likely that the bulk of any population will swim for less than 24 hours, the mean activity period being possibly 6–10 hours, which at observed swimming speeds of 90–160 $\mu m\ s^{-1}$ would give zoospores the potential to move several metres before encystment (Carlile, 1986). Obviously, this distance cannot be realized because of the high frequency of spontaneous changes in direction which, in *Phytophthora cinnamomi* for example, may occur at an average rate of $8°\ s^{-1}$. Redirection also occurs following collision, further reducing maximum displacement pathways, especially within particulate substrata; even in coarse, wet sand, zoospores may travel at most 6 cm from their starting point, the journey for the great majority being considerably shorter (Newhook *et al.*, 1981). High soil water potentials not only favour zoospore discharge, but also give rise to water flow through the pore system. If flow rates exceed swimming speed, then zoospores will be carried in the direction of the current; if flow rates are the same as, or lower than, swimming speed, zoospores regain control over their movement, and are thus able to respond to appropriate environmental stimuli (Young *et al.*, 1979). It would seem that, in perhaps the majority of situations, long-distance movement of zoospores is an essentially passive process. Then, having arrived at a destination in random fashion, swimming and the numerous directional changes associated with it allow zoospores to thoroughly explore a limited amount of the space surrounding them.

General irritability. In laboratory conditions, zoospores released into stable diffusion gradients of non-toxic organic solutes commonly exhibit positive chemotaxis, and this undoubtedly reflects similar behaviour in nature as a response to diffusates from living or dead cells (Bean, 1979; Carlile, 1983, 1986; Lange & Olson, 1983). Selected aspects of chemotaxis are detailed later, but some general comments might be made here. As perhaps would be expected, potential nutrients (mainly carbohydrates and amino acids) frequently guide swimming. However, non-nutrients can also fulfil this role. For example, zoospores of the rumen flagellate *Neocallimastix frontalis* show positive chemotaxis to a range of soluble carbohydrates, including mannose, sorbose and sorbitol, none of which can support its growth (Orpin & Bountiff, 1978; Mountfort & Asher, 1983). Negative chemotaxis has also been observed, notably in response to hydrogen ions and other cations. This occurs irrespective of the presence of counter-ions, and in conditions close to the threshold for cation-induced damage to the zoospores, thus enabling them to avoid highly acidic or cationic environments (Byrt *et al.*, 1982; Pommerville & Olson, 1987).

Whatever the peculiarities of their eventual homing behaviour, zoospores will also experience the force of gravity, so that in static water they will tend to sediment out. Amongst Oomycetes, this effect is counteracted by negative geotaxis which, in soil water films, may additionally move zoospores from the stationary boundary layers into the flowing layers where passive, long-distance dispersal can then take place. The ability for negative geotaxis may reside in zoospore conformation, being brought about through a combination of cell rotation and a propulsive force located anterior to the centre of gravity pushing the spore upwards (Bean, 1984). In Chytridiomycetes, similar upward swimming is brought about by positive phototaxis. For instance, zoospores of the marine species *Rhizophydium littoreum* are strongly attracted to white light, but at low intensities only to blue light of approximately 400 nm wavelength. This part of the spectrum is the most effective in penetrating oceanic waters, and in their normal environment of turbid coastal areas *Rhizophydium* zoospores could receive sufficient blue light to stimulate phototaxis at depths of up to 13 m. Positive phototaxis would facilitate their positioning within the photic zone inhabited by sessile algae on which they could then settle, this being aided finally by positive chemotaxis towards carbohydrates and amino acids (Muehlstein *et al.*, 1987, 1988).

Host location. Research on chemotaxis has concentrated almost exclusively on its role in host location by parasitic species, its great survival value being evidenced by the wide ecological diversity of those fungi that demonstrate it, amongst which are mycoparasitic and nematode-destroying chytrids and oomycetous pathogens of fish (Sayre & Keeley, 1969; Held, 1973; Smith *et al.*, 1984). It is, however, root-infecting species that have received most research attention, particularly the responses of *Phytophthora* and *Pythium* zoospores to host exudates.

There are numerous examples of chemotaxis towards roots *in vitro*, with encystment occurring mainly on the elongation region, which suggests that the necessary solute gradients originate from this zone (see Mitchell, 1976). However, there may also be attraction to, and encystment upon wounds and, in some instances, root cap cells (Goldberg *et al.*, 1989). For example, zoospores of *Pythium dissotocum* show a chemotactic response specifically to root cap cells of cotton (*Gossypium barbadense* and *Gossypium hirsutum*) with accumulation, encystment and germination occurring within minutes in cell–zoospore suspensions (Fig. 9.1). The process of directed swimming is quite distinct from those of attachment and encystment, and each is controlled by a different mechanism (Callow, 1984). Many compounds, including sugars, organic acids, amino acids, peptides, purines and pyrimidines, nucleotides, vitamins, and plant growth regulators, are capable of inducing zoospore accumulation, but most act by trapping randomly swimming spores rather than by directing them to the diffusion source. In *Phytophthora* and *Pythium*, chemotaxis *sensu stricto* seems to be largely brought about by non-specific responses to certain amino acids, principally arginine, aspartate, glutamate and methionine (Royle & Hickman, 1964; Khew & Zentmeyer, 1973). Ethanol can also act as an attractant, which is of interest since it is

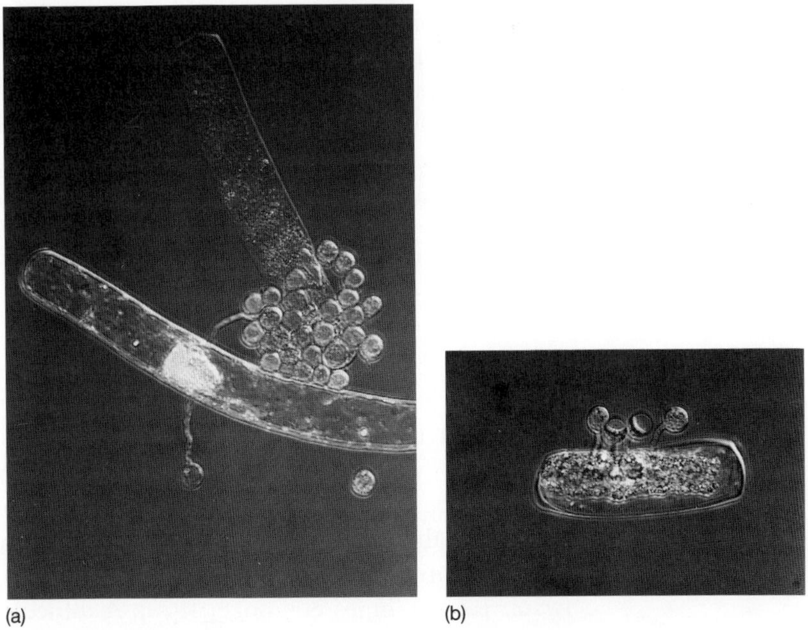

Fig. 9.1 Colonization of cotton root cap cells by *Pythium dissotocum* zoospores. (a) Accumulation of zoospores around an attractive root cap cell 1 minute after cells were added to a zoospore suspension; (b) germination of encysted zoospores and penetration of a cap cell (from Goldberg *et al.*, 1989, © National Research Council of Canada).

secreted by roots in anaerobic conditions arising from waterlogging; the latter also favours asexual reproduction and zoospore release (Allen & Newhook, 1973; Halsall, 1976). Chemical attraction may be augmented by a mechanism of autoaggregation (Thomas & Peterson, 1990). For instance, zoospores of *Achlya* species produce heat-stable attractants which cause them to congregate in the absence of exogenous chemotactic signals. This behaviour might serve to amplify initial attractant stimuli from the substratum via the induction of swarming.

Claims have been made for specificity of response. For example, it has been reported that *Phytophthora cinnamomi* zoospores rarely show positive chemotaxis to roots of non-hosts and that, similarly, those of *Phytophthora megasperma* respond only to roots of susceptible alfalfa cultivars, resistant cultivars evoking at best a weak response (Zentmeyer, 1966; Chi & Sabo, 1978). However, in view of the non-specific chemical composition of root exudates, such observations should be treated with caution.

If positive chemotaxis to roots is non-specific, then its end-point of zoospore adhesion and encystment may not be so, and it is probably here that host recognition takes place (Callow, 1984). Evidence suggests that adhesion is mediated by carbohydrate determinants of the root mucigel, the zoospore carrying receptors with specificity for complementary saccharide-containing ligands of the root slime (Hinch & Clarke, 1980; Irving & Grant, 1984; Grant *et al.*, 1985). There may also be recognition at the root cell level, although its mechanism is not clear. For instance, it has been observed that during infection of onion roots by *Pythium coloratum*, zoospores encyst preferentially on parts of the epidermis lying above short, but not long, hypodermis cells (Shishkoff, 1989).

Identification of stomata. Most, if not all, zoosporic leaf pathogens can enter their hosts only through stomata; success rates are high when encystment takes place upon or close to guard cells, or above the vestibule, or within the substomatal cavity, but are drastically reduced when it occurs on epidermal cells. Although there are suprisingly few detailed studies on swimming behaviour, there is strong evidence for the preferential encystment of zoospores over stomata (Arens, 1929a,b; Iwata, 1957; Whipps & Cooke, 1978). A difficulty arises here in distinguishing between chemotaxis in response to diffusion gradients originating at stomata, and the random trapping by stomata of zoospores that are moving feebly at the end of their motile phase. For instance, *Plasmopara viticola* zoospores on illuminated grapevine leaf discs encyst gregariously in groups of up to 10 over stomata. In the dark they do so in groups of up to 28, but over only those stomata that remain open, which indicates chemotaxis in response to diffusates from the leaf interior. By contrast, *Pseudoperonospora humuli* zoospores on illuminated hop leaf discs become associated singly with each stoma, but in the dark encyst randomly on the leaf surface. Chemotaxis is not involved here. What appears to happen is that a chemical stimulus linked to photosynthesis in guard cells considerably reduces the motile period; a physical stimulus, imparted by the configuration of the open stoma, then encourages settling (Royle & Thomas, 1973). Presumably the dimensions of the stoma cannot accommodate more than one zoospore.

Directional growth

Establishment of positional primacy is not only a question of arrival at the right time in the right place, but also depends on rapid consolidation on or in the new substratum. Speed is essential here, together with economic expenditure of endogenous energy reserves so as to maximize inoculum potential, the latter being favoured if germ tubes can react directionally with respect to colonization sites. Whilst germ tube behaviour will mainly be discussed here, the kinds of responses involved are shared, as will be seen later, by established hyphae and organized mycelia during guided habitat exploration. Such orientations are usually referred to as tropisms, but this terminology is eschewed here in favour of directional growth. Tropism carries the connotation of a repositioning of an organ via differential growth of one side of an intercalary extension zone. In fungi, with the notable exception of some sporophores, this does not take place; orientation is achieved through the repositioning of an apical extension zone rather than by the bending of an existing structure.

General responses. Directional growth of germ tubes towards diffusible factors appears to be widespread, although rarely has this been attributed to specific agents. For example, in *Achlya ambisexualis* there is a positive response to a variety of amino acids and peptides, and root-infecting higher fungi show similar behaviour towards root exudates (Schroth & Hildebrand, 1964; Musgrave *et al.*, 1977; Wall & Lewis, 1980; Manavathu & Thomas, 1985; Jansson *et al.*, 1988). However, it has been pointed out that in terrestrial habitats, especially within soil, continuity of the water phase is normally interrupted by a vapour phase, so that whilst concentration gradients of gases can occur across the vapour phase, those of non-volatile compounds cannot (Carlile & Matthews, 1988). At least in the rhizosphere soil environment, both the amount and diversity of volatile root exudates exceed those of water-soluble exudates. Furthermore, water-soluble compounds must pass through the microbially rich rhizosphere, and so may be removed or inactivated before being able to travel very far; volatiles differ in diffusing rapidly and distantly, and being less susceptible to microbial depletion (Gemma & Koske, 1988). It would therefore seem likely that terrestrial fungi have sensitivities more attuned to volatiles or gases than to non-volatiles. There is some evidence to support this view. Volatiles induce germ tubes of *Chaetomium globosum* and *Gigaspora gigantea* to grow towards wood and roots respectively, and a similar response to roots by *Phytophthora citricola* is attributable to growth either down an O_2 or up a CO_2 gradient (Koske, 1982; Carlile & Matthews, 1988; Carlile & Tew, 1988; Gemma & Koske, 1988).

Contact responses. Whilst adaptations to diffusates are to be expected amongst occupants of habitats where gradients can occur frequently enough, and with sufficient duration, to direct growth successfully, even those of soil volatiles are susceptible to disturbance during natural, minor perturbations of the general environment. This problem is most severe in aerial habitats, and is exemplified during location of stomata by some leaf pathogens. On sugar beet leaves incubated at high RH, but in the absence of a surface water film, conidial germ tubes of *Cercospora beticola* exhibit directional growth towards stomata and penetrate them, probably in response to a water vapour gradient. Similarly, if a conidium arrives close to a stoma, then the germ tube emerges at that point along the spore which is nearest to the opening. These responses do not take place if the leaf is wet; instead, germ tubes arise at random, wander over the leaf surface, and do not deviate towards stomata even when a close approach is made by chance (Rathaiah, 1977). Thus, although directional growth up the water vapour gradient allows precise location of stomata, the system is highly vulnerable to disruption by stomatal closure, rainfall or dew deposition. This is probably one reason why other leaf-infecting fungi, in particular rusts, have acquired an alternative means of stomatal location, in which germ tubes are guided towards pores through contact with leaf surface features rather than by stomatal emanations (Wynn, 1981; Callow, 1984). Contact guidance is brought about through the action of exogenous mechanical stimuli that fix the alignment of the axis of germ tube growth but do not determine the direction of growth along that axis. It therefore lacks the precision of chemically directed growth but is probably much less prone to the effects of microclimatic changes.

In most rust species, urediniospore germ tubes can enter hosts only via stomata, one exception being *Phakospora pachyrhizi* on soybean where there is direct epidermal penetration. When growing across the leaf, germ tubes tend to orient themselves at right angles to linear structural features, for instance epidermal cell wall ridges, or become aligned in response to regular, repetitive surface patterns. If surface topography is suitably related to stomatal position, then this behaviour results in a much increased chance of locating the pore.

During infection of wheat by *Puccinia graminis tritici* and maize by *Puccinia sorghi*, germ tubes extend in a direction parallel to the short axis of the leaf (Fig. 9.2). Stomata are arranged in longitudinal rows in which they alternate with epidermal cells, each stomatal row being separated from the next by several epidermal cells, and a stoma in one row tending to correspond with an epidermal cell in the adjacent stomatal row (Fig. 9.3). This distribution pattern is such that germ tubes growing transversely across the leaf have a high probability of contacting a stoma, and this axis is determined by the fine surface structure of the cuticle (Lewis & Day, 1972). In wheat, the latter is covered by a regular lattice of wax crystals, with a repeat pattern of approximately 800 nm along the transverse leaf axis and 1000 nm along

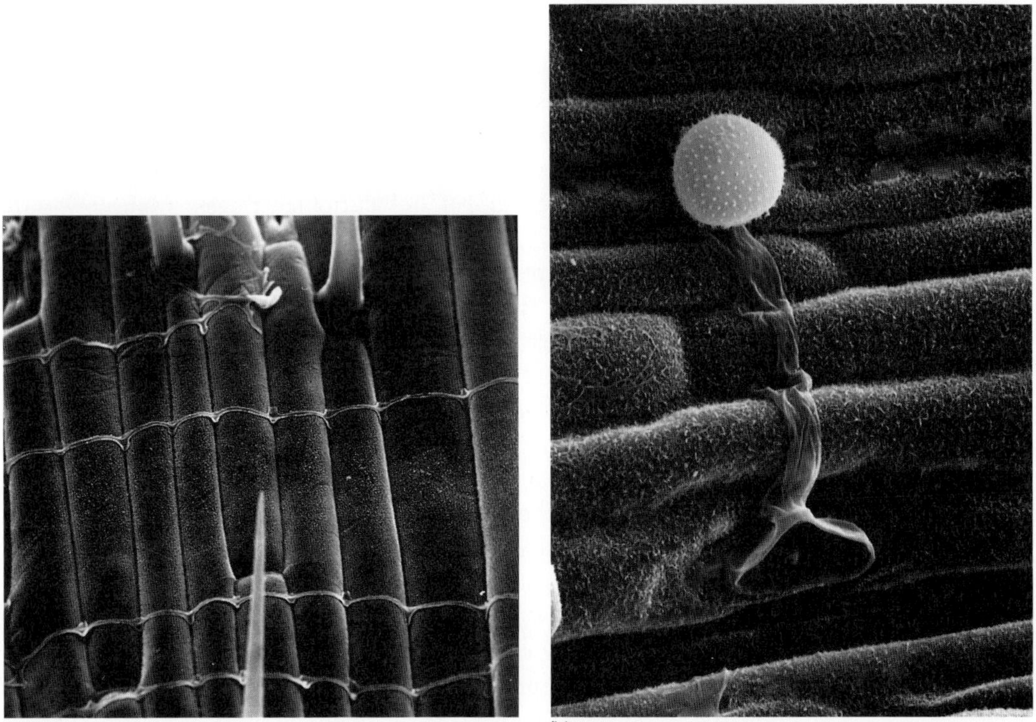

(a) (b)

Fig. 9.2 Directional growth of urediniospore germ tubes. (a) *Puccinia graminis tritici*, four germ tubes growing across a wheat leaf at right angles to the long axis of the leaf (from Lewis & Day, 1972, © British Mycological Society); (b) *Puccinia sorghi*, urediniospore on a maize leaf with the germ tube extending at right angles to the long axis of the epidermal cells. A terminal appressorium has formed over a stoma (photograph by W.K. Wynn, by courtesy of V.A. Wilmot).

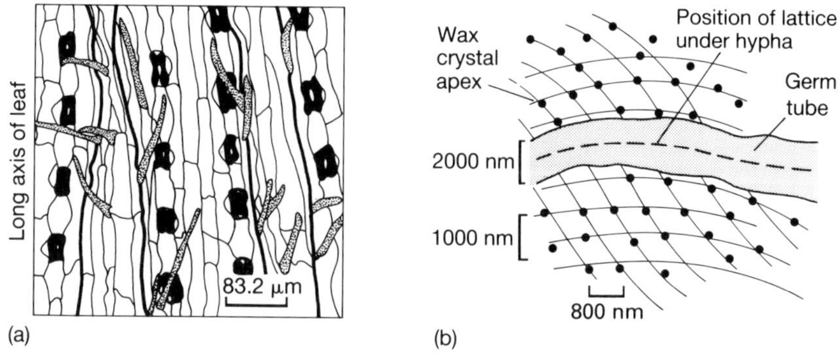

Fig. 9.3 Morphology of the wheat leaf surface in relation to directional growth of urediniospore germ tubes. (a) Arrangement of epidermal cells (outlined), stomata (in black) and trichomes (speckled) relative to the long axis of the leaf; (b) alignment of the germ tube relative to the crystal lattice which occurs on each epidermal cell surface (from Lewis & Day, 1972, © British Mycological Society).

the longitudinal axis. At germination, the germ tube tip adheres to the crystals and then proceeds to extend along a transverse row in the lattice. Germ tube width is about 2000 nm, so that as it grows it remains bounded on each side by a transverse row of crystals (Fig. 9.3). Stomata of *Antirrhinum majus* are also linearly arranged, and favourable orientation of *Puccinia antirrhini* germ tubes is achieved by their growth at right angles to parallel cuticular ridges (Maheshwari & Hildebrandt, 1967).

Many leaf- and stem-infecting species that normally penetrate the cuticle directly, rather than via stomata, also exhibit contact guidance, but respond only to gross features of leaf surface topography; germ tubes and hyphae travel, for example, along the grooves between epidermal cells. Irrespective of whether a fungus is responding to coarse or fine structure, contact guidance should be seen as a prelude to appressorium formation, which in itself appears to be mainly a contact response. It is from the appressorium, or more elaborate structures such as infection cushions, that the final orientation necessary for establishment takes place – the growth of penetrant hyphae into the substratum.

There is uncertainty concerning both reception and transduction of stimuli that trigger appressorium differentiation, and the nature and intensity of the signal required vary widely (Bourett *et al.*, 1987; Hoch & Staples, 1987; Staples & Hoch, 1987). For instance, in necrotrophs it may be a response to a hydrophobic surface or to intercell grooves, whilst amongst rusts some respond to smooth surfaces and others to minute surface elevations (Maheshwari & Hildebrandt, 1967; Lewis & Day, 1972; Wynn, 1976; Lapp & Skoropad, 1978).

Responses to microorganisms. Directional growth is crucial to the establishment of some interfungal and fungus–microorganism associations (see Chapter 10), although documented examples of this largely relate to antagonistic rather than mutualistic symbiosis. For instance, when growing in low nutrient conditions, vegetative hyphae of the basidiomycetous species *Agaricus bisporus*, *Coprinus quadrifidus*, *Lepista nuda* and *Pleurotus ostreatus* are strongly attracted to bacterial microcolonies (Barron, 1988). On reaching them, hyphae produce coralloid structures which bring about lysis of the bacterial cells thus, presumably, releasing nutrients for absorption (Fig. 9.4). *Pleurotus ostreatus* also produces droplets on its aerial hyphae which can paralyse nematodes on contact (Barron & Thorn, 1987). Subsequently, additional hyphae home towards the animal, penetrate it through one of its orifices and then, having killed it, digest the body contents (Fig. 9.5).

Similar behaviour frequently occurs amongst mycoparasites, and germ tubes of haustorial biotrophs, for example *Piptocephalis* species, commonly grow towards host hyphae, being particularly directed to their apices (Fig. 9.6). This has survival value, since axenic development of mycoparasitic biotrophs is frequently severely limited, with failure to contact a host resulting in rapid lysis and death of germlings. However, hosts do not invariably elicit a directional response. *Circinella mucoroides* is highly susceptible to *Piptocephalis fimbriata* but germ tubes of the latter do not show directional growth

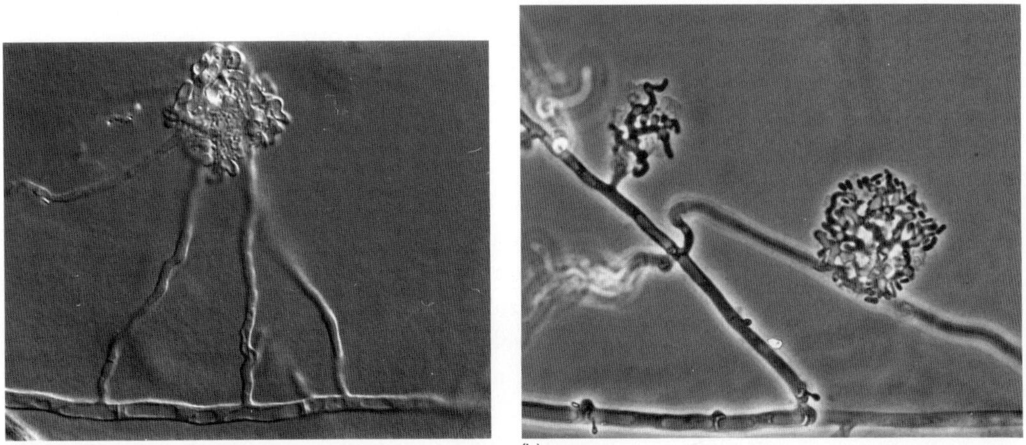

(a) (b)

Fig. 9.4 Directional growth of basidiomycete hyphae to bacterial colonies. (a) *Pleurotus ostreatus*, lateral hyphae growing into and attacking a microcolony of *Pseudomonas*; (b) *Agaricus bisporus*, branching absorptive hyphae within lysing microcolonies (from Barron, 1988, © National Research Council of Canada).

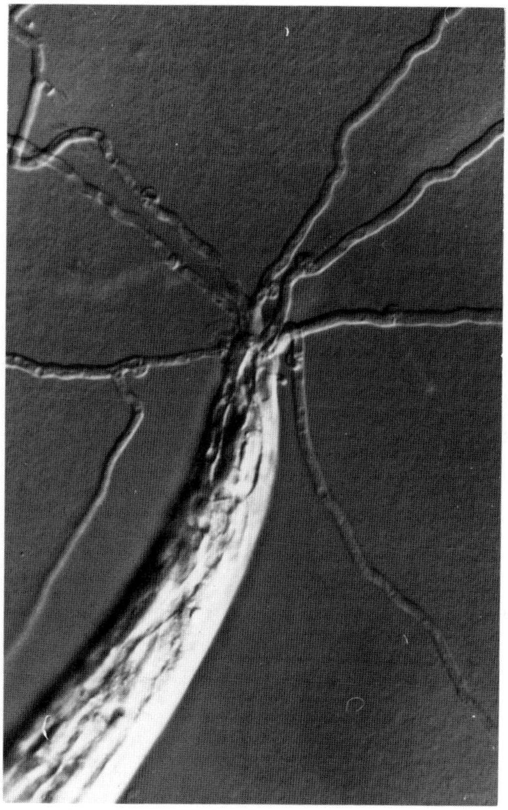

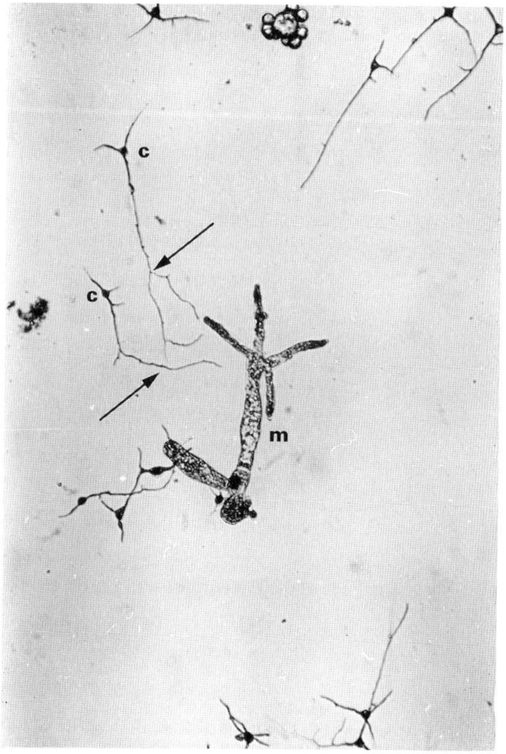

Fig. 9.5 (*left*) Directional growth of basidiomycete hyphae to narcotized nematodes. Hyphae of *Pleurotus ostreatus* homing to the buccal cavity of an immobilized nematode (from Barron & Thorn, 1987, © National Research Council of Canada).

Fig. 9.6 (*right*) Conidia (c) of the mycoparasite *Piptocephalis fimbriata* with germ tubes (arrowed) showing directional growth towards a germling mycelium (m) of *Mycotypha microspora* (from Evans *et al.*, 1978, © *New Phytologist*).

toward its hyphae (Evans *et al.*, 1978). Converse behaviour has been observed during mycoparasitism by contact biotrophs. Germinating spores of *Calcarisporium parasiticum* and *Gonatobotrys simplex* induce directional growth of host hyphae towards them, whilst in *Gonatobotrys fuscum* hyphal branches of the host grow towards those of the mycoparasite, to finally make end-to-end contact (Shigo, 1960; Whaley & Barnett, 1963).

In lichen synthesis, where directional growth of germ tubes or hyphae might be anticipated, evidence for this is lacking, and it may well be that chemotaxis of algal zoospores is of greater importance (Slocum *et al.*, 1980; Garty & Delarea, 1988). In this regard, hyphae of some ectomycorrhizal and vesicular–arbuscular mycorrhizal species also fail to respond directionally to host roots, even though proximity of the latter elicits greater hyphal branching.

Trophic and social orientations

In filamentous species, the established germling provides a springboard for habitat exploration by the expanding mycelium, which has an inherent capacity for the ordered growth necessary for it to adequately carry out its trophic function. Efficient sequestration of resources requires direction of leader hyphae towards nutrients and, subsequently, an appropriate disposition of their branches for maximum nutrient acquisition. Following this, in septate fungi, consolidation of the colony, co-ordination of its activities, and organogenesis are achieved via anastomoses that are brought about by oriented branch growth. Finally, as will be seen in Chapter 10, the ability to defend captured territory against invaders, or to oust occupants from their domains, often involves oriented growth responses. The interrelationships between hyphal orientations and colony form have been intensively studied using agar-grown cultures, the two-dimensional, physicochemically homogeneous system provided by the Petri dish being ideal for detailed, continuous observation; much is known of the fundamental processes that are involved. This information is frequently valuable for the interpretation of growth patterns in natural habitats, but it is much less so as a basis for predicting how fungi might behave in particular ecological situations, especially where development is three-dimensional and the substratum is physicochemically and biotically heterogeneous (Cooke & Rayner, 1984; Rayner & Coates, 1987).

Colony form

The history of the typically circular, radially expanding colony has been mapped numerous times (Prosser, 1983; Trinci, 1984; and see Chapter 3). Those hyphal orientations that confer this form become established soon after germination of the initiating spore via modification of the angles which branches subtend to their parent hyphae. For instance, in young mycelia of *Neurospora crassa*, the branching angle is 90° but becomes reduced to 63° as development proceeds, so pointing new branches away from the colony centre (McLean & Prosser, 1987). Such colonies characteristically exhibit a high degree of spatial organization, with regularly spaced, parallel marginal hyphae which rarely cross one another.

Maintenance of this state appears to involve avoidance responses, mutual repulsion occurring between neighbouring hyphae if they approach one another within a critical distance. Avoidance reactions also occur between germ tubes over distances of 10–30 μm, repulsion being attributable to growth away from zones of O_2 depletion (Robinson, 1973b). However, there is no firm evidence to suggest that the same mechanism operates in colonies, particularly since leader hyphae are usually separated by much greater distances both from one another and from the tips of their nearest branches. In fact there is reason to doubt whether mutual repulsion is essential for generation of circular colonies (Hutchinson *et al.*, 1980). Using *Mucor hiemalis*, data for hyphal extension rate, interbranch distance and branching angle have been used in computer simulations of colony development. A circular morphology was rapidly obtained solely as a consequence of randomly varying growth within these parameters, and without the necessity of avoidance responses. Yet, although the computer-bred colonies were certainly circular, their constituent 'hyphae' lacked the high degree of organization typical of real colonies of the same age (Fig. 9.7). This implies that whilst circular expansion may take place independently of avoidance reactions, the latter are important for the spatial arrangement of leader hyphae and their branches.

Whatever mechanisms underlie their generation, in nature circular colonies can be formed only on suitable solid or liquid surfaces, for instance, on freshly exposed foodstuffs or newly cut wood. In most other circumstances, colonies experience various degrees of physical confinement, interruption of nutrient supply, and conflict with co-inhabitants; all of these, either singly or in combination, cause departures from regular radial expansion. Therefore, natural colony forms are of infinite variety. Furthermore, in many situations, exploratory growth is effected by hyphal aggregates, such as cords and rhizomorphs, rather than by a diffuse mycelium. This level of organization requires hyphal fusion, a process which is also essential for other functions of the trophic mycelium.

Guided fusions

A characteristic of septate fungi is that, at some point beyond the germling stage, the potential emerges for fusions between vegetative hyphae (Watkinson, 1978; Gregory, 1984). Anastomoses create a nexus through which metabolites, cytoplasm and nuclei can travel, so minimizing the effects of localized trauma by providing alternative routes for translocation and transmigration. They also facilitate co-ordination within the colony, and can unite somatically compatible individuals of the same species, thus reducing intraspecific competition for resources. More importantly, as will be discussed in the following chapter, fusions mediate self and non-self recognition systems during intraspecific interactions, with subsequent far-reaching ecological consequences.

Irrespective of their role, fusions result from accurately guided hyphal growth, the mechanisms for which remain largely unknown. Interhyphal attraction may occur between two hyphae that have achieved a close approach initially by chance, or between a hypha and

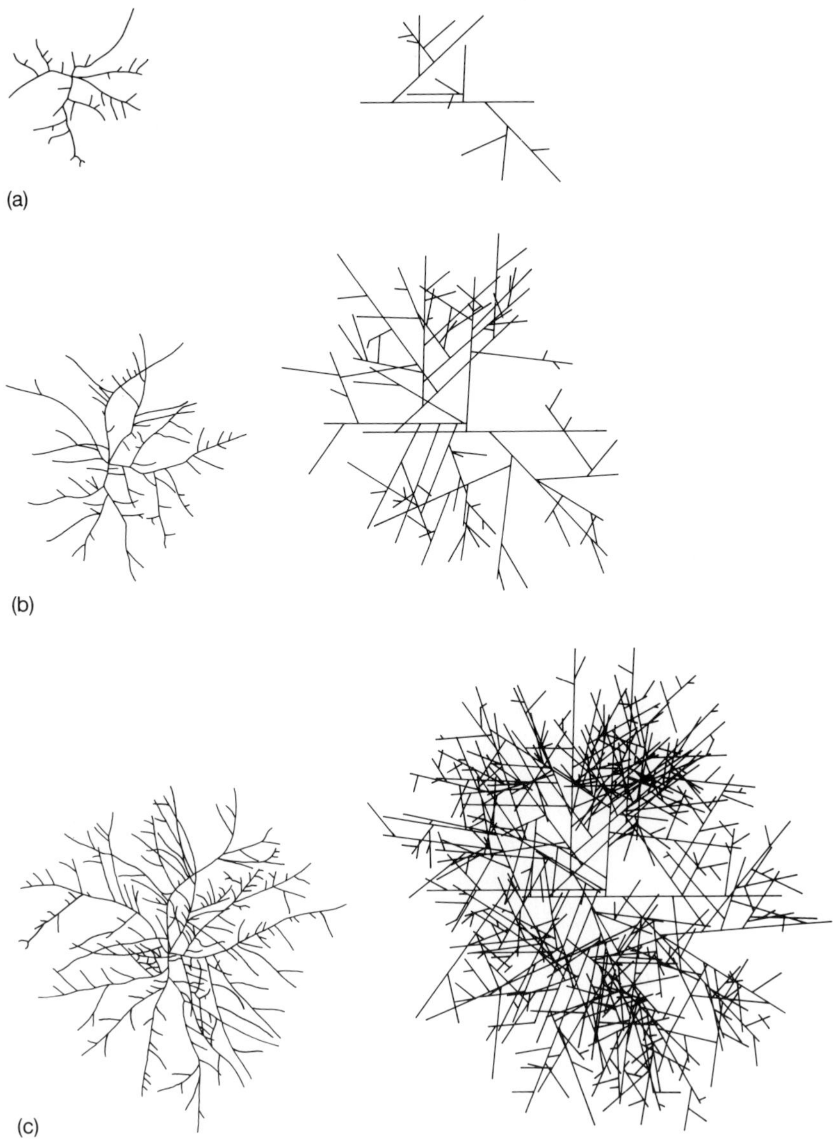

Fig. 9.7 Colony development in *Mucor hiemalis*. Left-hand diagrams are tracings of a colony arising from a single spore at (a) 5 hours; (b) 7½ hours; and (c) 10 hours after germ tube emergence. Right-hand diagrams are the corresponding computer simulations using data for branching variation and rates of hyphal elongation (from Hutchinson *et al.*, 1980, © British Mycological Society).

a short lateral branch arising from an adjacent hypha in response to its approach, or between two laterals emerging opposite each other from adjacent hyphae, or between a lateral and the side of an adjacent hypha. The maximum distance over which directing influences operate is approximately 10–15 μm, which suggests a mechanism that involves a steep concentration gradient of the guiding factor or factors. Lateral branch attraction to the same parent hypha is the basis for clamp connection formation, but here there is no evidence for chemically directed growth, the turning of the branch being rather under genetic control.

In some Basidiomycotina guided fusion is crucial to heterokaryon formation, hypha–spore homing being of particular interest (see also chapters 5 and 7). For example, oidia of *Clitocybe truncicola* attract hyphae of compatible homokaryons and induce the formation on them of lateral pegs which then fuse with the spores (Bistis, 1970). Similarly, both mono- and dikaryotic hyphae of *Schizophyllum commune* exhibit directional growth towards basidiospores over distances of up to 15 μm. The factor responsible is water soluble, and basidiospores release it only when stimulated to do so by the proximity of hyphae (Voorhees & Peterson, 1986).

Trophic responses

Given the behavioural versatility of the established mycelium, it is reasonable to assume that where nutrient depots are spatially separated, both in nature and in laboratory systems, hyphal growth will be directed towards them. Although there is a widespread tacit assumption that this is so, supporting evidence is remarkably sparse and, until recently, was confined to aseptate fungi, notably aquatic species. For instance, hyphae of *Achlya* and *Saprolegnia* are attracted to sources of amino acids, as are the rhizoidal components of *Blastocladia emersonii* thalli (Fischer & Werner, 1955; Harold & Harold, 1980; Robinson & Bolton, 1984; Manavathu & Thomas, 1985). This has resulted in the undoubtedly erroneous impression that such responses are common amongst lower fungi but are rare, if not unknown, amongst higher fungi (see Carlile & Matthews, 1988).

It has been pointed out emphatically that in nature, where commonly nutrients are discontinuously supplied both in space and time, there will be pressure on the mycelium for economic discovery of available resources, and that this is bound to be reflected in growth patterns at large, irrespective of the evolutionary status of the fungi concerned. At the same time, it has been proposed that two kinds of behaviour might facilitate efficient location of nutrient depots: first, the production of diffuse exploratory mycelia which, on contacting a resource unit, then develop further; second, the directional growth of hyphae, mycelia or mycelial aggregates towards resource units (Dowson *et al.*, 1986). Strongly directional growth has been demonstrated in *Phanerochaete velutina*, a basidiomycete causing decay of wood on the forest floor. When growing through soil from colonized wood blocks this species produces a sparse, symmetrical system of hyphal cords which, on approaching an uncolonized block, curve towards it and produce an effuse mycelium which spreads over it. This is followed quickly by a thickening of the cords connecting the two blocks.

A striking feature of directional responses by aggregated hyphae is that they can occur over large distances, and are thus presumably facilitated by volatiles rather than water-soluble compounds. Wood volatiles have been shown to elicit directional growth in a number of fungi, as have those from germinating seeds and organic debris in soil (Punja & Grogan, 1981; Mowe *et al.*, 1983; Norton & Harman, 1985).

Sporophore positioning

During dissemination, unimpeded spore fall or ejection and, in the case of projectiles, achievement of maximum trajectory, require appropriate positioning of spore-bearing structures relative to their substratum, and often also to surrounding objects, including neighbouring sporophores. Generally, this involves an, at least initially, negative response to gravity by the developing sporophore, or a positive response to light, or a combination of the two. It should, however, be noted that a vast number of fungi appear to be indifferent to these stimuli, their spore-bearing organs growing perpendicular to the substratum whatever its orientation or condition of illumination. By contrast with directional growth of vegetative hyphae, sporophore positioning often involves differential growth of one side of an intercalary extension zone, so that the responses can properly be called tropisms. The fundamental features of such behaviour have been long recognized, and early investigations on them began a major phase in experimental mycology (see Buller, 1909, 1934). Despite this, the underlying mechanisms are known in only a very few instances.

Multicellular structures

Several kinds of fruitbody can make environmentally cued adjustments during growth which guarantee freedom of spore dispersal. For instance, pilei of Agaricales and Boletales become oriented horizontally in response to gravity, so that their gills or pores are arranged vertically for free spore fall; coremia, perithecial necks and apothecial stipes show positive phototropism to unilateral illumination, thus directing released spores towards gaps between surrounding obstacles. As will be seen later, the physiological bases for similar responses by unicellular sporophores are broadly understood, but for fruitbodies the pathways from stimulus reception to the co-ordinated behaviour of whole tissues are unknown.

With regard to basidiomata, coarse adjustment of the pileus is brought about by negative geotropism of the elongating stipe; fine adjustment of the hymenial surfaces is then made via positive geotropism of the gills or pore dissepiments (Buller, 1909). It might, nevertheless, be noted that basidiomata of *Polyporus ciliatus* have been observed to form normally in microgravity conditions aboard the Salyut-5 and Salyut-6 orbiting satellites (Kasatkina *et al.*, 1980). However, photoresponses may also be involved, and positioning may result from an interplay between gravitational force and light incidence. This has been demonstrated in *Polyporus brumalis*, in which a subapical growth zone is responsible for stipe elongation (Plunkett, 1961). In darkness, the stipe is negatively geotropic and extends vertically, but if illuminated unilaterally, positive phototropism overrides the influence of gravity and extension becomes horizontal. When the stipe is in this position, the young pileus develops so as to be oriented vertically. However, its subsequent expansion screens the receptive growth zone of the stipe from light, so that negative geotropism becomes reasserted, and the pileus is adjusted to its normal horizontal position.

Amongst Basidiomycotina an apparently unique tropism is exhibited by the 'gemmifer' of the coffee leaf parasite *Omphalia flavida* (*Pseudoclitocybe*). This vegetative dispersal structure consists of a long pedicel bearing an apical, detachable 'gemma' (Buller, 1934). On emerging from the substratum the pedicel grows vertically by means of an intercalary extension zone, but as it increases in length its upper part turns away from the vertical as a negative response to neighbouring pedicels. This mutual repulsion results in the gemmae, which are adhesive, avoiding contact with one another, and so they remain available for wide dissemination.

Unicells Whilst responses to gravity and other influences by unicellular sporogenous structures are not unknown, it is light that is apparently the most widespread and important orienting factor. Amongst apothecial Ascomycotina, positive phototropism is commonly exhibited by the tips of ripe asci, which tend to protrude from the general hymenial surface just prior to spore discharge (Buller, 1934). It is also shown by the sporangiophores of a number of Zygomycotina, the oft-quoted examples being *Phycomyces* and *Pilobolus* species. In these latter genera, and probably in all other cases as well, the cell acts as a cylindrical lens which focuses unilateral light on the distal wall (Tsuru *et al.*, 1988). Detection of the increased illumination by a photoreceptor is then transmitted and translated in such a way that extension of the distal wall becomes greater than that of the proximal wall, resulting in bending of the whole structure towards the light. In nature, this allows *Pilobolus* to aim its sporangiophores towards the sun, which not only aids sporangium discharge through gaps in surrounding vegetation but also, since this occurs at approximately midday, ensures the correct sporangiophore elevation to achieve maximum flight trajectory. Detailed investigations of the phototropic mechanism have centred almost exclusively upon the giant sporangiophore of *Phycomyces blakesleeanus*, mainly because of its interest to sensory physiologists as a simple model system with which to study how cells translate signals into output, in this instance growth rate. As a consequence, a great deal has been revealed of the biophysics of this unique cell, but for all its intrinsic interest it is difficult to relate this to the natural history of *Phycomyces* which, despite its occurrence on a variety of substrata in nature, remains suprisingly obscure (Bergmann *et al.*, 1969). Furthermore, the sporangiophore also responds tropically to both gravity and solid objects, yet how – in nature – orderly growth is maintained during simultaneous input of all three major stimuli is not clear.

With regard to phototropism in *Phycomyces*, there is a large literature on the nature of the photoreceptors, their location within the elongating cell, and the nature of some of the early physiological events that accompany the light response (see Dennison, 1979). In summary, the action spectrum for the stimulus shows maxima within the blue region at 485, 455, 385 and 280 nm which, when taken with absorption spectrum data, indicates that the receptors are flavins (Delbrück *et al.*, 1976). The photosensitive zone is from 0.5 to 2.0 mm below the sporangium, and so lies within the 0.1 to 2.0 mm region that is responsible for cell wall extension. For ordered growth it would seem necessary for the flavin units to be fixed to the cell wall. In addition, flavins are dichroic, that is they have preferred electric vector orientations for light absorption, and the units are oriented parallel to one another in a helical arrangement around the cylindrical sporangiophore, and at an angle to its transverse axis (Jesaitis, 1974; Medina & Cerdá-Olmedo, 1977). Light stimulation rapidly produces a 50% increase in cyclic AMP, subsequent increased growth being marked by an increase in chitin synthase activity (Cohen, 1974; Jan, 1974). This is accompanied by softening of the cell wall on the stimulated side and subsequent bending of the cell. Extension growth, however, has a spiral component, the elongating region of the cell at 0.7 mm below the sporangium rotating at approximately $10^\circ \mathrm{min}^{-1}$, thus carrying the photoreceptors around with it.

When extending *Phycomyces* sporangiophores are placed horizontally in the dark, they exhibit negative geotropism and bend upwards until oriented vertically. There is a latent period of between 30 and 180 minutes before response, and approximately 12 hours elapse before horizontal cells achieve the vertical position (Bergmann *et al.*, 1969). The gravity receptor has yet to be identified, but there seems to be no redistribution of particles within the cell when it is placed horizontally as, for instance, is seen with statoliths in plants. If a sporangiophore is irradiated with a horizontal beam of blue light, it orientates itself at an angle of 20–30° above horizontal, having then reached an equilibrium between negative geotropism and positive phototropism. This equilibrium angle departs less and less from the angle of irradiation as the latter moves from the horizontal towards the vertical. Hence, in natural situations, sporangiophore orientation will not be determined solely by the direction of incident light.

Final orientation of the *Phycomyces* sporangiophore is additionally mediated by an avoidance response, which causes it to turn away from solid barriers and vertical cylindrical structures including, presumably, other sporangiophores. In darkness, avoidance occurs in response to glass rods of 150 μm diameter, but is inhibited by light, which has an overriding effect (Bergmann *et al.*, 1969; Harris & Dennison, 1979). No firm conclusions have been reached as to the avoidance mechanism, but the sporangiophore may detect modulations in the concentration of a self-emitted volatile chemical caused by its close approach to neighbouring objects (Johnson & Gamow, 1971; Cohen, 1976).

PART FOUR

Interactions with other Heterotrophs

10 Microorganisms

For most natural habitats it is difficult to express, except in relatively crude terms, the complex qualitative and quantitative effects of interactions amongst fungi, and between them and other microorganisms. Largely, interaction physiology has been studied using Petri dish culture in which community components, usually only two, are opposed, and the outcome assessed. Whilst such simple ecological models provide information relating to fundamental niche, they may not be even a coarse guide to the dimensions of realized niche. For example, species A may be aggressive towards species B in dual culture, and so restrict or reduce its domain. However, in nature a further species, C, may enter into combat only with species A, thus enabling species B to retain or expand its territory at the expense of both A and C. Using the agar plate has certainly enabled a great deal to be discovered concerning, for instance, antibiosis and other antagonistic mechanisms of fungi, yet their ecological importance often remains unproven. The gap between what is known of interactions at the individual trophic level in the laboratory and the significance of this for community processes is likely to remain for some time. In addition, as has occurred in the past, it is probable that emphasis will continue to be placed on antagonism as an important determinant of ecological success. Although the validity of this view cannot be denied, the steps involved in community development amount to much more than a catalogue of confrontations; neutralism and mutualism also have important parts to play.

It is, of course, impossible to ignore the salient features of individual encounters either between fungi or with other microorganisms, but equally it is important to view these within a community context wherever this can be done, even though information may be fragmentary. This is the approach taken here with regard to the interactive role of the fungal mycelium within selected microbial communities; the fate of propagules in this respect has been considered elsewhere (see chapters 6 and 8).

Precision in terminology has not always been the hallmark of fungal ecology, despite attempts to lay down a strict definitive scheme of usage (Cooke & Rayner, 1984). The basic ecological groupings of organisms are the population and the community. A population is an assemblage of individuals of the same species existing in the same proximity in space and time; a community is an assemblage of diverse species occupying the same, functionally discrete, environment. The community has a more complex level of organization than the population in having a distinctive structure, and its own activities and laws, which depend on the relationships between its constituents. A convenient ecophysiological framework within which to examine interactions is that of the microbiome, defined as a characteristic microbial community occupying a well-delimited habitat that has distinctive physicochemical properties. Microbiome refers not only to the microorganisms involved, but also encompasses their theatres of activity; major examples are the phylloplane, rhizoplane, rhizosphere and numerous kinds of uncomminuted plant residues.

Two organisms may be considered as interacting when the presence of one in some manner affects the performance of the other. With strict reference to two interacting mycelia, or a mycelium interacting with another microorganism, performance of either or both components is modified through one of three broad phenomena: antagonism, mutualism or neutralism. All are multifaceted, the types of specific mechanisms acting within each being dependent, as will be shown later in this book, on the particular mycelium–interactant combination in question. Unilateral antagonism leads to the exploitation or replacement of one interactant by another; bilateral antagonism may result in deadlock, in which performance of both interactants is impaired, but where some kind of equilibrium between them is maintained. Mutualism arises from situations in which each interactant benefits from the activities of the other. Neutralism occurs either where one interactant benefits from the activities of the other without conferring benefit or harm in return, or where there is passive co-existence with no discernible positive or negative effect on either organism.

Bacteria

For fungi developing within most microbiomes, frequent encounters with bacteria and Actinomycetes are the norm. Yet, despite their obvious potential importance, the ecological scale of such interactions largely remains unquantified, a notable exception being the consequences of lichenization with cyanobacteria. It is of course difficult, if not impossible, to assess the mediation by neutralism of community development, and since the possibilities for mutualism with non-photosynthetic bacteria are only now being recognized, there has been a resultant focus on the nature and outcome of antagonism.

There exist numerous qualitative descriptions of unilateral and bilateral antagonisms between fungi and bacteria, with a clear emphasis on the deleterious effects of bacteria on fungal growth. However, in some situations, antagonism of bacteria by fungi may be of equal ecological importance. For example, under low nutrient conditions *in vitro*, hyphae of the lignicolous basidiomycete *Pleurotus ostreatus* grow towards and branch intensively within colonies of *Agrobacterium* and *Pseudomonas* which subsequently lyse, so possibly acting as supplementary sources of nitrogen compounds (Barron, 1988). Other

Basidiomycotina behave in a similar manner, and *Agaricus bisporus* can utilize dead *Bacillus subtilis* cells as a sole supply of carbon and nitrogen (Fermor *et al.*, 1991). Destruction and exploitation of bacterial biomass may therefore be a significant factor in nutrient acquisition by wood- and litter-inhabiting species.

The antifungal activities of bacteria and Actinomycetes have been most closely studied with regard to their application to the biological control of plant pathogenic fungi on the phylloplane and rhizoplane, and within the rhizosphere and spermosphere. All these microbiomes contain rich bacterial communities, some components of which are potent antagonists. Principal amongst these are fluorescent *Pseudomonas* species that, when introduced appropriately, have been shown to reduce disease levels (Blakeman & Fokkema, 1982; Weller, 1988; Osburn *et al.*, 1989). However, although leaves, roots and seeds may be naturally densely colonized by bacteria and Actinomycetes, the frequency of antagonists can be very low in the absence of artificial enhancement of their numbers. For instance, only 4–7% of rhizosphere isolates may adversely affect fungal growth *in vitro* (Turhan & Grossman, 1986; Elad & Chet, 1987; Becker & Cook, 1988; Handelsman *et al.*, 1990). Thus, in nature, although antagonism is widespread it is probably only locally intense.

Antagonism is exerted by a variety of means, encompassing nutrient competition, direct parasitism and antibiosis. With their short generation times, bacteria are capable of subjecting fungi to severe nutrient competition, especially where resources are in short supply. On and around seeds and roots, bacteria may act as large nutrient sinks for carbon and nitrogen compounds, and may also come to occupy and exclude fungi from sources of nutrient-rich exudates, for instance cell junctions and points of root emergence (Weller, 1988; Parke, 1990). Competition can also be facilitated by attachment to hyphae (Fig. 10.1). *Enterobacter cloacae* possesses agglutinins which bind it to *Pythium ultimum* and *Rhizopus stolonifer*, subsequent inhibition of mycelial growth not being attributable to either antibiosis (for example via ammonia production) or enzyme or toxin production (Nelson *et al.*, 1986; Howell *et al.*, 1988; Wisniewski *et al.*, 1989). Iron limitation is another important factor. Some *Pseudomonas* species growing under low iron conditions produce yellow-green fluorescent (pyroverdine type) siderophores (see Chapter 2) which facilitate sequestration of iron. Fungi within the same microbiome may then suffer iron limitation, provided that: they themselves produce no siderophores; or are unable to use siderophores of other microorganisms; or produce too little siderophore, or a siderophore with a lower affinity than that of the antagonist; or produce a siderophore which can be used by the antagonist (Weller, 1988).

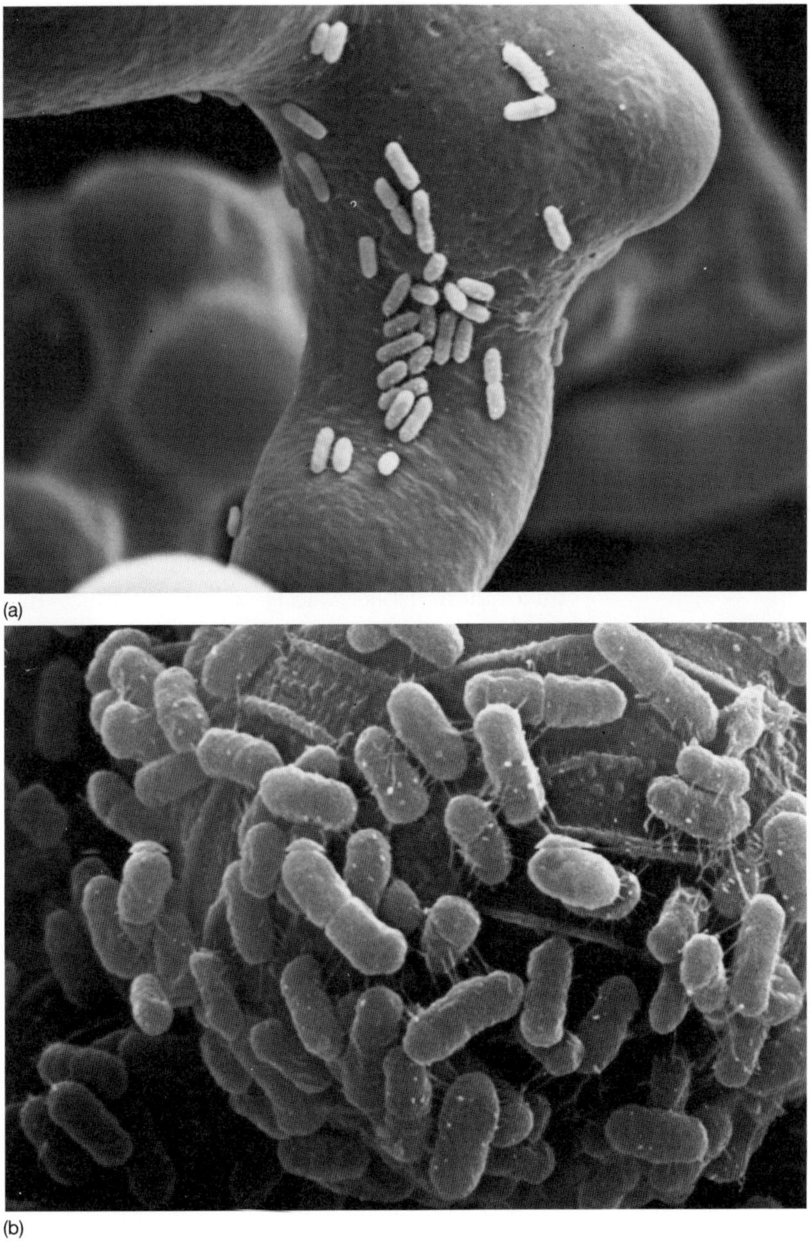

Fig. 10.1 Attachment of *Enterobacter cloacae* cells to *Rhizopus stolonifer*. (a) Adhesion to a hypha; (b) production of fimbriae by bacterial cells attached to a sporangiospore (from Wisniewski *et al.*, 1989, © National Research Council of Canada).

More direct intervention also occurs, with attachment by species of *Aerococcus*, *Alcaligenes*, *Pseudomonas* and *Xanthomonas* being followed by rapid lysis of hyphae (Nesbitt *et al.*, 1981). Filaments of *Streptomyces albus* show directional growth towards hyphae of *Nectria inventa* and form appressoria on contact, host cells collapsing and being invaded (Tu, 1986). It would, however, seem that it is through antibiosis that bacteria are most damaging to fungi and, while not disregarding the antibiotic propensities of other genera, fluorescent pseudomonads appear to be particularly well equipped in this regard. For example, *Pseudomonas fluorescens* and *Pseudomonas aureofaciens* produce phenazine-1-carboxylic acid *in situ* within the wheat rhizosphere, this being an effective antibiotic *in vitro* against a wide range of fungi (Brisbane *et al.*, 1989; Thomashow *et al.*, 1990). *Pseudomonas fluorescens* also releases pyoluteorin which suppresses *Pythium ultimum*-induced damping-off, whilst *Pseudomonas cepacia* produces pyrrolnitrin and two forms of pseudane (Fig. 10.2) which are effective against *Pyricularia oryzae*, *Rhizoctonia solani* and *Verticillium dahliae* (Howell & Stipanovic, 1980; Homma *et al.*, 1989).

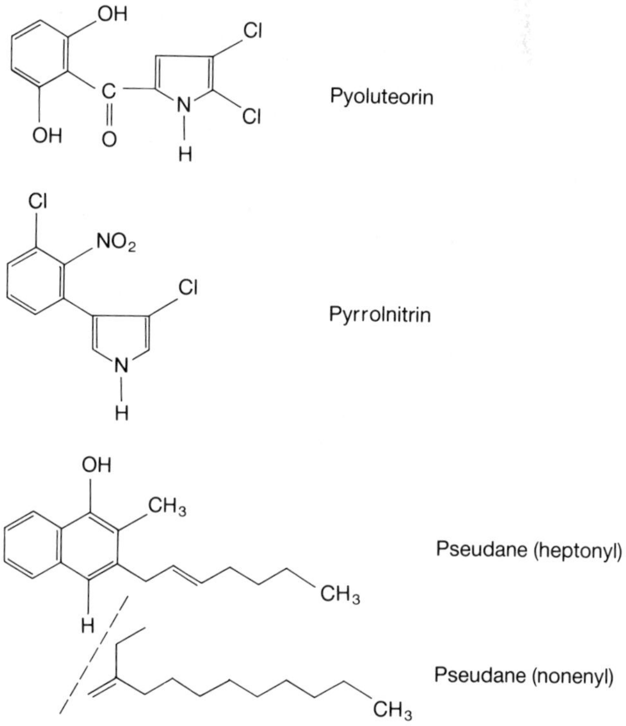

Fig. 10.2 Some antifungal antibiotics produced by *Pseudomonas* spp.

The effects of bacterial antagonism on selected plant pathogens, both *in vitro* and *in vivo*, are often quite clear, but predicting the broader outcome at the microbiome level is more difficult because of the influence of other ecological factors. Paradoxically, for example, inoculation of leaves or fruit with siderophore-producing bacteria may stimulate infection by some fungi (Blakeman & Fokkema, 1982). There is no immediate explanation, the bacteria presumably reducing constraints imposed on these pathogens by other microorganisms, including other, more antagonism-sensitive fungi. This raises the important, and frequently overlooked, point that bacterial antagonism is selective and, where it operates with sufficient strength, total fungal community structure inevitably will be changed.

Evidence for bacterially mediated shifts in community development tends to be circumstantial, but it has been demonstrated to occur in herbivore faeces (Safar & Cooke, 1988a,b). When grown together in sterile faecal resource units (copromes) the coprophilous ascomycetes *Ascobolus crenulatus*, *Chaetomium bostrychodes* and *Sordaria macrospora* interact such that fruiting of *Ascobolus* is almost entirely suppressed. Addition of either *Flavobacterium*, or *Methanobacterium*, or *Pseudomas* isolates (all faecally derived) to the microbiome facilitates fruiting of *Ascobolus* whilst depressing that of the other two fungi. The presence of all three bacteria, together with a *Staphylococcus* isolate, results in severe inhibition of fruiting by *Chaetomium* and *Sordaria* and enhancement of that by *Ascobolus* (Fig. 10.3). This seems to be related to the fact that, on agar, all bacteria inhibit mycelial extension of the former two species but stimulate that of the latter. Introduction of an aggressively competitive basidiomycete, *Coprinus*, to the community leads to fruiting by *Sordaria* alone. However, the presence of the bacteria, either singly or in combination, alleviates this effect to various degrees, although the mechanisms involved are not known (Fig. 10.4).

Reference has been made above to the possibility of mutualism with non-photosynthetic bacteria. Examples are few and uncertain, but this might reflect lack of study rather than a natural paucity of them. Sporocarps of many kinds house bacteria, and presence of the latter may contribute to normal morphogenesis in some instances. *Tuber* ascomata contain cocci which are also found inside mature asci, and it has been suggested that these bacteria aid development of sporocarp hyphae via metabolite interchange, and that they may also have a role in ascospore activation (Pacioni, 1990). A key event for basidioma initiation in *Agaricus bisporus* is thought to be the presence of *Pseudomonas putida* in the substratum. This bacterium may remove mycelial self-inhibitors of fruiting, so facilitating fruitbody formation (Rainey *et al.*, 1990). More indirect mutualisms occur between

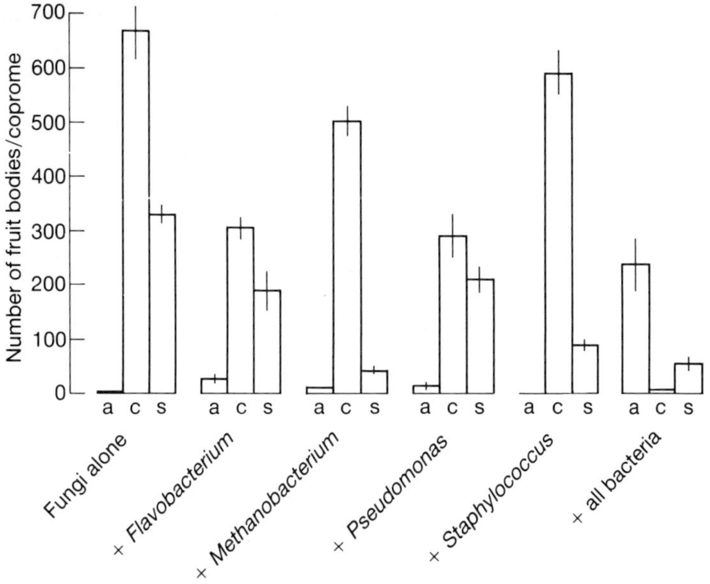

Fig. 10.3 Fruitbody production by three ascomycetous coprophilous fungi: a, *Ascobolus crenulatus*; c, *Chaetomium bostrychodes*; s, *Sordaria macrospora*, developing either together in faecal resource units (copromes) or together in the presence of faecally derived bacterial isolates (from Safar & Cooke, 1988b, © British Mycological Society).

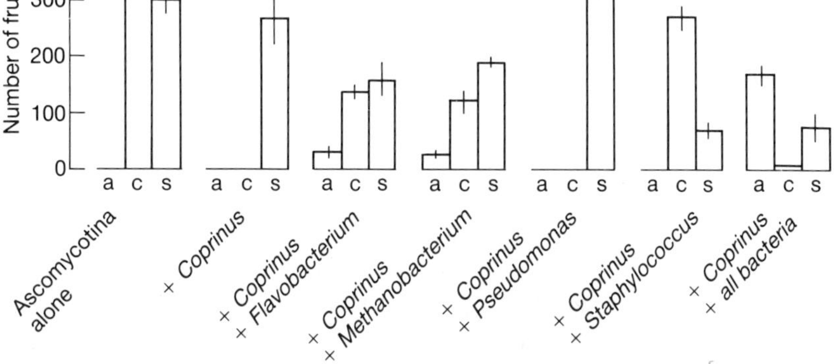

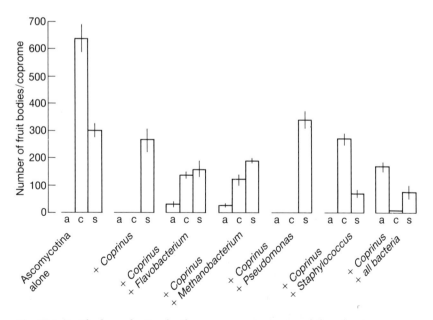

Fig. 10.4 Fruitbody production by three ascomycetous coprophilous fungi: a, *Ascobolus crenulatus*; c, *Chaetomium bostrychodes*; s, *Sordaria macrospora*, developing in the presence of a *Coprinus* species and bacteria (from Safar & Cooke, 1988b, © British Mycological Society).

rhizosphere bacteria and vesicular–arbuscular mycorrhizal fungi. For example, *Pseudomonas putida* populations increase mycorrhizal infection in *Trifolium subterraneum*, which then leads to an increase in numbers of other rhizosphere bacteria, but with fewer fluorescent pseudomonads and chitinoclastic Actinomycetes than occur in non-mycorrhizal rhizospheres (Meyer & Linderman, 1986a,b). A reduction in these two antifungal components of this microbiome must have consequences for both the success of further mycorrhizal infection and the structure and activity of the non-mycorrhizal fungal community.

Amoebae, Myxobacteria and myxomycetes

Some motile protists have voracious appetites for fungi, and in soil or litter may bring about large-scale destruction of spores (see Chapter 6). Equally serious damage can be inflicted on mycelia; the vampyrellid amoeba *Leptomyxa reticulata* is capable of perforating and emptying hyphae of their contents within 40–90 minutes of pseudopodial contact. The mechanism involved, removal of a disc of cell wall to permit entry of a pseudopodial branch, is identical to that used against spores (Homma *et al.*, 1979). Somewhat similar behaviour is exhibited by Myxobacteria, for instance streaming colonies of *Polyangium*, although penetration and protoplast lysis require a much longer period of 6–12 hours. *Cytophaga* and *Sorangium* species possess powerful extracellular chitinases and proteases, are active within the rhizosphere, and can exert partial control of damping-off in conifers caused by *Fusarium solani*, *Pythium intermedium* and *Rhizoctonia solani* (Hocking & Cook, 1972).

There is also a body of evidence for the consumption of mycelia *in vitro* by plasmodia of a large number of myxomycetes. For the most part the bulk of observations were made during the first half of this century and, in general, have not been expanded upon in more recent times (see Madelin, 1984). However, a remarkable feature of some interactions is the total dissolution of susceptible hyphae within a few seconds of their being touched by the plasmodial margin due, presumably, to the rapid secretion and action of extracellular chitinases (Stirling *et al.*, 1979). In nature, myxomycetes co-exist with wood-rotting Ascomycotina and Basidiomycotina, although it is not clear whether they feed upon them from the onset of decay, and so retard it, or appear later when decay columns have become fully established. Plasmodia are also commonly found on basidiomata, and undoubtedly contribute to their eventual disappearance, but to what degree they impair the processes of basidiospore production and dissemination is not known.

The roles of predator and victim between amoeboid microorganisms and fungi are commonly reversed, with fungi which capture and consume amoebae and testaceous rhizopods being common within, and readily isolated from, soils, leaf litter and aquatic habitats. These species are apparently nutritionally obligately dependent on the predaceous habit, and the majority belong within the Zoopagales (Zygomycotina). Prey become attached to undifferentiated hyphae and, having become immobilized, are then invaded by lateral branches which effect their digestion (Drechsler, 1936, 1937).

Nematodes Many fungal habitats also support thriving populations of relatively large microscopic animals, including rotifers, copepods, tardigrades and nematodes. All are potential prey for fungi that have become adapted to capture and consume them but, at least in terrestrial habitats, it is nematodes which are most important in fulfilling this role, since they usually predominate in biomass, and in numbers of both individuals and species. However, there are also possibilities for bilateral fungus–nematode antagonism, some genera of the latter are fungus-feeders with mouthparts in the form of a stylet, by which hyphae are penetrated and their contents sucked out. The gross effects of this on free-living fungi are not known, but the development and functioning of mycorrhizal species may be disrupted by such activity within the rhizosphere (Riffle, 1967; Shafer *et al.*, 1981; Finlay, 1983).

Whilst nematodes and other animals constitute a rich and abundant nutrient source for predaceous fungi, these commonly vigorously moving organisms must first be entrapped. This is achieved in a variety of ways, although mechanisms have been explored in detail only amongst nematode-destroying species (Barron, 1976, 1990). The vast majority of them capture prey by means of adhesive traps consisting of modified lateral branches, which can either be short and simple, or be elaborated into loops or three-dimensional networks. A constant feature of all adhesive cells is the presence, towards their periphery, of numerous prominent electron-dense inclusions, which may tend to reduce in number or disappear after capture has taken place. Their function is unclear, but they are possibly involved in secretion of adhesive materials to the cell surface and, subsequently, in the release of lytic enzymes for penetration and dissolution of the prey (Schenk *et al.*, 1980; Wimble & Young, 1983; Dowsett *et al.*, 1984). Trapping is efficient and usually reaches a rapid conclusion; wandering nematodes become quickly and firmly attached, are rendered immobile after a period of 1–2 hours, and are thereafter penetrated by a fine infection peg which gives rise to assimilative hyphae within the body.

As is indicated by these events, traps have a complex physiology. First, although some species produce them spontaneously, even in axenic culture, others do so only in response to the presence of nematodes; morphogenesis is possibly triggered by the recognition of anal excretory compounds. Following induction, trap formation may continue to be non-random in that it may exhibit an endogenous rhythm. For instance, in the network former *Arthrobotrys oligospora*, there is a post-triggering cycle with a mean period of approximately 42 hours (Lysek & Nordbring-Hertz, 1981). Second, encounters between nematodes and traps may not be entirely at random. A wide range of nematodes, especially fungivorous species, but also some plant parasites, is drawn to mycelia. In some predaceous fungi, such general attractiveness is enhanced by the presence of traps, although in others this has no effect (Field & Webster, 1977; Jansson & Nordbring-Hertz, 1979, 1980). The number and nature of attractants remains obscure, there probably being more than one with, possibly, small polypeptides being involved (Monoson *et al.*, 1973; Balan *et al.*, 1976). Finally, the swift death of trapped nematodes, well in advance of cuticular penetration by the fungus, suggests the action of a toxin, but this has not yet been convincingly demonstrated (Olthof & Estey, 1963). It might be noted at this point that hyphae of the non-predaceous lignicolous basidomycete *Pleurotus* produce secretory droplets that are capable of narcotizing nematodes which are then invaded, and there is increasing evidence for similar exploitation of these animals by other nonpredaceous litter- and wood-inhabitants (Barron & Thorn, 1987).

An additional aspect of trap physiology, and one which is central to an understanding of the influence of predation on nematode populations, centres on prey recognition and the possibility of discrimination between different nematode species. Some fungi are neither host specific nor even organism specific. For example, *Zoophagus pectosporus* (*Acaulopage pectospora*) captures both loricate rotifers and nematodes on lateral adhesive pegs arising from its sparse mycelium (Figs 10.5 & 10.6). Trap cells exude adhesive material to their surface via extrusion pits in the cell wall, and this has an affinity for both rotifer cilia and nematode cuticle (Whisler & Travland, 1974; Saikawa & Morikawa, 1985). By contrast, exclusively nematode-trapping species do exhibit specificity, at least *in vitro*. For instance, *Monacrosporium ellipsosporum* captures species in the Class Secernentea but not those in Adenophorea, this being apparently related to differences in cuticular structure between the two groups (Gaspard & Mankau, 1987). In relation to this, nematode adhesion is mediated by specific carbohydrate-binding proteins on the trap cell surface, and several such lectins have been identified (Zuckerman & Jansson, 1984).

For *Arthrobotrys oligospora* the lectin is specific for *N*-acetylgalactosamine, and binds with a glycoprotein receptor on the nematode

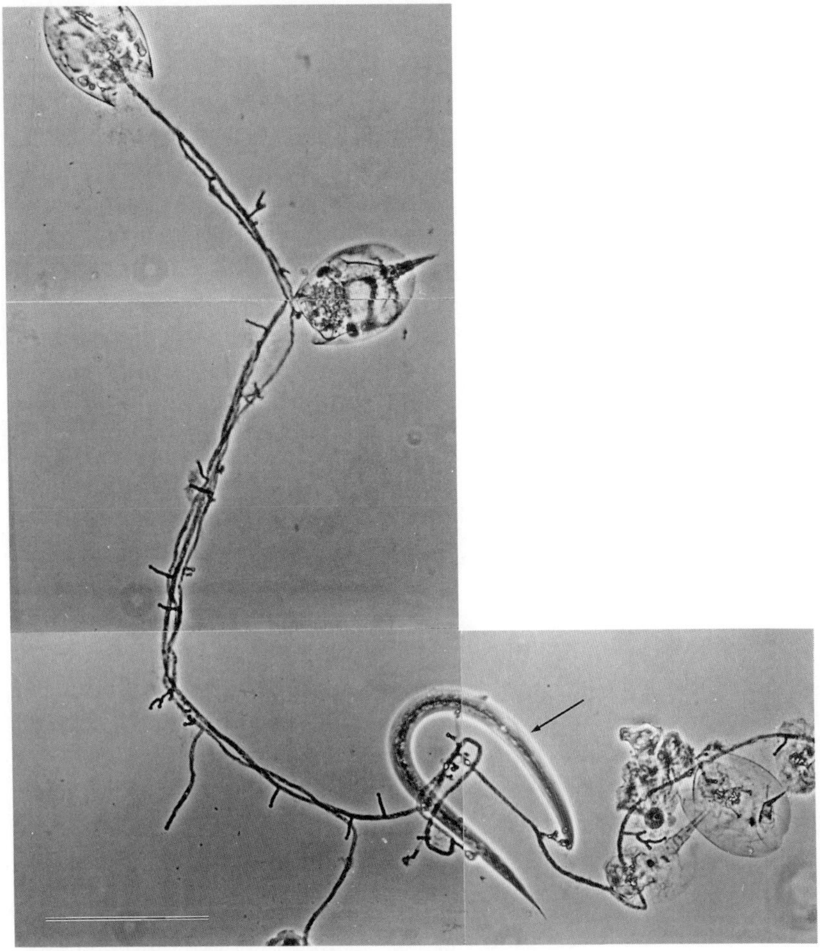

Fig. 10.5 Capture of rotifers by *Zoophagus pectosporus*. Several rotifers caught on the same hyphal system as a single nematode (arrowed). Bar = 100 μm (from Saikawa *et al.*, 1988, *Mycologia*, **80**, © 1988 New York Botanical Garden).

surface (Borrebaeck *et al.*, 1985). Nematode cuticle commonly contains D-galactose and related galactose residues but these are absent from some species; *Arthrobotrys oligospora* might therefore be expected to trap the former but not the latter type of nematode. The related species *Arthrobotrys ellipsospora* does not possess a sugar-specific lectin, but its traps produce a mucin-specific haemagglutinin with affinity for animal mucopolysaccharides (Yamanaka *et al.*, 1988).

Counter to these observations, it has been suggested that lectins may not be the sole binding agent, and that their efficiency *in vitro* may not match that expressed *in vivo*. Some tests have failed to demonstrate selective trapping, indicating that lectins may exhibit

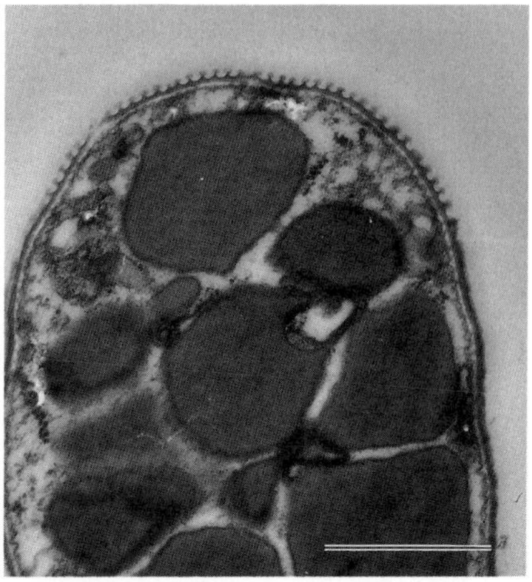

Fig. 10.6 Distal portion of an adhesive trapping branch of *Zoophagus* showing large inclusions possibly associated with the secretion of the adhesive fluid. Bar = 1 μm (from Saikawa *et al.*, 1988, *Mycologia*, **80**, © 1988 New York Botanical Garden).

'best-fit' specificity which, when adhesion is weak, is supplemented by other – and non-specific – binding forces, for instance bonding with the hydrophobic cuticle surface (Rosenweig *et al.*, 1985). Furthermore, *Dactylaria candida* and *Monacrosporium rutgeriensis* have lectins specific for 2-deoxyglucose, a saccharide that does not appear to occur in nature (Rosenweig & Ackroyd, 1984; Zuckerman & Jansson, 1984). With regard to the functioning of lectins in nature, it has been shown that *Saccharomyces cerevisiae*, which has a mannan-rich cell wall, binds to traps of *Arthrobotrys conoides* and blocks nematode capture, which is normally effected by a glucose–mannan-specific lectin (Rosenweig & Ackroyd, 1984). This has obvious implications for trapping efficiency in circumstances where microorganisms other than nematodes, but with similar lectin-interacting properties, are present.

The nutritional requirements of nematode-trapping fungi *in vitro* do not differ in any significant respect from those of strict saprotrophs, which suggests that the predaceous habit has been adopted as a means of easing competitive stresses, by facilitating utilization of an additional nutrient source unavailable to other fungi. This view is supported by studies on microcosms of various kinds, for which it has been shown that nematode trapping increases in the presence of saprotrophic competitors, and that it is not directly related to the numbers of nematodes present (Cooke, 1977; Quinn, 1987). Despite this density independence, trapping can markedly reduce nematode populations both in laboratory microcosms and in the field (Gaspard & Mankau, 1987; Grønvold *et al.*, 1987; Stirling, 1988).

Other fungi A salient feature of fungal community processes is the development of individuals of the same or different species in either close proximity or contact. Inevitably interactions arise, either as a consequence of common exploitation of a resource or, perhaps less frequently, because one individual serves directly as a nutrient source for another.

Clearly, it is possible for partitioning of resources between individuals to be mediated through various shades of neutralism and mutualism, but too few examples have been identified outside the laboratory to allow broader assessments of ecological significance to be made. Studies *in vitro* do, however, give an indication as to how neutralism and mutualism might be realized in nature. When mycelia of two species meet or intermingle, a frequent result is stimulation of sporulation in one or both fungi. Accelerated or enhanced reproduction can be viewed as being beneficial, but is obviously achieved at the expense of curtailed vegetative development. For any particular ecological situation, therefore, the advantages which accrue from interaction-induced reproduction must be judged against concomitant deleterious changes in the growth mode before assigning this phenomenon to either neutralism or mutualism. Interaction can also give rise to other kinds of morphogenetic change. For instance, formation of cords and rhizomorphs by Basidiomycotina may be stimulated by contact with, or proximity to, a wide variety of non-basidiomycetous fungi, this switch to a foraging mode facilitating rapid exploration of available resources.

In some circumstances, physiological complementarity between neighbouring individuals may provide a basis for mutualism. Cellulase production by a cellulolytic species could be enhanced if catabolite repression is relieved through removal of reducing sugars by an associated non-cellulolytic species. Furthermore, since cellulase is an enzyme complex, mixtures of cellulolytic species – each with its own spectrum of cellulases – may effect cellulolysis more efficiently than a single species acting alone, and to the advantage of all.

Competition, combat and antagonism Whilst remaining sensible of the above possibilities, it is clear that the strongest and most frequent determinant of fungal distribution and activity is competition for resources, during which – for any two interactants – the outcome is detrimental to one or both. With strict reference to mycelial species, two fundamentally different aspects of competition must be distinguished, namely, primary resource capture and combat (see also Chapter 1 in relation to strategy theory and niche determinants). Primary resource capture is restricted to early stages of community development, when vacant resources are wholly or still partially available, and describes the process of gaining access to, and achieving initial primacy within, unoccupied domains. Here, success depends on characteristics which give the opportunity for rapid arrival and establishment, such as: prolific production of xenospores, quick germination of these, high mycelial extension rate, possession of appropriate extracellular enzymes for available substrates and, in some cases, tolerance of any stress factors associated with the resource. These are features of fungi with ruderal primary or secondary strategies that depend for survival on priority of arrival and unhindered exploration.

As colonization of a resource proceeds, accompanied by recruitment of later arrivals, the territories of different individuals will come into contact, so raising the prospect of direct physiological challenge between individuals, and ensuing combat. This can be in the form of either defence of captured resources or secondary resource capture, that is the wresting of a resource from one individual by another. Successful defence results in mutual exclusion of individuals from each other's domains; secondary resource capture leads to the replacement of one individual by a more powerful combatant. With the exception of those species with defence capability, ruderals will thus be excluded by fungi with combative strategies, which then come to dominate. In this context, all mechanisms underlying combat are encompassed by the term antagonism (Cooke & Rayner, 1984).

Two basic forms of antagonism can be distinguished: in the first, combative interactions are mediated at a distance by means of diffusible or volatile metabolites, so giving rise to antibiosis; in the second, combat is initiated upon contact between individual hyphae or mass mycelia, and involves direct physical, rather than chemical, interventions of various kinds. Studies *in vitro* have demonstrated amply that septate fungi, especially, are able to secrete a wide range of antibiotic compounds inhibitory to mycelial development. The possible significance of antibiosis in nature remains, however, a matter of debate (Faull, 1988). On the one hand there is no evidence that antibiotic producers are ecologically more successful than non-producers, but this may merely reflect the fact that antibiosis is only one of several available options for combat. It can also be argued that, *in vivo*, antibiotics may not persist in sufficient quantity to have any marked effect, particularly in oligotrophic conditions, and that additionally they may become bound to substrates, or be decomposed by other microorganisms. Counter to this, excessive antibiotic production could be disadvantageous in that it might create too large a competitor-free zone to be exploited immediately, the result being selective invasion of this by further competitors with possibly greater antibiotic resistance. However, given that antibiosis can operate in nature, it may be particularly important during the ousting of ruderal fungi following completion of primary resource capture by them (Wicklow, 1981). In many habitats, Zygomycotina typically are the first colonizers, and these are particularly prone to the effects of antibiotics while not, in general, producing their own. Their replacement by combative, septate species is therefore quite likely to be achieved, at least in part, via antibiosis.

Of more certain ecological significance are some of the wide range of direct antagonisms that may follow contact between fungi, where there seem to be interfungal recognition systems which determine the nature and extent of any interaction. There are three broad forms of the latter: parasitism, non-parasitic interactions between different species, and non-parasitic interactions between individuals of the same species.

Mycoparasitism and hyphal interference

Bearing in mind the pitfalls of too strictly defining any symbiotic association, a mycoparasite is a fungus existing in intimate association with another, for either an extended or brief period, and from which it derives some or all of its nutrients while conferring no benefit in return. It is then axiomatic that there must be contact between the two fungi involved, which eventually may lead to death of all or part of the antagonized individual.

Mycoparasitism can affect all stages of development, so that not only vegetative hyphae are killed, but also spores and other propagules (Spencer & Atkey, 1981; Fries & Swedjemark, 1985; Whipps *et al.*, 1988; McClaren *et al.*, 1989). In addition, ascomata, including those of lichenized species, and basidiomata may be either destroyed or exhibit mycoparasite-induced abnormalities. This suggests that severe, but probably highly localized, constraints can be imposed at the establishment and reproductive phases of resource exploitation, as well as during resource capture. The probable influence of mycoparasitism on fungal communities has encouraged its study in relation to possible biological control of plant pathogens, and there is a great deal of evidence which shows that a number of important diseases are susceptible to such measures, imposed via either stimulation of resident mycoparasites in the relevant microbiome, or the introduction to it of non-residents (Burge, 1988; Whipps & Lumsden, 1989). Despite this interest, all too frequently an empirical approach has been employed, with total disregard to basic ecological principles, so that little progress has been made in understanding mycoparasitic behaviour in nature (see Whipps *et al.*, 1988). However, it must be acknowledged that the bulk of what is known, or surmised, concerning the ecophysiology of mycoparasites has evolved from the examination of potential biocontrol fungi.

Mycoparasites can be divided into two groups depending on whether they have a biotrophic or necrotrophic nutrition, some species showing shifts between these modes. They are largely, but by no means totally, mycelial in habit; for instance, there are many chytrids which attack either other chytrids or aquatic and soil-borne Oomycetes (Willoughby, 1956; Pemberton *et al.*, 1990). In ecological terms, even less is known of these than of filamentous mycoparasites, so that here consideration will be limited almost entirely to the latter.

Disregarding a handful of seldom-encountered or enigmatic species, filamentous biotrophs form a discrete assemblage of merosporangial Zygomycotina characterized by their almost total confinement to, and ability to form haustoria within, mucoralean hosts (Benjamin, 1959; Jeffries, 1985). Experimental studies have centred virtually exclusively on *Piptocephalis* species. These are obligate symbionts which exhibit only restricted development in monoculture with, at one extreme, spores of some species swelling but failing to germinate and, at the other, spores giving rise to minute germling mycelia of limited growth. Sporulation may then occur, but the resultant spores will not germinate in the absence of a host. The nature of the probably nutritional basis for obligate dependence is not known.

By contrast, when spores are sown on agar in proximity to either germinating host spores or mycelia, rapid development of the mycoparasite ensues. Spores germinate quickly and, in some mycoparasite–host pairings, germ tubes show directional growth towards host hyphae, with pronounced homing towards their tips (Fig. 10.7). Where growth-directing factors occur, for instance during guidance of *Piptocephalis fimbriata* germ tubes towards those of *Mortierella vinacea* and *Mycotypha microspora*, these appear to be proteinaceous in character (Evans, *et al.*, 1978; Evans & Cooke, 1982). Available evidence suggests that attachment to host hyphae is through recognition of specific glycoproteins of the host cell wall by germ tube receptors.

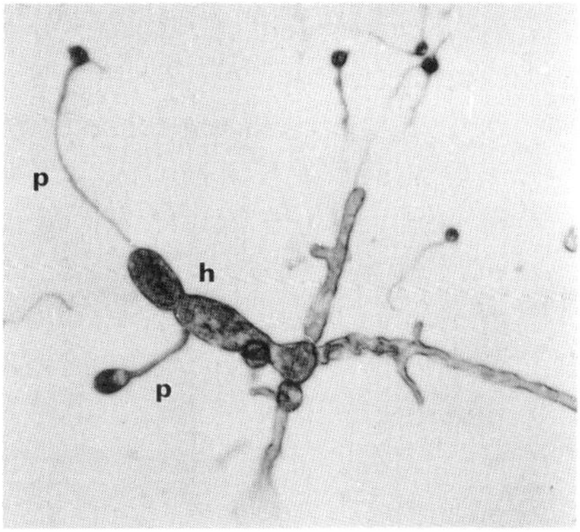

Fig. 10.7 *Piptocephalis fimbriata* attacking hyphae of *Mortierella vinacea*. Mycoparasitic hyphae (p) arising from conidia homing in and causing hyphal swellings (h) on the host (from Evans *et al.*, 1978, © *New Phytologist*).

In addition, penetration is accompanied by the production, by the mycoparasite, of factors which prevent incorporation of chitin precursors into the host cell wall, so that wall repair – and hence defence – is prevented (Manocha, 1985; Manocha & McCullough, 1985). The measurable effects of mycoparasitism on host development vary according to the *Piptocephalis*–host combination in question (Curtis *et al.*, 1978). Hyphal extension may be reduced, unaffected or even increased. Similarly, hyphae may retain a normal appearance or may show striking morphological abnormalities, ranging from atypical branching to massive, multilobed swellings, and in the case of the dimorphic species *Mycotypha microspora*, complete suppression of the mycelial phase, with persistence of the normally transient yeast phase (Figs 10.8 & 10.9).

Although biotrophic mycoparasites can severely debilitate their hosts *in vitro*, the extent to which they are able to do so *in vivo* is not known. But it is possible that, in certain situations, their activity may influence patterns of resource allocation between host and non-host species. During colonization of herbivore faeces by coprophilous fungi, the early appearance of saprotrophic Zygomycotina, which participate in primary resource capture, is frequently accompanied by that of dependent biotrophs, especially *Piptocephalis* species (Harper

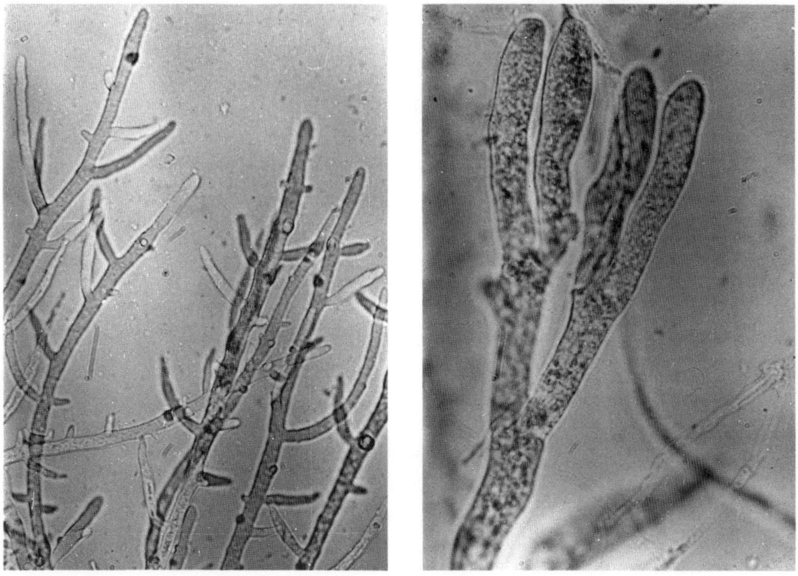

(a) (b)

Fig. 10.8 *Piptocephalis fimbriata* attacking a mature colony of *Mycotypha microspora*. (a) Marginal hyphae of an uninfected host colony; (b) swollen and abnormally branched marginal hyphae of an infected colony (from Curtis *et al.*, 1978, © *New Phytologist*).

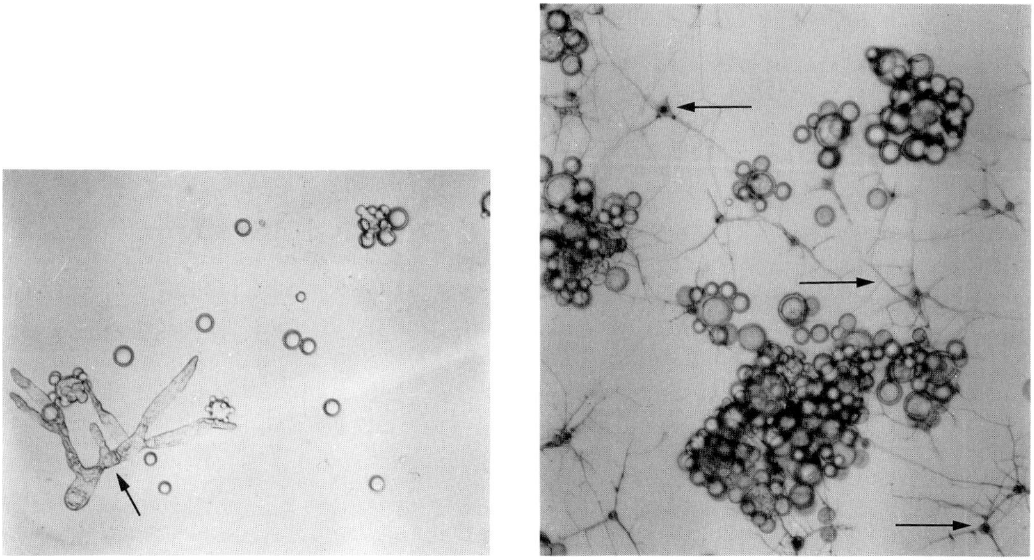

(a) (b)

Fig. 10.9 Suppression of the mycelial phase in *Mycotypha microspora* by *Piptocephalis fimbriata*. (a) Axenic development of *Mycotypha* with transient yeast phase colonies and the beginnings of persistent mycelial phase growth (arrowed); (b) development of *Mycotypha* in the presence of germinating *Piptocephalis* spores (arrowed). Large, persistent yeast phase colonies are formed with total absence of mycelial phase growth (from Evans *et al.*, 1978, © *New Phytologist*).

& Webster, 1964). It is possible that the ability of infected mycelia to establish domains is reduced, so permitting either greater colonization by non-host ruderals or, if resource defence is also impaired, the quicker ingress of combative fungi. Sporulation may also be depressed, and this may lead to a significant reduction in inoculum for establishment in new faecal habitats (Wood & Cooke, 1986).

In sharp contrast to biotrophy, necrotrophy is widespread and, whilst a few species which attack sclerotia and fruitbodies are ecologically obligate necrotrophs, a great number of saprotrophs exhibit facultative necrotrophic mycoparasitism. In agar culture, generally this has three broad phases: host recognition, contact growth and penetration, and nutrient acquisition (Whipps *et al.*, 1988; Lewis *et al.*, 1989). A mycoparasitic hypha in proximity to that of a potential host may show directional growth towards it, presumably in response to some kind of stimulus gradient. Reorientation may be accompanied by increased branching of the mycoparasite, reduction in the extension rate of the host hypha, and production by the latter of laterals with limited growth and abnormal morphology, this behaviour being a clear indication of metabolite exchange between the protagonists (Fig. 10.10). After meeting the host hypha the mycoparasite grows over

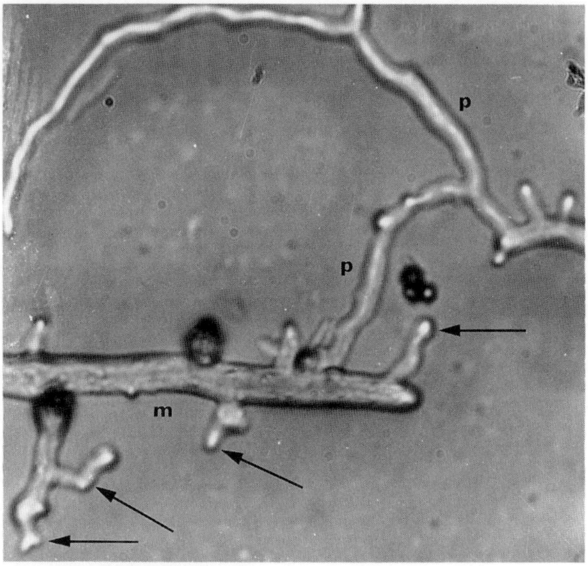

Fig. 10.10 Contact between the mycoparasite *Pythium oligandrum* (p) and a host species *Mycocentrospora acerina* (m). The contacted hypha of the latter has ceased to elongate and is producing abnormal lateral hyphae with refractile tips (arrowed) within the apical zone (from Lutchmeah & Cooke, 1984, © British Mycological Society).

or along it and commonly coils around it, contact growth probably being lectin mediated (Elad *et al.*, 1983a; Barak *et al.*, 1985). Penetration is then effected by fine invasive hyphae, sometimes preceded by appressorium formation, with the production of an array of inducible wall-lysing and cytoplasm-degrading enzymes. Subsequent death of the parasitized hypha, or some of its compartments, may be followed by extensive growth of the necrotroph within it (Elad *et al.*, 1983b; Elad *et al.*, 1985).

Necrotrophy may not be confined to filamentous species. Cells of the yeast *Pichia guilliermondii* attach to hyphae of *Botrytis cinerea* causing pitting and collapse (Fig. 10.11). These effects are brought about through tenacious adhesion and secretion of β-1,3-glucanases. Yeast cells are thus in a position not only to intercept exogenous nutrients destined for uptake by hyphae, but also to utilize sugars released from hyphal walls by enzyme action (Wisniewski *et al.*, 1991).

An outstanding feature of some necrotrophic interactions is the rapid death of host hyphae following contact and before there are any clear signs that penetration is taking place, with, for example, *Pythium oligandrum* inducing lethal disorganization of host cytoplasm within a few minutes of encounter (Fig. 10.12). This phenom-

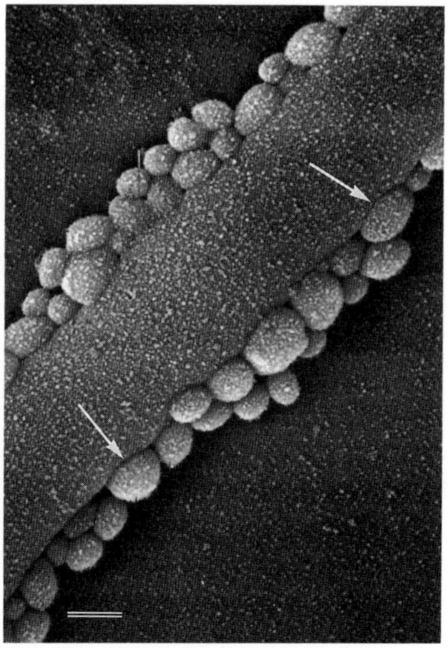

Fig. 10.11 Attachment of cells of the yeast *Pichia guilliermondii* to a hypha of *Botrytis cinerea* showing embedding of the cells (arrowed) via the wall-softening action of glucanase. Bar = 2.5 μm (from Wisniewski *et al.*, 1991, © Academic Press).

enon closely resembles, and may be identical to, that of hyphal interference, which is an antagonistic attribute of some lignicolous and coprophilous fungi that does not involve subsequent penetration of host hyphae (Ikediugwu *et al.*, 1970; Ikediugwu & Webster, 1970a,b).

No evidence has yet been obtained to indicate the degree to which necrotrophy contributes directly to the nutrient demands of mycoparasites. It seems more likely that rather than exploiting other fungi as hosts *sensu stricto*, this kind of antagonism is directed towards removing competitors so as to invade their partially exploited domains. Secondary resource capture, via either necrotrophic mycoparasitism or hyphal interference alone, is probably particularly important for territorial gains during the struggle for finite resources. Both provide an effective and precise device for the localized creation of space within the often crowded milieu of resource-limited communities. Access to residual nutrients would then be complemented by metabolite release from the moribund mycelia of the previous occupants.

Mycelial interactions and territoriality

The physicochemical constitution of many habitats precludes the development of, and hence meetings between, compact mycelia of the kind that are typical of agar cultures. In such natural situations, combat is confined to the encounter of either sparse hyphal systems or

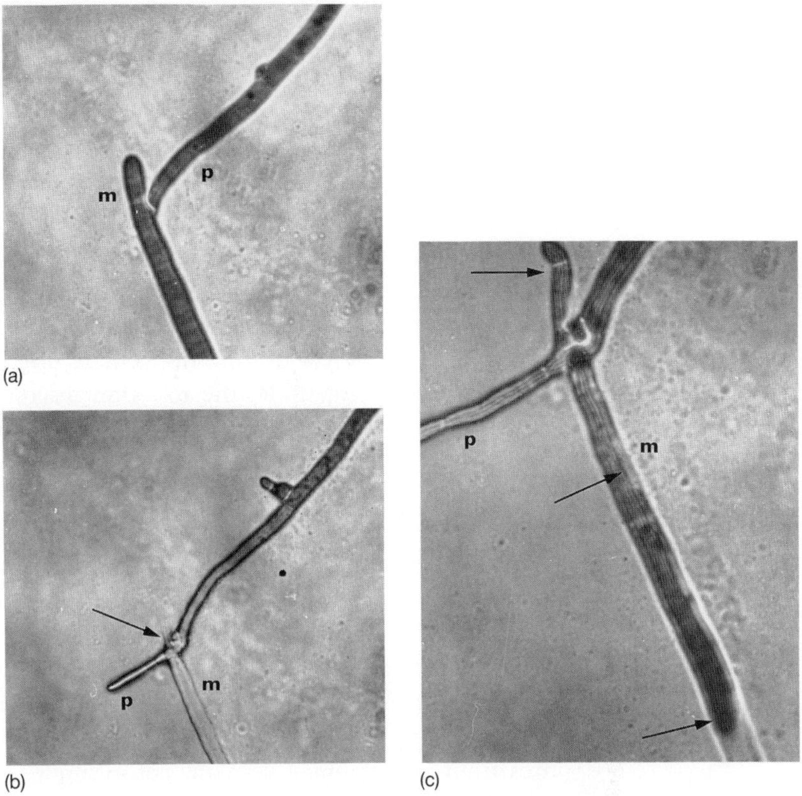

Fig. 10.12 Rapid killing of a hypha of *Mycocentrospora acerina* (m) by *Pythium oligandrum* (p). (a) Contact between host and mycoparasite; (b) same hyphae 14 minutes later with loss of opacity of host hypha (transparent tip of host hypha arrowed), and continued growth of the mycoparasitic hypha; (c) appearance 70 minutes after contact with *Pythium oligandrum* producing side branches (arrowed) within the host hypha (from Lutchmeah & Cooke, 1984, © British Mycological Society).

of individual hyphae. However, some resources provide conditions ideal for compact mycelial growth, and here the extent of primary and secondary resource capture is determined by collisions between individual mass mycelia. Typically, these interactions occur amongst ascomycetes and basidiomycetes that colonize standing and fallen timber; the participants usually exhibit reactions of a different order of magnitude from those so far outlined, in that their responses are often marked by macroscopic changes in colony morphology (Rayner & Webber, 1984; Rayner & Coates, 1987). Interaction is obviously not restricted to confrontation between individuals of different species, but also occurs between those of the same species, although the outcome may be the same, namely the demarcation of discrete territories.

Development of conspecific individuals in wood is commonly characterized by the appearance of rejection zones. These delimit adjacent mycelia, as a consequence of somatic incompatability which prevents the access of non-self nuclei. Hyphal fusion between individuals takes place, but is quickly followed by death of the fusion compartments via lethal activation of phenoloxidase and protease systems (Rayner *et al.*, 1984; Ainsworth *et al.*, 1990). This maintains the existence of individual mycelia by preventing physiological integration with others of differing genetic origin. Where individuals are somatically compatible, the rejection response is overriden, non-self nuclei obtain access, and there is the emergence of a secondary mycelium with different characteristics from its parents. In these cases, collectivism creates possibilities for advantageous growth modes, for instance greater somatic organization through increased hyphal fusion and aggregation.

Production of distinctive morphological forms, that is alternative phenotypes, also occurs as a response to the opposition of different species, assimilative mycelium being replaced by defensive or offensive non-assimilative mycelium. Since it is non-absorptive, the latter can be maintained in hostile territory by translocation of metabolites from the occupied domain. For example, when a mycelium of *Hypholoma fasciculare* meets that of another wood-decay fungus *in vitro*, the latter may deploy a barrage of aerial hyphae in the contact zone, which may result in successful defence of its territory. However, this reaction may be ineffective, in which case the mycelium of *Hypholoma* may give rise to a rhythmically extending replacement front that invades the opponent's domain, its advance being marked by bands of aerial mycelium emerging successively deeper within the invaded area. Alternatively, there may be production of linear mycelial aggregates or cords which spread radially into the opposing colony and rapidly replace it. In some pairings, invasion may be bilateral, cords of *Hypholoma* intrude into colonies of *Coriolus versicolor* and lyse them, at the same time its own assimilative mycelium is replaced by a dense mat-like growth of *Coriolus* hyphae. Whilst the recognition events leading to such responses remain uncertain, the importance of the resultant developmental switches is quite clear; the ability to assume an alternative functional mode allows for ecological versatility within an unpredictable, and frequently hostile, environmental setting.

11 Macroscopic animals

Although their physiology is relatively unexplored, in ubiquity and variety fungus–animal interactions have equal standing with those occurring between fungi and plants. Recent increased interest within this field has resulted in several substantial, detailed reviews of such interactions, with particular emphasis on arthropod–fungus symbioses (Anderson *et al.*, 1984; Wheeler & Blackwell, 1984; Wilding *et al.*, 1988). In these, the viewpoint has been predominantly zoocentric, with few attempts being made either to assess the evolutionary and ecological implications for the fungal participants, or to examine those physiological traits which fit fungi for association with animals (but see Dowding, 1984; Martin, 1984; Swift & Boddy, 1984; Pirozynski & Hawksworth, 1988). This imbalance probably reflects, in part, the difficulties inherent in a generalized approach to fungus–animal relationships. These arise from the fact that many outstanding examples are both unique and so bizarre as to defy synthesis, and that even at a simple level, for instance mycophagy, there exists an almost infinite number of subtly differing variations.

Accommodation of fungus–animal interactions within the broader context of fungal ecophysiology would seem to require adopting a stance which is at once mycocentric and only loosely based on previously emphasized, and copiously documented, considerations of symbiosis. One way by which this may be achieved is through confining discussion to basic features of habitat relationships, namely the mechanisms by which animal-associated fungi come to occupy their principal substrata, and the outcome with regard to their spatiotemporal distribution. When describing the niche characteristics of fungi that occupy habitats either shared with or provided by animals, it is inevitable that much available information relates to association with arthropods; there is no doubt that the terrestrial ascendancy of these invertebrates is closely linked to their co-evolution with fungi, and the importance of this is reflected in the abundance of relevant literature. However, in wider terms, co-habitation with, or inhabitation of, animals of all kinds is often made possible by the ability of the fungi in question to tolerate extreme physical environments, and to degrade complex polymers of both animal and plant origin, the result being unilateral or bilateral transfer of either nutrients or other benefits between fungi and animals.

Shared habitats

Seldom can animal habitats be fungus free but, at large, encounters between the two kinds of organism are casual, and carry few implications for the ecology of either. However, some animal groups, and arthropods in particular, have adapted to feed directly on mycelia and sporophores (Kukor & Martin, 1987). In addition, detritivores consume vast amounts of decomposing organic matter that contains significant levels of living fungal material. Such behaviour has potential for considerable impact on fungal distribution, biomass production and reproductive capacity, although any effects so produced are not always readily quantifiable. In some fungi, as might be expected, feeding pressure has resulted in the evolution of various deterrents against direct or indirect mycophagy. On conceptual grounds, a distinction should be made between mechanisms for protection of the soma *per se*, and those which prevent the exploitation by animals of a common resource, for example, decaying organic debris or fungally infected host plants. In some instances this is possible, but in others it is not easy to do so because an ability for habitat defence may be conferred by factors similar to, or identical with, those which preserve the fungus itself, the only difference being the level at which they operate.

One further point requires mention here, namely the often highly selective nature of mycophagy and detritivory, which may therefore have a strong effect on competitive encounters between fungal species, so influencing both community structure and, where applicable, decomposition processes (Wicklow, 1988b). Possession of defences can thus confer advantages beyond the immediate one of avoiding consumption.

Defence systems

It has been convincingly argued that fungal defences are distributed in direct proportion to the risk of attack on a particular structure, and to the value of that structure in terms of potential fitness loss should it be damaged or destroyed (Wicklow, 1988b). If this were the case then, in general, reproductive and organized survival structures should be better provided for than vegetative hyphae; the former are static and less capable of repair or renewal than is dynamic, and for the most part diffuse, trophic mycelium. Available evidence suggests that this might be so.

Protection of organs. The degree of organization of reproductive and survival structures gives scope for the elaboration of physical barriers to ingestion; as well as this, and perhaps more importantly, many secondary metabolites which accumulate during organogenesis are either unpalatable or noxious to animals (Wright *et al.*, 1982).

Mechanical defences have been little studied, but they can take many forms. For example, the exudation of wound-induced latex by basidiomata of *Lactarius* species may deter mycophagous arthropods; for other fungi the protection of unripe hymenia by a veil prevents oviposition by those insects whose larvae begin their life within the basidial layer (Hanski, 1989). In ascomata, sclerotia and rhizomorphs, a tough, melanized outer layer may prove strong enough to deflect the mouthparts of some mycophages; some ascomata and rhizomorphs are further protected by dense, ornamental hairs and deposits of oxalic acid crystals respectively (Brasier, 1978; Wicklow, 1979).

Basidiomata also commonly contain chemical deterrents and, in some instances, freedom from particular mycophagous insects can be attributed to the presence of specific compounds, for example, L-dopa in *Strobilomyces floccopus* and α-amanitin in toxigenic *Amanita* species (Bruns, 1984; Jaenike, 1985). Other defences may depend on rapid chemical reactions to injury. Amongst Boletales, the instantaneous staining which commonly occurs upon wounding the basidioma is due to the release of enzymically oxidized compounds, such as benzoquinones and diphenylcyclopentenones, that possibly deter further feeding by decreasing either the palatability or nutrient quality of affected tissues.

Antifeedant chemicals appear to be widespread amongst other reproductive or survival structures, and there is increasing evidence that, as has been implied above, these may differ in either quantity or quality from those found in parent mycelia. This suggests that, in some situations, feeding has been a selective force which has resulted in the preferential allocation of resources to the conservation of reproductive and survival potential (Wicklow, 1988b). Thus, in *Penicillium brevicompactum*, conidiophores produce the secondary metabolite brevianamide A, which is not found in the immersed mycelium, and which inhibits ingestion by larvae of the fruit-fly *Drosophila melanogaster* (Bird *et al.*, 1981; Paterson *et al.*, 1987). Similarly, whilst the mycelium of *Aspergillus flavus* contains aflatoxins B1 and B2, and cyclopiazonic acid, its sclerotia additionally have aflatoxins G1 and G2 together with aflatrem and dihydroxyaflavinine (Fig. 11.1). This results in the enhanced toxicity of sclerotial tissues to mycophagous insects (Wicklow & Shotwell, 1983; Wicklow *et al.*, 1988).

Preservation of resources. Detritivores are potentially far more destructive to fungi than are strict mycophagists, because they consume both fungal biomass and the substratum which supports its development, thus at once eliminating mycelia and their material resources which might not, at that time, have been fully exploited by their occupants. A valuable attribute of fungi subject to such pressures might be an ability to synthesize antifeedant chemicals, so preserving both life and resource. In this regard, the possible role of mycotoxins, especially those produced by *Aspergillus*, *Fusarium* and *Penicillium* species on stored cereal grains and other foodstuffs, has been reviewed exhaustively a number of times (see Wright *et al.*, 1982; Ciegler, 1983;

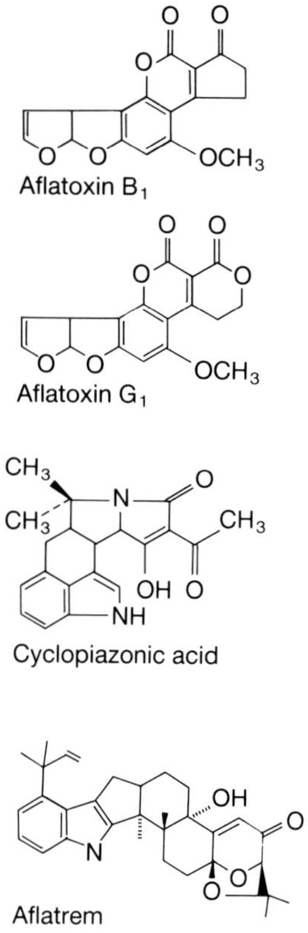

Fig. 11.1 Metabolites from the mycelium and sclerotia of *Aspergillus flavus* which have activity against mycophagous insects.

Frisvad, 1986). Despite the weight of this kind of evidence, the scale and importance of mycotoxin production in more natural ecological situations largely remains a matter for speculation (Whittaker & Feeny, 1971; Janzen, 1977; Wicklow, 1988b).

Herbivorous insects pose a similar threat to the ostensibly defenceless exhabitants and inhabitants of aerial plant parts. However, in the case of some mycophylla, the presence of the endophyte may result in the production of chemicals that deter feeding, so conserving the habitat not only for itself but also for any attendant fungal assemblages. Although mycophylla are widely distributed throughout the plant kingdom, defence systems have been studied in detail mainly in grasses and sedges, whose ascomycetous endophytes are anamorphic states of Clavicipitaceae (Siegel *et al.*, 1987; Clay, 1988). Their mycelia are generally sparse, hyphae ramifying throughout leaves, stems, ovules and seeds (Fig. 11.2). Occupation of the latter facilitates seed transmission of the symbiosis and may also deter granivorous

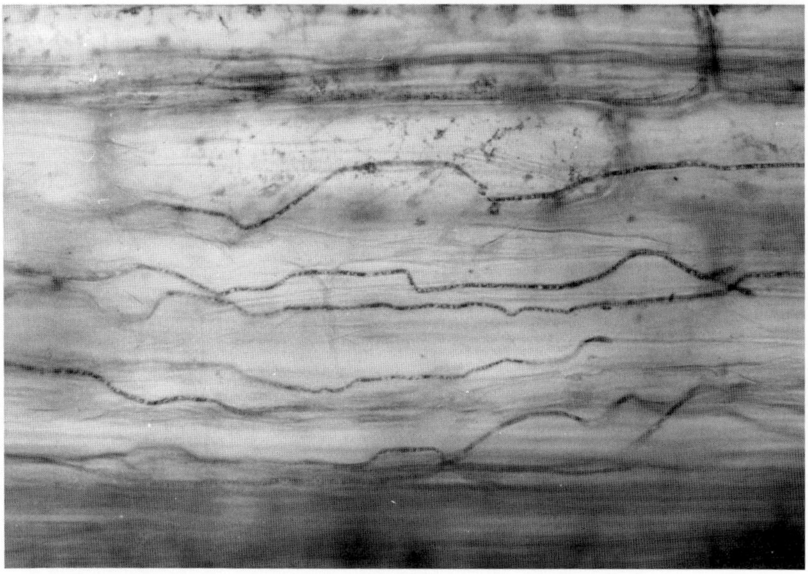

Fig. 11.2 Hyphae of an endophyte within the stem of *Festuca arizonica* (photograph by J.F. White).

arthropods. The systemic nature of infection precludes selective feeding on particular tissues, and although levels of deterrent chemicals may vary between different plant parts, and with season, they are always present (White & Cole, 1985; Hardy *et al.*, 1986; Lyons *et al.*, 1986).

Herbivore responses to mycophylla depend on the insect involved and on the host–endophyte combination, but include depression of feeding, significantly reduced survival and discrimination against infected plants (Table 11.1). These effects may arise either directly from the production of fungal toxins, or indirectly through endophyte-induced synthesis of noxious secondary plant metabolites. Individual mycophyllas usually contain a spectrum of such compounds, and it is often difficult to establish whether particular compounds are of fungal origin, or plant origin, or are endophyte induced in plant tissues, or are induced in the endophyte by plant tissues (Clay, 1988). Most deterrents appear to be alkaloids, yet few have been identified as being of fungal origin. For example, protection of the *Festuca arundinacea–Acremonium coenophialum* mycophylla has been attributed to the presence of the pyrrolizidine alkaloids *N*-acetyl and *N*-formyl loline, but these compounds have not been isolated from axenic cultures of the endophyte (Johnson *et al.*, 1985). However, in *Lolium perenne–Acremonium lolii*, the protectant alkaloid peramine together with lolitrem neurotoxins, which are also active, are possibly fungal metabolites (Rowan *et al.*, 1986).

Table 11.1 Examples of endophyte-mediated effects on herbivorous insects of Gramineae.

Host plant	Endophyte	Herbivore	Response	References
Cyperus virens	*Balansia cyperi*	*Spodoptera frugiperda* (fall armyworm)	Reduced survival and weight gain	Clay *et al.*, 1985a
Dactylis glomerata	*Epichloë typhina*	*Agrotis segetum* (cutworm)	Reduced survival	Schmidt, 1986
Festuca arundinacea	*Acremonium coenophialum*	*Rhopalosiphum padi* (oat-birdcherry aphid) *Schizaphis graminum* (greenbug aphid)	Strong discrimination against infected plants	Johnson *et al.*, 1985
		Draculacephala antica (sharpshooter leaf hopper) *Chaetocnema pulicaria* (corn flea beetle)	Reduced numbers on infected plants	Kirfman *et al.*, 1986
		Spodoptera frugiperda (Fall armyworm)	Reduced weight gain	Clay *et al.*, 1985b
Lolium perenne	*Acremonium lolii*	*Acheta domestica* (cricket)	Total mortality of test population	Ahmad *et al.*, 1985
		Sphenophorus parvulus (bluegrass billbug) *Crambus* species (sod webworm)	Reduction in feeding damage	Funk *et al.*, 1983; Ahmad *et al.*, 1986
		Listronotus bonariensis (Argentine stem weevil)	Reduced feeding and frequency of oviposition	Mortimer & di Menna, 1983
		Spodoptera frugiperda (fall armyworm)	Reduced survival	Clay *et al.*, 1985b
Paspalum dilatatum	*Myriogenospora atramentosa*	*Spodoptera frugiperda* (fall armyworm)	Reduced survival	Clay *et al.*, 1985b

Effects of mycophagy and detritivory

Despite the existence of defences, and their undoubted importance for some fungi, it is probable that in a wider ecological context they are commonly either not present or are easily overcome. This is indicated by the large number of arthropod taxa which are either known, or presumed, to be reliant on fungi as a sole or major food source. Amongst insects, members of over 30 Orders are in this category, and in Coleoptera alone over 60 Tribes have mycophagous representatives

(Hammond & Lawrence, 1989). However, the outcome is not always simply one of mass destruction. First, as has already been mentioned, feeding preferences may exert a selective effect on fungal community structure. Second, the propagules of many fungi are dispersed by animals, often the initial step being attraction of the vector to edible sporophores. In such circumstances, it is clear that substantial ecological benefits to fungi can arise from mycophagy and detritivory.

Consumption of fruitbodies and propagules. Fruitbodies provide mycophages with a variety of resources, there being wide differences in their quality and spatial and temporal availability. What little is known of the effects of mycophagy upon them is confined almost totally to basidiomata, and even here there is a lack of quantitative data (Hanski, 1989).

Basidiomata have large numbers of arthropod associates, but it is doubtful whether, in most situations, mycophagy interferes significantly with spore discharge. For instance, the fleshy basidiomata of most agarics are relatively short-lived and probably have released the bulk of their spores by the time that any resident mycophages become established; the latter then consume exhausted tissues, but have little effect on reproductive capacity. Longer-lived, especially perennial, polyporous basidiomata may be more affected, although their low water content and physical toughness can discourage attack. However, brackets of *Polyporus adustus* may be destroyed by *Cypherotylus californicus* (Coleoptera: Erotylidae) before spore production takes place, and hymenial galls induced by *Agathomyia wankowiczi* (Diptera: Platypezidae) may prevent discharge in *Ganoderma applanatum* (cited by Hanski, 1989).

Location of agaric basidiomata by mycophagous insects may not be a random process. For example, 1-octen-3-ol has been implicated as a volatile attractant, and the same role might be played by luminescence (Pyssalo, 1976; Sivinski, 1981). By contrast, the attraction of truffle flies – *Suillia* species (Diptera: Heleomyzidae) – to the hypogeous ascomata of *Tuber melanosporum* is well documented. Truffles release a range of odorous volatiles, amongst which dimethyl sulphide can be selectively detected by truffle-hunting pigs and dogs; its involvement in host location by *Suillia* species is a distinct possibility (Talou *et al.*, 1990). It has been suggested that, in nature, attraction to and consumption of hypogeous fruitbodies by rodents indicates a co-evolved mutualism, in which food is exchanged for dispersal of those spores that pass through the gut unharmed (Pirozynski & Malloch, 1988). A similar relationship might be claimed for the attraction of slime-feeding flies by stinkhorns and related species, and of other insects to fungal honeydews.

Whilst spore dispersal via mycophagy and detritivory is widespread, for the most part it would seem to occur inadvertently. The most familiar example is that of coprophilous fungi, where herbivores are not only vectors but also act as vehicles for activation of dormant spores. However, other instances abound, and there is no doubt that some spores are remarkably resistant to digestion which, in many cases, additionally promotes subsequent germination. Oospores are notable in this respect (Salvatore *et al.*, 1973; Kueh & Khew, 1982; Gardiner *et al.*, 1990; Goldberg & Stanghellini, 1990). Those of *Phytophthora* and *Pythium* respectively survive passage through snails and larvae of the fungus gnat *Bradysia impatiens* (Diptera: Sciaridae), with germination levels in *Phytophthora megasperma* being raised from 5 to 93% following ingestion by the land snail *Helix aspersa*.

There is also evidence for the frequent ingestion by earthworms and macroarthropods of zygospores of vesicular–arbuscular mycorrhizal fungi, the result being redistribution within the soil profile, and hence an increased chance of encountering a new host root (McIlveen & Cole, 1976; Rabatin & Stinner, 1985). Within soil and litter, spores are probably a major dietary component for many microarthropods, and indeed some Collembola prefer spores to hyphae, given that both are edible. In these circumstances only a small proportion survive passage (McMillan, 1976; Ponge & Charpentie, 1981). Quantitative data relating to rates of spore destruction are rare, but in one study 80% of ascospores of *Coniochaeta nepalica* placed in soil were removed by microarthropods over a 3-month period. Germinability of those which reappeared in faecal material was reduced to 8–30%, whilst those remaining *in situ* were often damaged by perforating mites (Gochenaur, 1987). Survival of sclerotia can also be adversely affected by arthropods. For example, those of *Sclerotinia sclerotiorum* that have been damaged by larvae of the dark-winged fungus gnat *Bradysia coprophila* may germinate poorly, and their susceptibility to mycoparasites is greatly increased; reduction in germination, at least in part, is due to chitinase-rich salivary secretions (Anas & Reeleder, 1988; Anas *et al.*, 1989).

Destruction of mycelia. Biomass disappears in two ways: either by grazing, that is mycophagy *sensu stricto* in which hyphae are removed from their substratum without extensive destruction of the latter; or through detritivory, in which hyphae plus resource are consumed. The distinction between these processes may not always be clear cut, but where they are selective their ecological consequences – at least in theory – can differ markedly with respect to their influence on community structure. Grazing, through its elimination of preferred species, opens up the resource to colonization by incoming fungi or, possibly more frequently, leaves already-established, but less-preferred, occupants in command. On the other hand, detritivory destroys those resources that are occupied by preferred species, whilst leaving other materials which support less-preferred or repellent fungi; this results in the disappearance of habitats rather than a change in community development in existing habitats.

Destruction of fungal biomass by invertebrates occurs on a vast scale, and is a key step in nutrient flux pathways in many aquatic and terrestrial ecosystems (Anderson & Ineson, 1984; Bärlocher, 1985; Peterson *et al.*, 1989). However, in spite of its importance, the influence of selective feeding on habitat persistence and community structure has been suprisingly little studied, with most attention focusing on collembolan grazing within leaf litter (Visser, 1985). Mycophagous Collembola exhibit strong dietary preferences, and appear to be attracted to palatable fungi via the release from them of volatile odour compounds (Bengtsson *et al.*, 1988). For instance, within experimental microcosms, *Onychiurus subtenuis* avoided hyaline mycelia of basidiomycetes, but fed upon *Cladosporium* and a pigmented sterile mycelium, thus placing the less-preferred fungi at a competitive advantage (Visser & Whittaker, 1977; Parkinson *et al.*, 1979). Similarly, in Sitka spruce litter, in both the laboratory and the field, *Onychiurus latus* prefers mycelium of *Marasmius androsaceus* to that of *Mycena galopus*. In the field, *Marasmius* mycelium is found in the surface 3 mm of the L-horizon, whilst that of *Mycena* is located in the lower F_1-horizon; this is due in part to the drought tolerance of the former and drought sensitivity of the latter (see also Chapter 4). In laboratory conditions, *Marasmius androsaceus* colonizes L-horizon and F_1-horizon litter at twice the speed of *Mycena galopus* but, in the field, within the moister F_1-horizon this competitive edge is removed by grazing populations of drought-sensitive *Onychiurus latus*, thus producing spatial separation of the two fungi (Newell, 1984a,b). In this instance, the ecological outcome of selective grazing is the obviation of competitive interactions between fungi, and the consequent reallocation of resources between them. This may not be a unique example of intervention by grazers in competitive interactions between fungi. Zones of somatic incompatibility between individual mycelia of Ascomycotina and Basidiomycotina attract fungus gnats and are selectively fed upon by them (Boddy *et al.*, 1983). If this also occurred during combat between mycelia of different species, then it could play a major role in determining the result.

At the whole terrestrial community level mycophagy can also have a considerable effect on fungal symbioses through, for instance, selective grazing on mycorrhizal fungi, with consequent changes in the performance of host plants (Finlay, 1985; Shaw, 1985). Again, within communities of saxicolous lichens, the grazing slug *Pallifera varia* shows distinct preferences based, at least partially, on the antifeedant activity of stictic and protocetraric acid; the most palatable lichen, *Aspicilia gibbosa*, is able to maintain itself only by virtue of its rapid growth rate (Lawrey, 1980).

Moving to aquatic environments, seasonal input of terrestrial leaf litter is the starting point for energy flow and nutrient fluxes in many freshwater systems. In lotic (flowing) waters, it is thought that initial resource utilization is carried out primarily by hyphomycetous fungi, which condition leaves so that they become palatable to invertebrate detritivores. This involves both the production of nutritious fungal biomass, and the enzymic conversion of inedible leaf materials to a more acceptable form (Bärlocher, 1985; Anderson & Cargill, 1987). There are indications that consumption of conditioned tissues is selective, and so may have consequences for fungal community structure. However, the extent to which this occurs in nature is unknown, and the complexities of resource–fungus–invertebrate relationships are not fully understood. Difficulties arise because different detritivores exhibit different preferences that depend on the fungi involved in processing. The bases for these differences are not known, but have been related variously and tenuously to the degree of leaf softening, nitrogen content of the resource–fungus combination, the presence of lipids or growth factors, the removal of antifeedant chemicals and digestive specialization in the animal concerned (Arsuffi & Suberkropp, 1989).

Finally, it should be noted that, in terrestrial and aquatic spheres, the conversion of fungal biomass or fungally modified resources to faecal material generates a new range of habitats for suitably specialized fungi. For example, faecal pellets of mycophages and detritivores are often rich in fungal chitin, thus enabling the development of chitinoclastic species upon them.

Habitat provision Defaecation is only one instance of the many ways in which animals provide fungi with fresh habitats. In regard to these, broadly there are two possibilities. First, an individual or group of animals may, for a variety of reasons, either deposit or accumulate organic materials to form discrete resource units with a low indigenous fungal biomass. This kind of enrichment disturbance results in the development, from either within or without, of particular fungal communities or populations, whose nature is determined by resource composition and, in some instances, by utilization of developing biomass by the animals concerned. Second, an individual or group of animals may, by virtue of their breeding or feeding habits, introduce particular fungi into unoccupied or sparsely colonized habitats which, in other circumstances, they would be incapable of entering. Again, composition of the resultant flora is resource related, and may also be influenced by breeding or feeding activity of the vector within the new habitat. The degree of animal–fungus interdependence varies widely with, at one extreme, ecological benefits accruing only to the fungus, and at the other the existence of an ecologically obligate mutualistic association between the two organisms.

Nests and hoards

Many animals accumulate plant materials of various types to create nests, stores or feeding depots. Whilst some nests are temporary, and some food caches quickly become exhausted, others may be added to constantly, so maintaining the fungi that inhabit them. Such replenishment often reflects nutritional interdependence of the animal and the microbial occupants of the resource.

Vertebrates. Although they have been little studied, conditions within the nests of many vertebrates are conducive not only to fungal growth in general but also, more particularly, to the development of specialized communities. For example, keratinophilic species are common inhabitants of feather-lined birds' nests, and thermophilic fungi permeate the large, self-heating, compost-rich incubators constructed by some Crocodilia, and the Mallee fowl, *Leipoa ocellata* (Pugh & Evans, 1970; Tansey & Brock, 1978). Similarly, the woodland agaric *Hebeloma radicosum* fruits above, and is connected to, the leaf-filled nests of moles, where there may also be nitrogen enhancement because of their use as middens (Sagara, 1989; Sagara *et al.*, 1989). Nests may also contain hoarded food and this too can encourage particular fungi to appear. In North American deserts, the underground seed caches of banner-tailed kangaroo rats (*Dipodomys spectabilis*) support assemblages of *Penicillium* species which are notable for their relatively low production of mycotoxins. This suggests that, in this animal, evolution of the seed-hoarding habit has taken place in such a way as to select for contamination by less harmful fungi than might otherwise occur (Frisvad *et al.*, 1987).

Termites and ants. Outstanding examples of resource accumulation are to be found amongst fungus-dependent termites and ants (Batra & Batra, 1979; Martin, 1984; Cherrett *et al.*, 1989; Wood & Thomas, 1989). Within their nests virtual monocultures of fungi, which are incapable of living at large, are maintained on massive, long-lived food depots that are constantly supplied with foraged plant materials. Although achieved in different ways, the result with both termites and ants is similar, in that nutrients released from plant tissues by fungal activity are tightly cycled and efficiently partitioned between the fungus and its animal symbiont, although the direct contribution made by fungal biomass to the diet of termites is far greater than that for ants.

Termites of the subfamily Macrotermitinae are widespread in tropical Africa, and are also distributed through Arabia and Indo-malaya, where they, and associated species of the agaric *Termitomyces*, have a major role in the processing of wood and leaf litter. For example, in a West African savannah, an estimated 27% of annually produced woody and leaf litters may be consumed by Macrotermitinae with other termites accounting for only 8%. Whilst details of this operation vary with the particular termite–fungus combination involved, its general features are as follows.

The termites construct nests which are either in the form of large, conspicuous mounds or are entirely subterranean (Fig. 11.3). Within these they build convoluted, sponge-like combs that may be located centrally in a hive or brood chamber, or in smaller chambers distributed throughout the nest (Fig. 11.4). Combs consist of faeces derived from ingestion of plant litter and are permeated by mycelium of *Termitomyces*, there being strong indications of specificity of association between *Termitomyces* species and different termite genera. The comb surface bears nodules of fungal material consisting of spherical conidiomata; termites feed on combs, at the same time ingesting mycelial biomass and conidiomata (Fig. 11.5). As older parts of combs

Fig. 11.3 Mound of *Macrotermes bellicosus* in the Guinea savannah (by courtesy of the Natural Resources Institute).

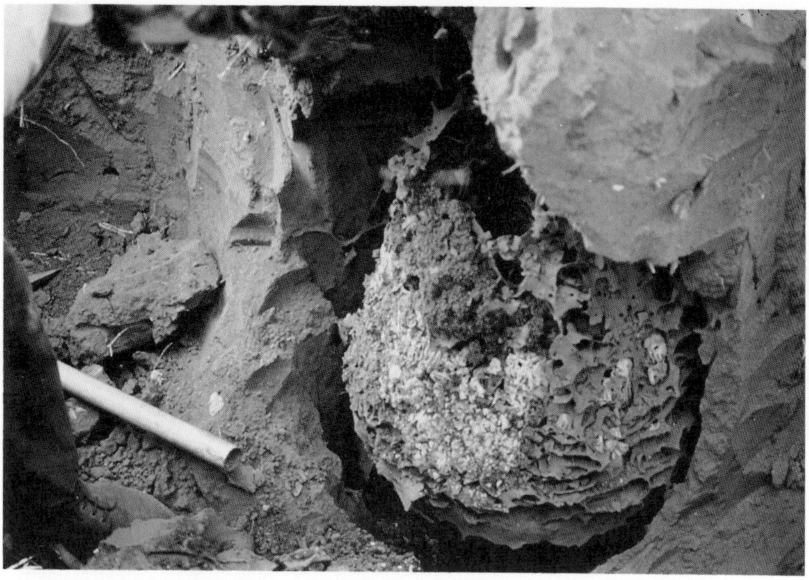

Fig. 11.4 Interior of mound of *Macrotermes bellicosus* showing detail of its chambered structure and a fungus-bearing comb (by courtesy of the Natural Resources Institute).

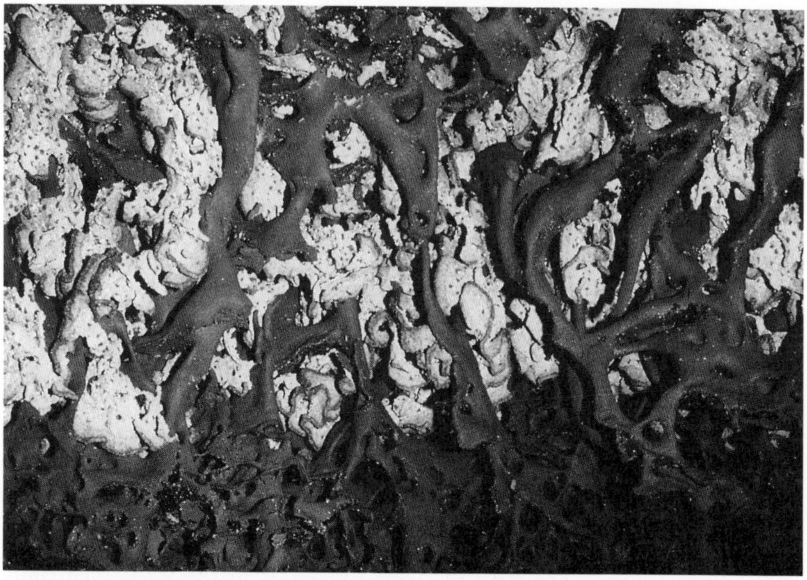

Fig. 11.5 *Macrotermes bellicosus*. Comb material removed from the nest with the food fungus *in situ* (by courtesy of the Natural Resources Institute).

are eaten away, new faeces are continually added which, after a period of 5–8 weeks, are sufficiently fungally conditioned to be eaten. Digestion of lignocellulose by the termite–fungus combination takes place in two phases. Fresh faecal deposits, although finely comminuted, are of structurally unaltered plant remains. During the first, conditioning, phase *Termitomyces* degrades lignin, cellulose and polyphenol–protein complexes; breakdown products are then consumed by termites which also obtain nitrogen compounds from mycelium and conidiomata (Garnier-Sillam *et al.*, 1988). During the second, digestion, phase there is further cellulolysis within the gut, this being effected partly by enzymes acquired from conidiomata and partly by those secreted by midgut epithelium and salivary glands (Martin, 1984).

The central feature of this symbiosis is the microbiologically unimpeded colonization of non-sterile faeces by *Termitomyces* mycelium, which is slow growing and has poor competitive ability. Its almost total command of comb resources is achieved through termite activity (Thomas, 1987a–c). First, the fungus is provided with a finely divided substratum which favours hyphal permeation. Second, contaminating fungi are reduced in number when plant remains are initially masticated and ingested and also, subsequently, when comb material is digested. In particular, termite saliva would seem to be fungistatic, as are the oral defensive secretions and queen exudates of some species. *Termitomyces* has, presumably, developed tolerance of these factors.

The importance of termite secretions to the maintenance of *Termitomyces* is indicated by the fact that, if combs are removed from the nest, they become overgrown quickly by other saprotrophs. This raises the question of how the symbiosis is continued when old nests are deserted and new ones founded. In some termite species, alates ingest conidial nodules before swarming, so that spores are carried as a bolus in the gut and then pass out to inoculate the primordial comb. In others, there is evidence for a mutually adapted life cycle between fungus and termite. When the comb is abandoned, basidioma primordia replace conidiomata, and fruiting then takes place outside the nest, each basidioma being connected to the comb via a mycelial cord. Foraging workers from new colonies then consume basidiospores and carry them back to establish comb systems within the nests.

Ecological parallels can be drawn between Macrotermitinae and the ant tribe Attini (Formicidae: Myrmicinae) which, by contrast, is confined to South and Central America and some regions of North America. All but one of nearly two hundred attine species cultivate agarics in fungal gardens within subterranean nests. The identity of the fungi remains obscure, but they are probably mostly species of

Leucoagaricus (Powell & Stradling, 1986; Muchovej *et al.*, 1991). Some attines grow them on collected caterpillar faeces and decaying plant litter, others on vegetable matter containing fallen flowers or fruit. However, what are considered to be the most evolutionarily advanced species, those of *Acromyrmex* and *Atta*, provide their fungi with fresh leaves, fruit and flowers, gathered from living plants, and it is this aspect of the attine–fungus relationship that has been studied most intensively, and which has been written of in detail on numerous occasions (see Cherrett *et al.*, 1989). If the intricacies of individual and colonial ant behaviour are set to one side, the most important mycological facet of the latter, more developed symbioses, relate to the nature and volume of the resources made available to the fungi, and the manner in which they are processed.

The harvesting of plant material by leaf-cutting attines occurs on a vast scale; in tropical rain forests their activities may account for loss of 17% of total leaf production, and in grasslands single colonies of *Atta* species may utilize up to 250 kg of dry matter per hectare annually. Whilst the range of plants utilized can be wide, within it ants exhibit preferences due to the presence of deterrent factors, including secondary plant metabolites (Waller, 1982; Hubbell *et al.*, 1983; Febvay *et al.*, 1985; Howard *et al.*, 1988). The collection of relatively unpalatable, as well as preferred, leaf material is a reflection of the differing dietary requirements of ant adults and larvae. The former obtain the bulk of their nutrients from plant sap, and so actually feed on a narrower spectrum of plant species than that from which they gather leaves. On the other hand, larvae are probably fed almost exclusively on the fungus, which is grown on the leaves and so, indirectly, can utilize a wider variety of them provided that they can be suitably conditioned (Quinlan & Cherrett, 1979).

Leaf conditioning is a complex procedure that is still imperfectly understood; its end-point is a monoculture of the agaric which, like *Termitomyces*, probably has no free-living existence outside the nest. Swollen hyphae, gongylidia, are produced which aggregate into nodular staphylae that are cropped by the ants and fed to the larvae. Various ant activities preserve the monoculture, maintain its unique morphology, and aid its saprotrophic activity.

On arrival within the nest, plant fragments are freed of cuticular waxes by licking, and then are crushed to expel sap and to soften tissues prior to fungal colonization. Fragments are incorporated into the fungal garden and are anointed with faecal fluid. Ants then remove hyphae from well-colonized parts of the garden and use these to inoculate treated tissues. Faecal fluid contains amylases, cellulases, chitinases and pectinases, all of which are possibly acquired by the ants via the ingestion of fungal hyphae. In addition, a serine proteinase and two metalloendoproteinases are present, which have been established as being of fungal origin (Martin, 1984). The fungal symbiont does not secrete these three enzymes to any extent, and grows only slowly if the nitrogen source provided is in the form of polypeptides. The application of faecal fluid to freshly cut leaves augments the capacity of the inoculum to degrade protein, with acquired pectinases perhaps additionally macerating tissues to facilitate their rapid penetration by hyphae.

Fungus gardens are continuously contaminated by microorganisms imported on new leaf material and on the bodies of foragers yet, as with termite comb, foreign fungi do not overrun the resident monoculture until gardens become abandoned. Suppression of microbial, and especially fungal, competitors appears to be effected by a number of different, but interacting, chemical factors (Iwanami, 1978; Hervey & Nair, 1979; Nair & Hervey, 1979). First, the garden occupant may produce antifungal antibiotics, one such being lepiochlorin (Fig. 11.6). Second, attines secrete acidic compounds from their metathoracic glands. Two of these, β-hydroxy-decanoic acid and phenylacetic acid, have antibiotic properties, whilst a third, indolylacetic acid, enhances activity of β-hydroxy-decanoic acid by depressing the pH of the mixture (Fig. 11.7). In addition, application of faecal chitinases to uninoculated leaves by the ants may destroy any spores or hyphae that are present upon them. Acidic ant secretions also reduce substratum pH to 4.5–4.8, the optimum for gongylidium production; the latter is also possibly favoured by the suppression of basidioma formation by as yet undetermined factors.

By contrast with Macrotermitinae, transmission of the food fungus is not via ingestion and defaecation by workers. Before swarming, founder queens compress hyphae into a pellet which is then stored in the infrabuccal pouch, a small cavity under the oesophageal entrance. After a queen has excavated the first subterranean chamber of the new nest, the pellet is expelled and tended by her, faecal fluid being used to nurture the fungus until emergence of the first workers, which then forage for leaves.

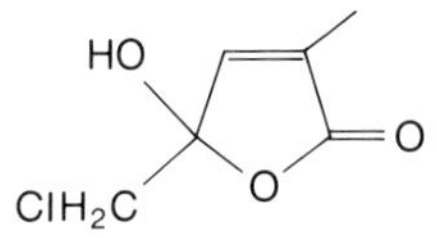

Fig. 11.6 Structure of lepiochlorin, an antifungal antibiotic produced by the food fungus of attine ants.

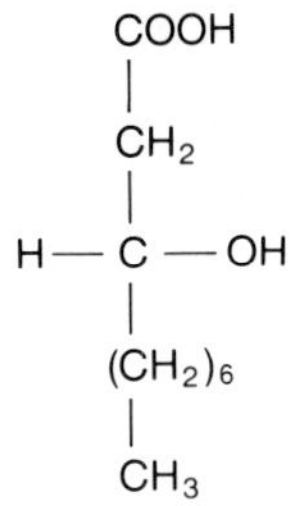

Fig. 11.7 Structure of β-hydroxy-decanoic acid, an antibiotic secreted by attine ants.

Intact plant tissues

Bulk tissues of plants offer fungi a variety of abundant, renewable, virtually universally distributed, resources that are, initially, either free from, or only lightly colonized by, microorganisms. Yet intact living tissues repel most potential invaders, and some dead tissues, for example wood, are not easily penetrated. For a great many fungi, such problems are solved when tissues are mechanically damaged by animal activity; wounds permit the ingress of fungi which would otherwise be denied entry. The range of fungi that exploit this kind of opportunity is vast, as is that of the animals which facilitate it and, for the most part, there is only a casual relationship between the two kinds of organism. There are, however, some specific associations between insects, in particular, and either single fungal species or well-defined fungal assemblages; here fungal life cycles have become fully integrated with various insect traits, including not only feeding patterns but also reproductive behaviour. Such relationships are characteristic of, but are not entirely confined to, insects that either inhabit wood or are closely associated with it (Swift & Boddy, 1984).

It is difficult to visualize the total volume of woody substrata available, but in both temperate and tropical forests 92–99% of above-ground biomass is in the form of wood. It is, however, a low quality resource, due to the physical and biochemical consequences of lignification, the presence of allelopathic compounds, and the additional protection afforded by enveloping bark. Utilization of unmodified lignin-rich wood components is confined largely to Basidiomycotina and, within the insects, to some Isoptera and Coleoptera; but many non-basidiomycetes can live within wood, on cell contents and unlignified tissue components, after being introduced there by an insect associate.

By far the most numerous examples of such associations occur within various groups of bark-inhabiting and wood-boring beetles that are obligately mutualistic with ascomycetous fungi. In detail these symbioses appear to be almost infinitely diverse, with each beetle species having unique features with respect to the manner in which the fungi are transmitted, and sometimes also to the identity of the fungi concerned (see Beaver, 1989; Berryman, 1989). Furthermore, many problems remain unsolved, for instance regarding the precise nutritional relationships between beetle and fungus and the degree of specificity involved. Hitherto, attention has mostly been focused on species of *Xyleborus* (Scolytidae: Xyleborini) which is one of the largest insect genera, and thus has ecological prominence amongst wood-boring Coleoptera. Other genera share some of the major features of *Xyleborus*–fungus relationships, and the latter may represent the fundamental pattern of these symbioses.

Xyleborine adults and larvae inhabit tunnel systems mainly excavated within sapwood, although some species may also enter hardwood. Fungi are transported to fresh wood principally by female adults, which carry them within mycetangia. These are fluid-filled organs in which the fungi grow in a yeast-like form. In many *Xyleborus* species, mycetangia consist of paired mandibular pouches, but in others they occur elsewhere on the body, for instance as a single pouch extended below the mesonotum. In other genera there is wide variation in their location, but all have a similar basic structure, being tubes, sacs, fissures or pits, with associated glandular cells. The fluid that they contain is probably fungistatic, which precludes alien fungi from growing in it and also, perhaps, imposes the yeast form on the symbiotic species.

As adult females burrow, secretion of mycetangial fluid increases, enabling them to smear the tunnel walls with fungal cells from which a mycelium then develops. Available evidence suggests that the principal genus transmitted is *Fusarium*, although this is associated with other ascomycetous fungi, including yeasts. Hyphae penetrate the phloem and other unlignified tissues; the beetle galleries become covered with an ambrosial layer, consisting of an erect mycelial palisade bearing chains of moniliform cells, conidia or chlamydospores, which is grazed by adults and their larvae. This ambrosia form appears to be maintained by the browsing process and also, perhaps, by adult or larval secretions; the latter seem to minimize development of non-ambrosial fungi. When larvae pupate and leave the tunnels, and the occupying brood females have died, the resident fungi are quickly overgrown and eliminated by alien species.

Bark beetles, which are essentially phloem feeders rather than mycophages, are frequently found in constant association with a characteristic flora of ascomycetous fungi (Wingfield & Gibbs, 1991). These are often Ophiostomaceae, principally *Ceratocystis* species, the ascospores of which are carried either on the body surface, or via passage through the gut, or in simplified mycetangia. They may also be transmitted by mites that are hyperphoretic on adult beetles (Moser *et al.*, 1989). There are also indications that ascomycetous communities may contain cryptic growths of ligninolytic basidiomycetes, which may enhance the colonizing capabilities of the complex as a whole (Whitney *et al.*, 1987).

Access of fungi to uncolonized wood is also facilitated during egg-laying by woodwasp species (Siricidae and Xiphidriidae), the larvae of which are wood-inhabiting. Adult females deposit eggs in the moist wood of weakened trees or freshly cut timber by means of a long, fine ovipositor. At the base of this are paired, fluid-filled mycetangial pouches which contain oidia of basidiomycete species, notably *Amylostereum areolatum* and *Stereum sanguinolentum*. Oidia are deposited with the eggs and germinate before the latter hatch, so that larvae develop within galleries or cavities lined with mycelium, their diet being a mixture of wood fragments and fungal hyphae. Digestion of the former is made possible by the acquisition, within the larval gut, of cellulases and other polysaccharidases released from ingested hyphae (Martin, 1984).

Similar insect–fungus interdependence in succulent plant tissues appears to be rare, but is found amongst gall midges (Diptera: Cecidomyiidae) that induce leaf, stem or fruit neoplasms, containing a larval chamber; this, at some stage, becomes lined with fungal mycelium on which the larva feeds (Bissett & Borkent, 1988). In species of *Asphondylia*, the female midge pierces tissues with its ovipositor to create a small cavity containing one egg and a few conidia. The latter are carried in paired mycetangia associated with the ovipositor, and are usually those of *Macrophoma* species, which are probably anamorphic states of *Botryosphaeria*. The egg hatches within a few days, and the presence of the larva in some way induces meristematic activity in the plant, resulting in growth of gall tissue around it. Only at this stage does the fungus develop, eventually producing a compact mycelium occupying most of the gall chamber; the larva then grows rapidly using the fungus as its sole food source. The fungus is not, however, totally destroyed, and when the gall becomes uninhabited, pycnidia may form on its exterior. The means by which new, and presumably fungus-free, females acquire fresh inoculum remains unknown, as does the ecological status of the fungus beyond the bounds of gall tissues.

An interesting feature of most of these insect–fungus associations is the existence of specialized transmission organs. Within them, fungi may not be merely housed, but rather pass through a distinct developmental phase determined by host secretions, which also deter the growth of contaminating fungi. The yeast form commonly adopted by otherwise filamentous species probably reflects the stresses involved, but nevertheless ensures that, given the ability to adapt, fungi can successfully exploit living animals. The nature of animal–fungus relationships when the animal itself becomes a habitat is the subject of the next chapter.

12 Animals as habitats

Unlike bacteria, fungi have generally failed to overcome the severe problems involved in exploiting the potential microbial habitats that exist upon and within living animals. However, although the mycofloras of animals are neither as rich nor as diverse as their bacterial equivalents, the range of habitats occupied is wide. With regard to behaviour, at one extreme there are highly specialized, stress tolerant, neutral, or (more rarely) mutualistic exhabitants and inhabitants, which take up residence for the natural lifetime of their hosts. At the other is a broader spectrum of casual or opportunistic species, normally free-living, but which can become occupants if circumstances permit; many then become pathogenic. There also exists a large number of obligately parasitic species, whose presence may reduce host longevity considerably.

The literature of animal-inhabiting fungi is dominated by studies on pathogens and the aetiology of the diseases so caused, and there is no doubting the ecological, agricultural, veterinary and medical importance of mycoses. Acknowledging this, it can nevertheless be argued that, rather than being produced by mechanisms which have been evolved for that specific purpose, disease is often simply the inevitable consequence of the presence of a fungus on or in a living resource to which it may be imperfectly adapted. Indeed, in ecological terms, disease induction may be looked upon as being frequently disadvantageous, in that it commonly involves premature death of host tissues, or even of entire hosts. This is not only wasteful of resource, but also can expose the pathogen to competition from necrophilic microorganisms. From an ecological standpoint, there is virtue in reducing the emphasis on disease manifestation, and in isolating colonization and exploitation processes from those of disease causation. This serves to focus attention directly on the attributes of the fungi concerned, instead of on host responses to invasion, and also permits pathogenic and non-pathogenic species to be viewed together.

Physicochemical constraints

Habitat types are determined by a number of factors. At the simplest level these include body size, morphology, anatomy and physiological constitution; with the additional influences of poikilothermy, endothermy and the numerous behavioural patterns that different kinds of animals exhibit. If biochemical site characteristics are also taken into account, then it becomes impossible to briefly enumerate and define the multitude of habitats that may be occupied typically by zoophilic fungi. Customarily this difficulty is avoided by describing attributes of those fungi which are characteristically associated with particular animal groups, or which are restricted to specific locations on or in these animals. To a large extent this practice will be followed here, but first some general points might be made in relation to major problems that face all zoophilic species, irrespective of whether they are neutral, mutualistic or antagonistic towards their hosts.

Broadly, habitats can be divided into the external and internal. Excluding the majority of microfauna, most external body surfaces potentially can support at least limited fungal growth, provided that the substratum supplies appropriate nutrients, and is sufficiently stable physically to permit establishment. For instance, skin and its keratinized appendages, calcified collagenous scales and chitinous exoskeletons, all provide long-lived, carbon- and nitrogen-rich substrata for those species that are able to utilize their refractory, polymeric structural components. Since these materials are often dead, attempted colonization generally does not provoke massive mobilized defence reactions, although it may be hampered by desquamation, the presence of mucus, the release of fungistatic secretions, and low water potentials. The general harshness of external habitats, and the consequently relatively slow and restricted development of residents, virtually eliminates interfungal competition; the influence of bacteria then assumes possibly greater ecological importance.

Potential invaders of deeper regions have access to abundant supplies of water and readily available nutrients, but they also encounter severe stress and disturbance factors, including generally high bacterial numbers, which restrict regular occupation to suitably adapted species. For example, gut inhabitants are exposed to enzyme action, high solute concentrations, ensheathment by mucus and, in endotherms, elevated temperatures. At the same time, their retention within the animal can be endangered by the peristaltic flow of digestive fluids. Invasion of the blood or solid tissues may be countered by phagocytosis or immune responses.

Invertebrates

The huge variety of invertebrates has given fungi countless opportunities to exploit them, the antiquity and subsequently extended co-evolution of such interrelationships being reflected in their present ubiquity and often high degree of sophistication. Discussion will be confined largely to arthropods since these, and especially insects, are prone to many kinds of infection which have been much more adequately described than those occurring in other groups. Nevertheless, it should be noted that numerous fungi commonly inhabit Protozoa, rotifers and nematodes, and that infection features amongst microscopic animals may point to the origins of fungus–invertebrate interactions at large (Barron, 1989, 1991).

As with all fungal infections, the first problem to be overcome by fungi of microscopic animals is that of incursion. Potential hosts are mobile, often vigorously so, and this may tend to prevent spore attachment. Additionally, as typically with nematodes, there may be a strong cuticle which deters penetration. Many species have overcome both difficulties by infecting via ingestion, their spores being of an appropriate size and shape which allows them to be swallowed or otherwise taken in. Thus, conidia of *Cochlonema* species are engulfed by soil amoebae, and those of *Diheterospora* and *Harposporium* are swallowed by, and lodge within the guts of, rotifers and nematodes respectively (Fig. 12.1). Perhaps less commonly, conidia are produced which can attach firmly to the body surface, this often being achieved by means of an adhesive appendage borne either directly on the spore or on a short germ tube (Figs 12.2 & 12.3). The risk of detachment of *Nematoctonus* conidia from nematodes is reduced by the appendage producing a toxin which rapidly immobilizes the animals (Giuma & Cooke, 1971). Here, little or nothing is known in general of fungus–host recognition systems, the exception being *Drechmeria coniospora*, the conidia of which attach preferentially around the mouths of males and females of the nematode *Panagrellus redivivus* and, additionally, at the posterior region of the males (Fig. 12.4). These are the sites of sensory organs, and are also characterized by the localized presence of sialic acid (*N*-acetylneuraminic acid) as a cuticle component. This non-random distribution of spores is probably mediated by the binding of a specific fungal lectin with sialic acid, the result being that spores become located on those areas of the body with the weakest physical defences against invasion (Jansson & Nordbring-Hertz, 1984; Jansson *et al.*, 1985).

Once ingested or attached, spores quickly germinate, their penetrant germ tubes giving rise to either discrete thalli or a mycelium (Jaffee & Zehr, 1982). There is no evidence that microscopic animals can resist invasion and, for the most part, death swiftly ensues; this is followed by sporulation via externally produced fertile hyphae or exit tubes. The nutrition of endozoic fungi is almost exclusively necrotrophic, although putative biotrophy has been attributed to some amoeba-inhabiting species whose hosts survive for lengthy periods despite carrying well-developed infections (Cooke, 1977). However, it is clear that the occurrence of biotrophy in fungus–animal interactions is extremely rare, and indeed on closer examination may be found not to exist. The constitution of animals, and in particular their array of defensive reactions to invasive organisms, would seem to militate against the development of biotrophic associations, and therefore excludes the evolution of mutualisms based on this nutritional mode. This situation contrasts markedly with that found amongst autotrophs, where fungal symbioses are dominated by biotrophic mutualisms.

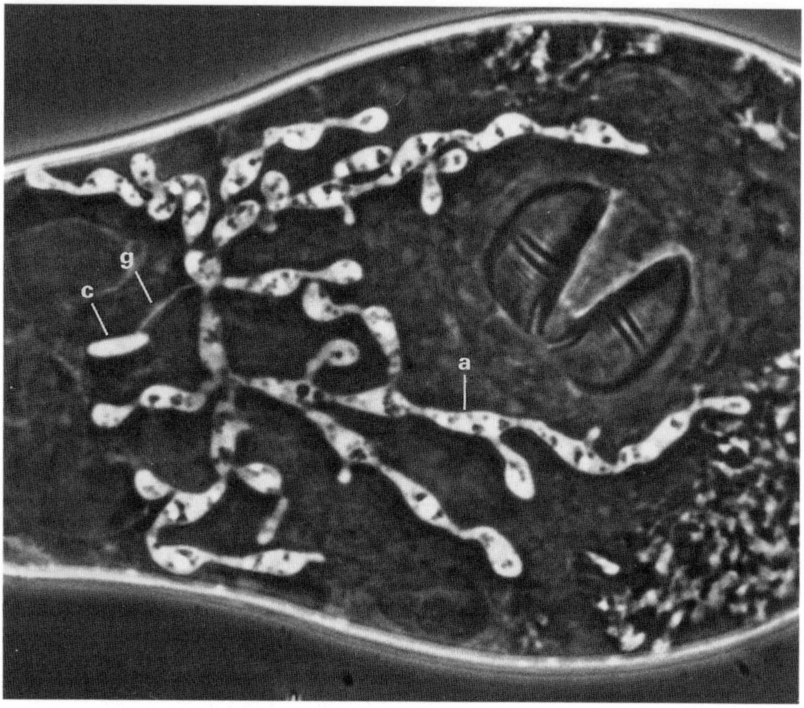

(a)

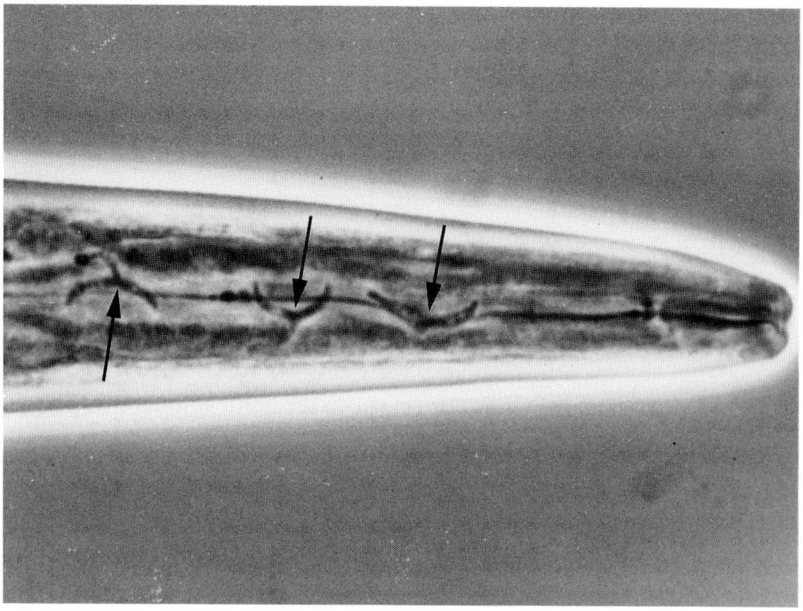

(b)

Fig. 12.1 (a) Development of *Diheterospora cylindrospora* within a rotifer. An ingested conidium (c) has given rise to a fine germ tube (g) which has then produced a series of branches (a) which are assimilating the body contents (from Barron, 1985, © National Research Council of Canada). (b) Curved conidia of *Harposporium* (arrowed) lodged within the oesophagus of a nematode (photograph by G.L. Barron).

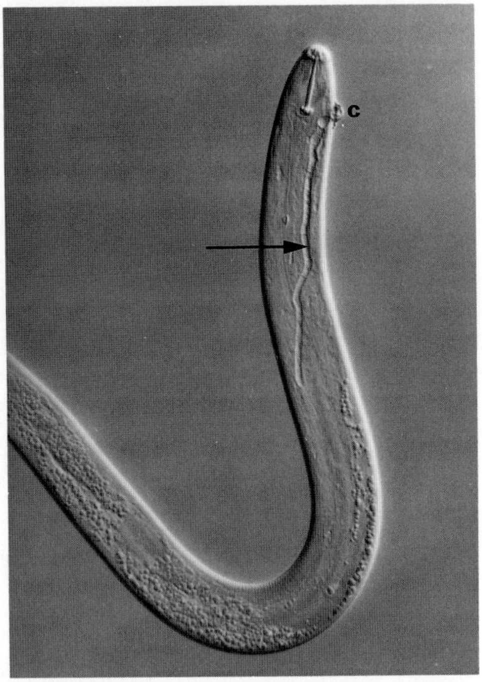

Fig. 12.2 Infection of the nematode *Heterodera schachtii* by *Hirsutella rhossiliensis*. An attached conidium (c) has effected penetration and an assimilative hypha (arrowed) is growing within the body (photograph by B.A. Jaffee).

Arthropods: general considerations

The rich variety of arthropod life styles generates complex combinations of biotic and abiotic determinants of infectivity which exclude most potential colonizers. However, during arthropod–fungus co-evolution, a number of fungal assemblages have emerged to successfully exploit the arthropod body. For the most part, these can be divided into three groups: entomopathogens, which rapidly invade and destroy the animal; ectosymbionts, the thalli of which are confined to the cuticle, although penetrant organs may pass into deeper, softer tissues; and endosymbionts, which inhabit the gut, with no apparent harmful effect on the host.

Before considering factors underlying specific habitat relationships, some comments of wider relevance are appropriate here. In nature, body surfaces may remain remarkably free from permanent fungal occupants, and this can, in some cases, be attributed to the presence of antifungal compounds. These may take the form of generally distributed epicuticular lipids or of defensive secretions from particular sites. For example, the bull ant, *Myrmecia nigriscapa*, releases a lipid-rich fluid from its metapleural gland which reduces mycelial development in a range of common soil-borne fungi, and in the two entomo-

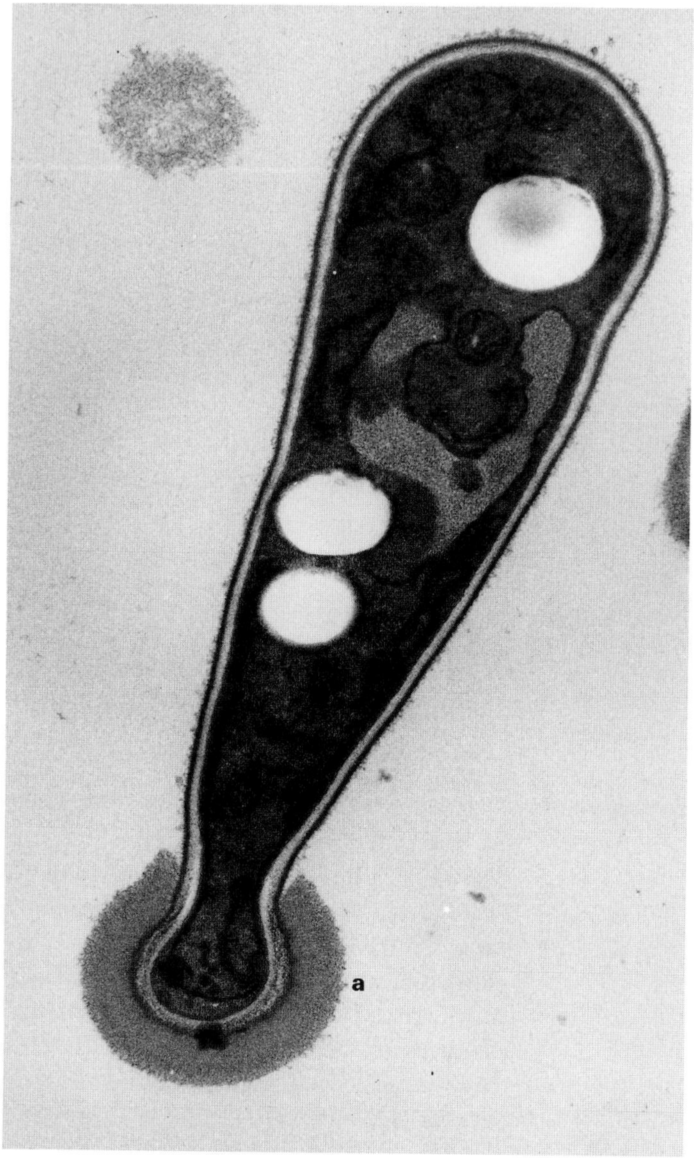

Fig. 12.3 Ultrastructure of the conidium of the nematode-parasitic fungus *Drechmeria coniospora* showing the basal pad of adhesive material (a) which facilitates attachment to the host (from Dijksterhuis *et al.*, 1990, © British Mycological Society).

pathogens *Beauveria bassiana* and *Paecilomyces lilacinus* (Beattie *et al.*, 1985). In this context, it has also been proposed that secretion of β-hydroxy-decanoic acid from the metathoracic glands of attine ants (which suppresses microbial contamination of their fungus gardens –

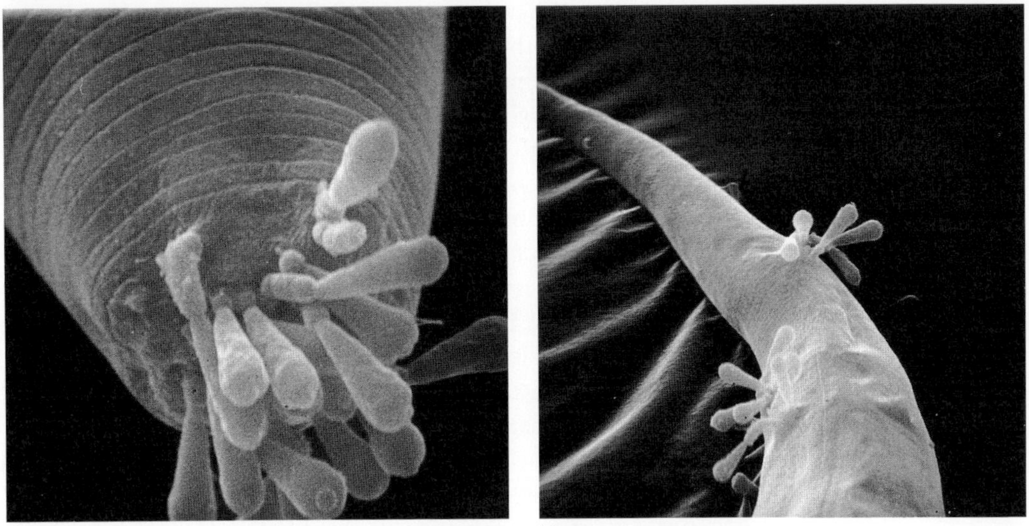

(a) (b)

Fig. 12.4 Attachment of conidia of *Drechmeria coniospora* to the nematode *Panagrellus redivivus*. (a) Mouth region with heavy infestation; (b) part of tail region of male with adhering spores (from Jansson & Nordbring-Hertz, 1984, © Society for General Microbiology).

see Chapter 11) first evolved as a protectant of naked larvae against potential soil-borne pathogens (Beattie, 1986). Similar antifungal mechanisms occur in certain millipedes (Xystodesmidae) which produce benzoic acid, benzoyl cyanide and phenol (Roncadori *et al.*, 1985).

As with the exterior, the body interior may remain uncolonized even where feeding involves the regular ingestion of large microbial inocula. It has already been noted that the propagules of many fungi can withstand passage through the gut but, in the main, conditions are hostile to their germination; contributing factors are anaerobiosis, adverse pH, enzyme activity and a range of antifungal phenolics produced by the bacterial gut flora (Charnley, 1989; Dillon & Charnley, 1991). Peristaltic and antiperistaltic movements also militate against spore attachment. Even spores of entomopathogens fail to infect here, and the normal invasion route for these species is via the cuticle. Given a suitable host, this leaves the gut habitat open to colonization by those relatively few endosymbionts which are adapted to do so.

Putting aside detail, for the moment, the arthropod body can be envisaged as providing two contrasting habitats: the exoskeleton and the organs contained within it. At the physiological level, detailed studies on colonization of these has been centred, almost exclusively, on insect pathogens; this emphasis is related to the economic importance of insect pests, and the potential of such fungi for their biological control (Charnley, 1984, 1989; Gillespie & Moorhouse, 1989). What follows is thus restricted in scope, but it seems likely that what is known concerning insect-pathogenic species can be more generally applied throughout the Arthropoda.

Occupation of the exoskeleton

The first terrestrial insects appear in the fossil record during the Devonian, at which time fungi were well established on the land. From that point, insect chitin would have been a frequent and increasing component of organic debris, so encouraging the emergence of ancestral chitinoclastic fungi and, arising from these, species capable of utilizing the exoskeletons of living insects (Evans, 1988). Present-day examples of the latter fall into two broad, overlapping ecological groups: one where somatic development takes place entirely or mostly on the exoskeleton, for example Laboulbeniales (Ascomycotina) and related fungi; and the other in which the exoskeleton is invaded as a first step in the colonization of the whole body, as occurs with entomopathogens. For both groups, success depends on the ability to utilize chitin and those compounds with which it may be chemically complexed. The problems involved in commanding such refractory resources are somewhat offset by lack of interfungal competition, although there may be some antagonism from any resident bacterial flora.

Cuticle structure and composition. Cuticle constitution varies according to species, and also goes through changes during development of the individual, but generally it consists of several non-cellular layers, with contrasting physicochemical properties, produced above an epidermis (Fig. 12.5). The outermost – epicuticle – is thin, lacks chitin, and is hard, hydrophobic and relatively inert to enzymic

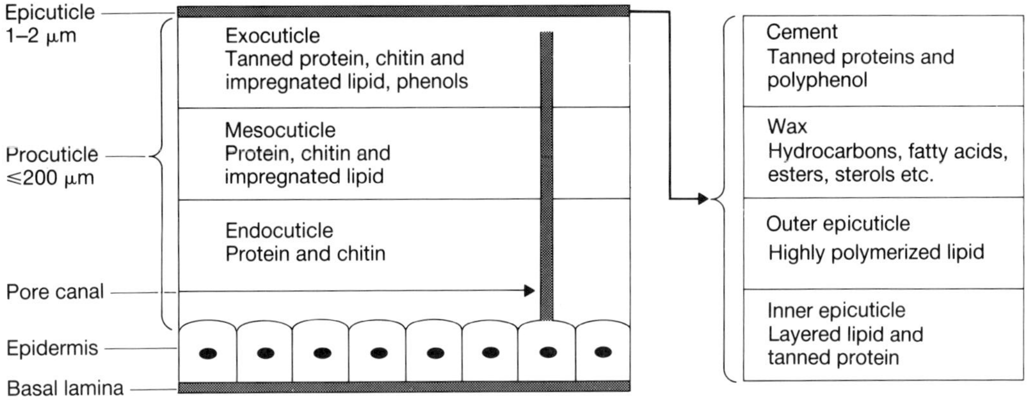

Fig. 12.5 Generalized structure of insect cuticle.

degradation due to its components of waxes, tanned proteins and polymerized lipids; some of its fatty acids may be antifungal. The underlying, laminated, procuticle is much thicker and is rich in chitin (for chitin structure see Fig. 2.8) which is complexed to varying degrees with proteins and lipids. The procuticle is pierced by minute canals containing waxes, lipids and other materials, and where it borders with the epicuticle it may be tanned. Irrespective of whether a fungus is preparing to traverse the cuticle or is attempting permanent settlement there, the initial problem is the same – attachment of the propagule to the inhospitable epicuticle for a period sufficient to allow establishment to take place.

Attachment and germination. The attachment process is poorly understood, and evidence for specific cuticle recognition mechanisms is scanty. In aquatic entomopathogens, zoospores may preferentially encyst on the intersegmental membranes of larvae, implying detection of particular substrates present in the flexible cuticle of these regions. In the mosquito pathogen *Coelomomyces psorophorae* a similar system may account for, respectively, frequent and rare encystment on larvae of susceptible and resistant species (Zebold *et al.*, 1979). For conidial fungi arrival is at random, with little difference in germination levels on different body sites. However, germ tubes may exhibit directional growth towards flexible cuticle, in particular that of intersegmental membranes (Boucias & Pendland, 1991). Attached spores are extremely difficult to dislodge, but there is as yet no evidence for lectin-mediated binding and, in some conidial species, firm adhesion occurs on both hosts and non-hosts (Charnley, 1989).

Given the great physicochemical variability of cuticle, particularly in its nutritional potential on the one hand and its antifungal properties on the other, it is not surprising that germination requirements are equally varied. Thus, it would appear that spores of relatively unspecialized fungi, for example *Beauveria bassiana* and *Metarhizium anisopliae*, will germinate on a range of insects, provided that the cuticle can supply suitable levels of water-soluble nutrients, mainly amino acids. By contrast, species with more restricted host ranges may be more fastidious. For instance, in *Nomuraea rileyi* germination is enhanced by diacylglycerols and polar lipids, which may be provided by its mainly lepidopteran hosts, whilst *Erynia variabilis*, which is restricted to small dipterans, has a requirement for oleic acid (Boucias & Pendland, 1984; Kerwin, 1984).

As a preliminary to cuticle invasion, germ tubes anchor firmly to its surface by secretion of glucan-rich mucilage. The cuticle beneath them may be degraded, as may that below ungerminated spores, suggesting that even during prepenetration it can be tapped as a nutrient source.

Penetration and utilization. Germ tube elongation ceases and, in some fungi, this is followed by appressorium formation; both are probably cued by a combination of cuticle substrates and contact stimuli arising from cuticle micromorphology. Incursion is then effected by either a relatively undifferentiated hypha, or a penetration peg, or by penetrant hyphae produced by appressoria. Invasion of cuticle is opposed by successive layers of material, each of which has its own distinctive properties. Passage through, and utilization of, these requires the sequential disposition of different physical and enzymic forces by the invader (Charnley, 1989; Charnley & St Leger, 1991).

The outer epicuticle is enzyme resistant, and its impermeability may prevent passage of enzymes into the inner layers. However, in insects it is usually fragile and is probably penetrated by the exertion of modest physical force. The inner epicuticle is toughened by polymerized lipoprotein stabilized by quinones, but it is nevertheless susceptible to endoproteases and lipoprotein lipases, which are known to be produced by some entomopathogens. Having breached the epicuticle by these means, the formidable barrier of the massive procuticle must then be passed. Its degree of resistance is determined by the relative proportions of its chitin fibrils and the proteinaceous matrix within which they are embedded, and further by the amount of protein–quinone cross-linking. The latter process is accompanied by extrusion of water, the result being toughening via sclerotization (tanning) particularly of the exocuticle. Hyphal growth within the outer procuticle often takes place laterally between the lamellae, as the fungus takes the path of least resistance, vertical penetration eventually being resumed. There is rarely widespread histolysis of the procuticle, but there is abundant evidence for localized enzyme action around hyphae.

Enzymic competence has been studied largely *in vitro* by growing entomopathogens on pulverized insect cuticle (Charnley, 1989; Charnley & St Leger, 1991). In such a system, *Beauveria bassiana, Metarhizium anisopliae* and *Verticillium lecanii* produce large amounts of endoprotease and variable levels of other enzymes, including chitinases and lipases. Enzymes appear in the sequence: esterase plus endoprotease, aminopeptidase, and carboxypeptidase (at approximately 24 hours) followed by *N*-acetylglucosaminidase and, after 3–5 days, chitinase and lipase. The exact role of these enzymes during cuticle exploitation *in vivo* is not yet known, but proteolytic enzymes are produced in quantity by appressoria and probably play a major part in solubilizing cuticle proteins, not only making these available for nutrition but also unmasking chitin. It is probably only then that chitinase, which is inducible, can be brought to bear. This explains the late appearance of chitinase *in vitro*, and the fact that marked enzyme activity *in vivo* occurs only when infection is well advanced.

Details of cuticle invasion so far described relate solely to entomopathogens, establishment of ectosymbionts having been observed much less frequently. There do, however, appear to be distinct differences in behaviour between these two groups. Infection by Laboulbeniales is via ascospores which become attached at one end to host cuticle. The attachment region usually darkens, and from it penetrant haustorial hyphae ramify into the cuticle to either remain confined there or, in some species, pass into the tissues beyond it (Tavares, 1985). In some related fungi, for instance *Pyxidiophora* on mites, there is no penetration; instead, attached ascospores differentiate into complex, cellular thalli that become large enough to support conidiation, which indicates that nutrients can be obtained from the cuticle without eroding it (Blackwell & Malloch, 1989).

Colonization of the interior

Depending on the nature of the host, invading hyphae may reach the haemolymph as quickly as 2 days after spore adhesion, with sporulation being completed by the fifth day. Massive colonization of the body and rapid assimilation of its abundant nutrients for reproduction can take place because, although defence mechanisms are triggered at break-in, these are usually ineffective.

On entering the haemocoel a few fungi continue filamentous development, but in the majority there is a morphic shift to the production of blastic unicells, hyphal fragments resembling arthroconidia or, in the case of *Entomophaga* species, naked protoplasts. Walled cells may be phagocytosed by haemocytes, which may also accumulate around them to form a capsule, but neither host response is necessarily lethal to the invader. With regard to rapid colonization and avoidance of host defences, the ability to produce naked protoplasts confers several advantages (Murrin & Nolan, 1987). They can assimilate nutrients directly, have greater energy efficiency, and hence can divide more rapidly than walled cells. In addition, and by contrast with walled cells, they do not elicit responses by haemocytes. There is evidence for the existence of apparently harmless haemolymph inhabitants (Jouvenaz & Kimbrough, 1991). The ascomycetous species *Myrmecomyces annellisae* occupies the blood of the fire ant *Solenopsis*, where it grows in a yeast form without inducing any measurable changes in either the histology or behaviour of its host.

As the host becomes terminally enfeebled by an entomopathogen, filamentous growth is re-established, hyphae pass to the exterior, and sporulation takes place. Death is due to a combination of factors, including haemocoel blockage, disruption of vital organs, and the release of proteases. Hosts may die after only limited fungal development, probably because of the action of toxins.

The gut Although the gut environment is hostile to spore attachment and germination, should a fungus be able to achieve this then, given tolerance of the stresses imposed within the digestive tract, it would have access to a rich and constantly replenished nutrient fluid. As with other complex animals, the digestive processes of arthropods are sequenced, with different parts of the gut being appropriately differentiated and, as a consequence, having varying potential to support microbial development (Bignell, 1984).

Conventionally, the gut is considered to be a feature external to the body, the cuticle of the body surface being for the most part continuous with that of the gut. The epicuticle of the latter confronts the gut lumen but lacks a wax layer, so facilitating partial or selective permeability. Also in this regard, gut cuticle is not tanned, that is sclerotized, except where it forms triturating surfaces or valves. The arthropod gut has three divisions: foregut, midgut and hindgut (Fig. 12.6). The midgut provides a digestive surface and therefore does not secrete a cuticle, this being replaced by a chitinous peritrophic membrane, which protects the underlying epithelium from direct contact with solid food material. It also acts as a selective filter, allowing digestive enzymes to enter the lumen, and products to pass in the opposite direction. The peritrophic membrane is elaborated continuously, and so moves constantly rearwards until it reaches the hindgut, where it disintegrates as it experiences muscular contractions or meets cuticular spines.

The gut provides the sole habitat for Trichomycetes, which have affinities with Zygomycotina, whose thalli become firmly attached to the gut wall but do not penetrate it (Figs 12.7 & 12.8). Most species

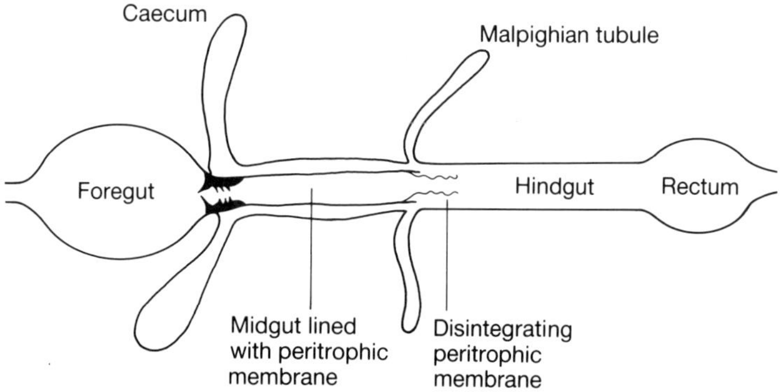

Fig. 12.6 Generalized organization of the insect gut.

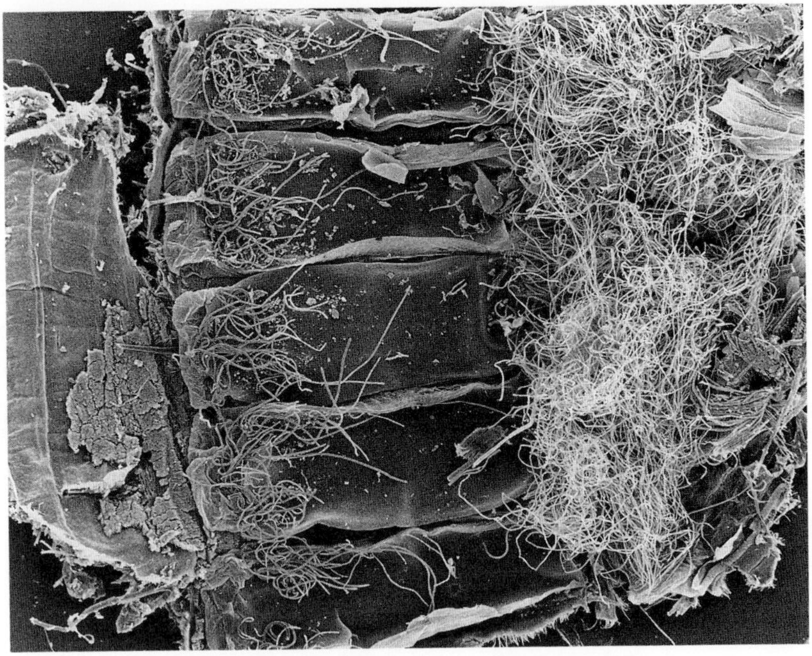

Fig. 12.7 Thread-like thalli of the endosymbiont *Enterobryus* attached to the cuticle lining the pyloric region of the hindgut of the millipede *Thyropygus* (photograph by S.T. Moss).

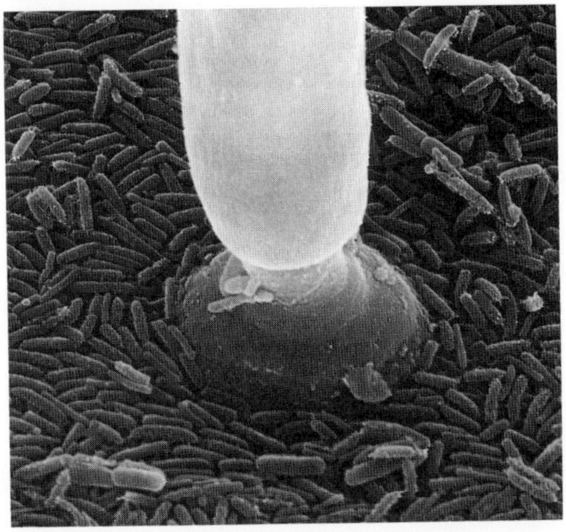

Fig. 12.8 The basal, secreted holdfast of *Enterobryus* which fixes the thallus to the gut cuticle. The cuticle has an attached bacterial community but the holdfast is largely bacteria free (photograph by S.T. Moss).

occupy the hindgut, some the foregut and others the peritrophic membrane of the midgut. The latter are restricted to freshwater dipteran larvae in which the peritrophic membrane is produced by a band of cells at the anterior end of the midgut. Trichomycetes are usually distributed amongst herbivorous and omnivorous mandibular arthropods but have not been found within carnivores (Table 12.1).

The location of most species within the hindgut, which contains only faecal residues, the absence of cuticular penetration by thalli, and the lack of discernible differences in fitness between infected and uninfected hosts, suggests that Trichomycetes are strictly neutral symbionts that are highly adapted to an oligotrophic habitat. Nothing is known in detail of their nutrition, although a number have been cultured axenically, but they possess some striking behavioural characteristics that ensure successful colonization and transmission (Moss, 1979).

Table 12.1 Distribution of gut-inhabiting Trichomycetes amongst Arthropoda Orders (after Moss, 1979).

Order	Habitat	Stage infected
Diptera (flies)	Freshwater, terrestrial	Larva
Ephemoptera (mayflies)	Freshwater	Nymph
Odonata (dragonflies)	Freshwater	Nymph
Plecoptera (stoneflies)	Freshwater	Nymph
Trichoptera (caddis-flies)	Freshwater	Nymph
Coleoptera (beetles)	Freshwater, terrestrial	Larva, adult
Collembola (springtails)	Terrestrial	Adult
Amphipoda (shrimps)	Freshwater, marine	Adult
Isopoda (woodlice etc.)	Freshwater, marine, terrestrial	Adult
Decapoda (shrimps)	Freshwater, marine	Adult
Diplopoda (millipedes)	Terrestrial	Adult

Infection is by means of ingested spores (Fig. 12.9). Attachment needs to be rapid since thalli can grow only in a restricted region of the digestive tract; furthermore, food may pass very quickly through the entire gut. To this end, spores possess either holdfast initials or mechanisms for swift secretion of holdfast material; the biochemical constitution of the holdfast probably also facilitates recognition of the appropriate gut surface. Hindgut specificity in some species appears to be determined via the stimulation of germination by potassium and high pH as spores pass through the midgut, followed by holdfast formation as a response to a fall in pH in the hindgut (Horn, 1989). Thalli are not only spatially confined but also have temporal limitations placed on their development due to disturbance. The cuticle of the fore- and hindgut is shed on ecdysis, so that growth and sporulation require completion during the intermoult period. Similarly, thalli attached to the peritrophic membrane must reach culmination before their travelling substratum passes into the hindgut and becomes fragmented (Moss, 1970).

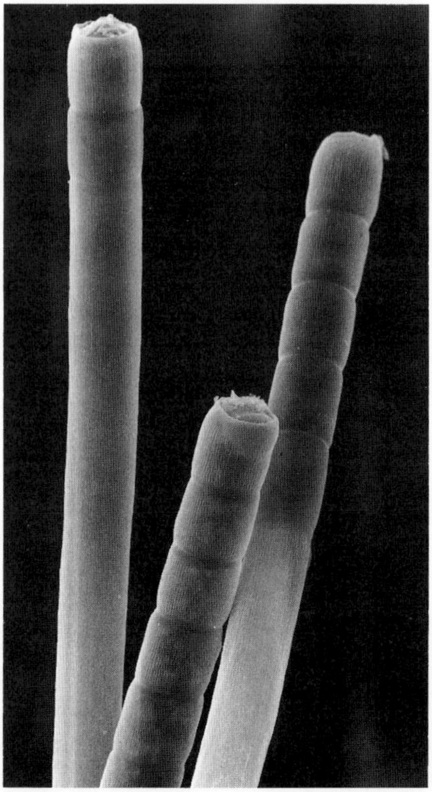

Fig. 12.9 Distal regions of three *Enterobryus* thalli which have differentiated to produce sporangia. Primary sporangiospores released from these pass from the gut to infect new hosts (photograph by S.T. Moss).

Another successful group of obligate gut inhabitants is associated with some anobiid beetles in which the digestive tract is specially modified to house populations of endosymbiotic yeasts, possibly species of *Torulopsis* (Jurzitza, 1979; Douglas, 1989). The forepart of the midgut consists of a number of evaginations (mycetomes), which have an epithelium of mycetocytes that contain large numbers of yeast cells. The latter may bud within the epithelial cells but are continually lysed, and dead individuals are expelled to the mycetome lumen. Their nutrition appears to be entirely saprotrophic despite being located inside living cells. These endosymbionts are possibly concerned in some way with the internal recycling of host-excreted nitrogen, which may additionally provide the animal with essential amino acids, and there is some evidence that hosts depend on them as a source of B-group vitamins. They are also found growing saprotrophically in the ovipositors of adult females, where they are located in fluid-filled tubules that connect with the ovipositor sheath. During laying, the egg surface is contaminated with fungal cells, the larval gut becoming infected when the larva hatches and eats the egg case.

Vertebrates Exterior and interior surfaces of the vertebrate body offer potential fungal colonists vast and immensely varied habitats. Throughout life, these are continuously exposed to fungal propagules but, although the latter are recruited in large numbers via ingestion, inhalation and deposition, most rarely survive for very long. Species that do achieve establishment are of two kinds. First, there is a relatively small number of fungi within which there is adaptation for colonization of specific habitats, for instance the skin and its keratinized appendages, and mucosal epithelia, and a concomitantly reduced ability – or total inability – for a free-living existence away from the host. Second, there are numerous fungi that, whilst normally existing as free-living saprotrophs, have a capacity to grow upon or within animals under particular circumstances.

Habitat exploitation can be cryptic, with lack of overt signs of fungal activity, but for the most part it is marked by fungus–host interactions which result in disease expression. Pathology has become the focus for most studies in this area, with understandable emphasis being placed on diseases of humans, or of laboratory animals used in human-related clinical research. Much of this is beyond the scope of fungal ecophysiology, but it is relevant to note that the spectrum of human fungal pathogens is constantly expanding, mainly because of an increased incidence of immunosuppression, due to both the administration of therapeutic drugs and the acquisition of certain viral infections, which leave the body open to invasion by common saprotrophs. Indeed, the hospital environment may often encourage colonization by numerous species of *Alternaria*, *Aspergillus*, *Curvularia*, *Paecilomyces*, *Fusarium* and Zygomycotina (Walsh & Pizzo, 1988).

The majority of fungi that are adapted to the body are confined to surfaces, the nature of which varies widely (Latham, 1979). For example, exposed skin presents potential occupants with problems of low water potential, poor nutrient supply, fungistatic secretions, and a generally antagonistic bacterial flora. By contrast, establishment on epithelia of the oral cavity, gastrointestinal tract and urinogenital system requires tolerance of low oxygen levels, adverse pH, action of digestive enzymes, elevated temperature and dense bacterial populations. However, some adapted species, and many opportunistic fungi, have potential for invasion beyond surfaces, which then leads to confrontation with activated host defence responses.

Skin and its appendages

Detailed studies on the mycoflora of dermal surfaces have been largely confined to mammals, with consequent neglect of habitats afforded by feathers or the mucus-coated collagenous scales of fish. With regard to humans, the capacity of healthy skin to support fungi depends on a number of factors, including climate, body location, age, and the quantity and quality of skin secretions. Normally, the mycoflora is sparse and is dominated by yeasts. Most of these, for instance species of *Aureobasidium*, *Geotrichum*, *Rhodotorula* and *Torulopsis*, would appear to be casual arrivals; but others, in particular *Pityrosporum* species, are lipophilic and seem to be native to sites of high sebum secretion (Mok & Barreto da Silva, 1984; Roth & James, 1988).

Whilst yeast populations, albeit scattered, can exist on nutrients supplied by secretions, growth of filamentous fungi requires access to the skin itself and utilization of keratin. This is accomplished by dermatophytes, which may exist cryptically on healthy skin but which come to prominence as incitants of skin diseases. Keratin is a highly insoluble fibrous protein made complex by unusually frequent cross-linking with disulphide bonds. Keratin is associated with other complex proteins and cementing substances to form the organized structures of hair, nails, hooves and horns. Dermatophytes probably satisfy their major nutrient requirements through exploitation of their keratin-rich substrata, but evidence for keratinolysis, as opposed to more general proteolysis of associated compounds, is weak (see also Chapter 2). However, *Trichophyton* species possess an array of serine proteinases which can unlock amino acids from keratinized proteins, single strains often producing multiple isoenzymes (Odds, 1991). Enzyme profile may be the basis for the observed host and positional specificity of many dermatophytes, in that it determines the kind of keratin–protein substratum on which any fungus can grow. For example *Trichophyton gallinae*, which is restricted to birds, will utilize feathers *in vitro* but not mammalian hair.

The gastrointestinal tract

Within the digestive tract, as may also be the case for the urinogenital system, there are three possible sites for fungal colonization: the luminal contents, the mucous gel overlying the epithelium, and the epithelium itself. Two groups of fungi have achieved success here: the yeast *Candida albicans* and related *Candida* species, and the anaerobic Chytridiomycetes.

Candida albicans is restricted *in vivo* to the gut, where it survives saprotrophically, in healthy animals, in low numbers associated with the mucosa. In compromised or debilitated animals it can become invasive and cause disease. Whichever of these life styles obtains, an essential feature is attachment to the gut walls, which prevents removal with the digesta. Cells first arrive on the mucosal surface at random through mixing of the gut contents. Here, the gel presents a barrier to penetration by yeast-size particles, but cells can pass through it, either slowly via less viscous channels or more rapidly with the aid of proteolysis, to then attach to the epithelium. Initial adhesion to the mucous blanket, and subsequently to the epithelium, involves specific binding mechanisms in which there is production of extracellular adhesins, possibly mannoproteins, that attach to mucosal and epithelial receptors (Kennedy, 1991).

Whilst general physicochemical conditions in the gastrointestinal tract preclude development of all fungi, except for a few yeasts, some regions may be so anaerobic as to exclude even these. In the rumen and caecum of herbivores, the headspace atmosphere consists largely of carbon dioxide and methane, any traces of oxygen being rapidly taken up by facultative anaerobic microorganisms. Yet fungi are found here in quantity, and form an important component of the fermentative microbial population which carries out pre- and postgastric digestion of plant material. These are obligately anaerobic Chytridiomycetes that have no capacity for a free-living existence in nature, and which are probably typical of all ruminants and non-ruminants that subsist on fibrous diets (Ho & Bauchop, 1991; Li *et al.*, 1991; Teunissen *et al.*, 1991). Ingested feed is quickly colonized by mono- or polycentric thalli, the rhizomycelia of which ramify over and within plant fragments. Their growth and subsequent zoo-sporogenesis require haem compounds, which occur in all vegetable material, and germination of encysted zoospores is dependent on acetic acid and fermentation carbohydrates with which the rumen fluid is enriched (Orpin & Greenwood, 1986a,b). It is probable that neonatal animals are fungus free, and so must acquire symbionts before weaning. Rumen fungi have been isolated from both saliva and faeces, being capable of survival in the latter for over 6 months. There is thus the possibility of transfer from adult to juvenile via grooming or saliva aerosols, or through ingestion of faecally contaminated herbage (Milne *et al.*, 1989; Theodorou *et al.*, 1990).

Rumen chytrids possess all the enzymes necessary to attack key fibre components, for instance cellulose, xylans, galactomannans and hemicellulosic arabinoxylans; they can additionally break down starch, but cannot hydrolyse pectic substrates. Cellulases may be catabolite-regulated by glucose and cellobiose until these become depleted; fermentation end-products are then principally formate, acetate, lactate and ethanol (Williams & Orpin, 1987a,b; Borneman *et al.*, 1989; Morrison *et al.*, 1990). These activities are not divorced from those of the other rumen inhabitants. For example, whilst the fungi can degrade lignified fibres and produce erosion zones in xylem, bacteria can attack only the periphery of fibres, and cannot erode xylem. Ruminal bacteria alone, therefore, bring about only slight tissue loss from resistant material; fungi render it more easily broken down during rumination, and give greater access to digesta (Akin *et al.*, 1989). The presence of bacteria also increases the effectiveness of the fungi. If the chytrids *Neocallimastix frontalis*, *Piromyces communis* and *Sphaeromonas communis* are co-cultured with the methanogen *Methanobrevibacter smithii* on xylan *in vitro*, there is a five- to sevenfold increase in the activity of fungal xylanase, β-D-xylosidase and α-L-arabinofuranosidase (Joblin *et al.*, 1990). This is correlated with much reduced xylose levels, which suggests that in the presence of the methanogen, fungi are better able to utilize this sugar. It is possible that *Methanobrevibacter* removes hydrogen, so shifting metabolism from the production of formate–lactate–ethanol towards that of acetate–methane. Lactate has been shown to adversely affect cellulolysis, and formate to inhibit fungal degradation of plant tissue. By contrast, acetate improves the potential yield of fungal ATP from 3 to 4 moles per mole of hexose fermented, with consequent enhancement of general metabolism.

Interactions with white blood cells

Fungi may enter blood or tissues in a number of ways, including passage through the mucosal epithelia of the digestive and urinogenital systems, and via wounds. However, the most common route is probably by inhalation of spores which, if small enough, can bypass upper respiratory tract defences and reach the distal lung regions. Whatever the detail of invasion, potentially dangerous spores, yeast cells or hyphal units, then become subject to the defensive activities of white blood cells. The relatively rare occurrence of, for instance, mycotic respiratory disease—despite constant inundation of the lungs by spores—indicates that propagules are probably cleared rapidly, with no unnecessary inflammatory reactions. But, if a greater threat is perceived, more aggressive and co-ordinated responses may be stimulated. However, whilst polymorphonuclear leucocytes, monocytes and macrophages have the capacity to kill, damage or inhibit development of infective units, some agents of serious fungal diseases are resistant to attack (Waldorf, 1991).

The process of phagocytosis involves four distinct phases: recognition, attachment, ingestion and killing. Phagocytes can discriminate not only between self and non-self, but also between self and altered states of self. During the recognition phase, contact with a fungus leads to the production of mediators, which mobilize phagocytes and stimulate them to produce chemotactic factors which attract more white cells to the invasion site. The mechanism of attachment to fungi is not known, but there is some evidence for binding to lectin-like receptors. During attachment, the phagocytic plasma membrane is perturbed, and this induces a respiratory burst, during which there is a dramatic increase in oxygen consumption linked to the NAD(P)H oxidase system. This results in the release of substantial quantities of superoxide anion (O_2^-) and hydrogen peroxide. These may then interact to produce short-lived oxygen species, for instance singlet oxygen and hydroxyl radicals, and longer-lived oxidants such as hypochlorous acid and monochloramine, all of which are cytocidal to microorganisms. Attachment also triggers ingestion, during which the products of respiratory burst are released, together with lysosomal enzymes, into the phagocytic vacuole where killing then takes place. In some circumstances cytocidal factors and enzymes are released to the cell exterior, which is important where the invader is too large to be phagocytosed, for example, in the case of hyphae or some yeast cells.

The propagules of some pathogens escape destructive phagocytosis. Yeast cells and spores of *Cryptococcus neoformans* possess surface polysaccharides that inhibit ingestion and also bind antibodies. *Histoplasma capsulatum* survives and multiplies within macrophages, in which it does not evoke a respiratory burst, which then become sites of primary infection. Similarly, conidia of *Aspergillus fumigatus* release compounds which inhibit phagocytosis, probably by reducing the ability of white cells to move, and also restrain release of superoxide anion and hydrogen peroxide (Robertson, 1991). It is of interest to note here that conidia of the non-pathogenic species *Aspergillus niger* may have similar defences against ingestion by the vampyrellid amoeba *Arachnula impatiens* (Old & Darbyshire, 1978).

Bibliography

Abawi, G.S., Grogan, R.G. & Duniway, J.M. (1985) Effect of water potential on survival of sclerotia of *Sclerotinia minor* in two California soils. *Phytopathology*, **75**, 217–221.

Abe, K., Kusaka, I. & Fukui, S. (1975) Morphological change in the early stages of the mating process of *Rhodosporidium toruloides. Journal of Bacteriology*, **122**, 710–718.

Abuzinadah, R.A. & Read, D.J. (1986a) The role of proteins in the nitrogen nutrition of ectomycorrhizal plants. I. Utilization of peptides and proteins by ectomycorrhizal fungi. *New Phytologist*, **103**, 481–493.

Abuzinadah, R.A. & Read, D.J. (1986b) The role of proteins in the nitrogen nutrition of ectomycorrhizal plants. III. Protein utilization by *Betula, Picea* and *Pinus* mycorrhizal associations with *Hebeloma crustiliniforme. New Phytologist*, **103**, 507–514.

Adler, E. (1977) Lignin chemistry–past, present and future. *Wood Science and Technology*, **11**, 169–218.

Adler, J.H., Gealt, M.A., Nes, W.D. & Nes, W.R. (1981) Growth characteristics of *Saccharomyces cerevisiae* and *Aspergillus nidulans* when biotin is replaced by aspartic and fatty acids. *Journal of General Microbiology*, **122**, 101–107.

Aggab, A.M. & Cooke, R.C. (1981) Self-stimulation and self-inhibition in germinating ascospores of *Sclerotinia curreyana. Transactions of the British Mycological Society*, **76**, 155–157.

Ahmad, S., Govindarajan, S., Funk, C.R. & Johnson-Cicalese, J.M. (1985) Fatality of house crickets on perennial ryegrasses infected with a fungal endophyte. *Entomologia Experimentalis et Applicata*, **39**, 183–190.

Ahmad, S., Johnson-Cicalese, J.M., Dickson, W.K. & Funk, C.R. (1986) Endophyte-enhanced resistance in perennial ryegrass to the bluegrass billbug, *Sphenophorus parvulus. Entomologia Experimentalis et Applicata*, **41**, 3–10.

Ainsworth, A.M., Rayner, A.D.M., Broxholme, S.J., Beeching, J.R., Pryke, J.A., Scard, P.R., Berriman, J., Powell, K.A., Floyd, A.J. & Branch, S.K. (1990) Production and properties of the sesquiterpene, (+)-torreyol, in degenerative mycelial interactions between strains of *Stereum. Mycological Research*, **94**, 799–809.

Akin, D.E., Lyon, C.E., Windham, W.R. & Rigsby, L.L. (1989) Physical degradation of lignified stem tissues by ruminal fungi. *Applied and Environmental Microbiology*, **55**, 611–616.

Akushie, P.-L. & Clerk, G.C. (1981) Effect of relative humidity on viability of *Rhizopus oryzae* sporangiospores. *Transactions of the British Mycological Society*, **76**, 332–334.

Alberts, B., Bray D., Lewis, V., Raff, M., Roberts, K. & Watson, J.D. (1983) *Molecular Biology of the Cell*, New York, Garland.

Al-Hamdani, A.M. & Cooke, R.C. (1987) Effects of water potential on accumulation and exudation of carbohydrates and glycerol during sclerotium formation and myceliogenic germination in *Sclerotinia sclerotiorum. Transactions of the British Mycological Society*, **89**, 51–60.

Alí, N.A. & Jackson, R.M. (1988) Effects of plant roots and their exudates on germination of spores of ectomycorrhizal fungi. *Transactions of the British Mycological Society*, **91**, 253–260.

Ali, N.A. & Jackson, R.M. (1989) Stimulation of germination of spores of some ectomycorrhizal fungi by other micro-organisms. *Mycological Research*, **93**, 182–186.

Allen, R.N. & Newhook, F.J. (1973) Chemotaxis of zoospores of *Phytophthora cinnamomi* to ethanol in capillaries of soil pore dimensions. *Transactions of the British Mycological Society*, **61**, 287–302.

Aluko, M.O. & Hering, T.F. (1970) The mechanisms associated with the antagonistic relationship between *Corticium solani* and *Gliocladium virens. Transactions of the British Mycological Society*, **55**, 173–179.

Amon, P.J. & Arthur, R.D. (1981) Nutritional studies of a marine *Phlyctochytrium* sp. *Mycologia*, **73**, 1049–1065.

Anagnostakis, S.L. (1984) The mycelial biology of *Endothia parasitica*. I. Nuclear and cytoplasmic genes that determine morphology and virulence. In *The Ecology and Physiology of the Fungal Mycelium*, eds Jennings, D.H. & Rayner, A.D.M., pp. 353–366. Cambridge, Cambridge University Press.

Anas, O., Alli, I. & Reeleder, R.D. (1989) Inhibition of germination of sclerotia of *Sclerotinia sclerotiorum by* salivary gland secretions of *Bradysia coprophila. Soil Biology & Biochemistry*, **21**, 47–52.

Anas, O. & Reeleder, R.D. (1988) Consumption of sclerotia of *Sclerotinia sclerotiorum* by larvae of *Bradysia coprophila*: influence of soil factors and interactions between larvae and *Trichoderma viride. Soil Biology & Biochemistry*, **20**, 619–624.

Anderson, J.G. & Smith, J.E. (1976) Effects of temperature on filamentous fungi. In *Inhibition and Inactivation of Vegetative Microbes*, eds Skinner, F.A. & Hugo, W.G., pp. 191–218. London, Academic Press.

Anderson, J.M. & Ineson, P. (1984) Interactions between microorganisms and soil invertebrates in nutrient flux pathways of forest ecosystems. In *Invertebrate–Microbial Interactions*, eds Anderson, J.M., Rayner, A.D.M. & Walton,

D.W.H., pp. 59–88. Cambridge, Cambridge University Press.

Anderson, J.M., Rayner, A.D.M. & Walton, D.W.H. (eds) (1984) *Invertebrate–Microbial Interactions*, Cambridge, Cambridge University Press.

Anderson, N.H. & Cargil, A.S. (1987) Nutritional ecology of aquatic detritivorous insects. In *Nutritional Ecology of Insects, Mites, Spiders, and Related Invertebrates*, eds Slansky Jr, F. & Rodriguez, J.G., pp. 903–925. London, Wiley.

Anderson, T.R. & Patrick, Z.A. (1978) Mycophagous amoeboid organisms from soil that perforate spores of *Thielaviopsis basicola* and *Cochliobolus sativus. Phytopathology*, **68**, 1618–1626.

Andrews, J.H. & Rouse, D.I. (1982) Plant pathogens and the theory of *r*-and *K*-selection. *American Naturalist*, **120**, 283–296.

Anikster, Y. (1988) Water potential as a regulating factor in teliospore germination of three rust fungi. *Mycologia*, **80**, 568–570.

Ann, P.J. & Ko, W.H. (1988) Induction of oospore germination of *Phytophthora parasitica. Phytopathology*, **78**, 335–338.

Arens, K. (1929a) Physiologische Untersuchungen an *Plasmopara viticola*, unter besonderer Berücksichtigung der Infektionsbedingungen (Physiological investigations of *Plasmopara viticola*, with special reference to the conditions governing infection). *Jahrbuch für Wissenschaftliche Botanik*, **70**, 91–159.

Arens, K. (1929b) Untersuchungen über *Pseudoperonospora humuli* (Miy. u. Tak.) den Erreger der neuen Hopfenkrankheit (Investigations on *Pseudoperonospora humuli* (Miy. and Tak.), the causal organism of the new hop disease). *Phytopathologische Zeitschrift*, **1**, 169–193.

Arora, D.K. (1988) Effect of microorganisms on the aggressiveness of *Bipolaris sorokiniana. Canadian Journal of Botany*, **66**, 242–246.

Arora, D.K., Filinow, A.B. & Lockwood, J.L. (1983) Bacterial chemotaxis to fungal propagules *in vitro* and in soil. *Canadian Journal of Microbiology*, **29**, 1104–1109.

Arora, D.K., Filinow, A.B. & Lockwood, J.L. (1985) Decreased aggressiveness of *Bipolaris sorokiniana* conidia in response to nutrient stress. *Physiological Plant Pathology*, **26**, 135–142.

Arpin, N. & Bouillant, M.L. (1981) Light and mycosporines. In *The Fungal Spore*, eds Turian, G. & Hohl, R.H., pp. 435–454. London, Academic Press.

Arsuffi, T.L. & Suberkropp, K. (1989) Selective feeding by shredders on leaf-colonizing stream fungi: comparison of macroinvertebrate taxa. *Oecologia*, **79**, 30–37.

Ayers, W.A. & Adams, P.B. (1985) Interaction of *Laterispora brevirama* and the mycoparasites *Sporidesmium sclerotivorum* and *Teratosperma oligocladum. Canadian Journal of Microbiology*, **31**, 786–792.

Ayers, W.A. & Lumsden, R.D. (1975) Factors affecting production and germination of oospores of three *Pythium* species. *Phytopathology*, **65**, 1094–1110.

Aylor, D.E. & Ferrandino, F.J. (1986) Germination of urediniospore clusters of *Uromyces appendiculatus* and *Puccinia recondita. Transactions of the British Mycological Society*, **86**, 591–595.

Ayres, P.G. (1983) Conidial germination and germtube growth of *Erysiphe pisi* in relation to visible light and its transmission through pea leaves. *Transactions of the British Mycological Society*, **81**, 269–274.

Ayres, P.G. & Paul, N.D. (1986) Foliar pathogens alter the water relations of their hosts with consequences for both host and pathogen. In *Water, Fungi and Plants*, eds Ayres, P.G. & Boddy, L., pp. 267–285. Cambridge, Cambridge University Press.

Azcón, R. (1987) Germination and hyphal growth of *Glomus mosseae in vitro*: effects of rhizosphere bacteria and cell-free culture media. *Soil Biology & Biochemistry*, **19**, 417–419.

Baard, S.W., van Wyk, P.W.J. & Pauer, G.D.C. (1981) Structure and lysis of microsclerotia of *Verticillium dahliae* in soil. *Transactions of the British Mycological Society*, **77**, 251–260.

Backhouse, D. & Stewart, A. (1988) Large sclerotia of *Sclerotium cepivorum. Transactions of the British Mycological Society*, **91**, 343–346.

Bacon, C.W. & Hinton, D.M. (1988) Ascosporic iterative germination in *Epichloë typhina. Transactions of the British Mycological Society*, **90**, 563–569.

Bahn, M. & Hock, B. (1973) Morphogenese von *Sordaria macrospora*: die Induktion der Perithezienbildung (Morphogenesis in *Sordaria macrospora*: the induction of perithecia formation). *Beriche der Deutschen Botanischen Gesellschaft*, **86**, 309–311.

Bajwa, R. & Read, D.J. (1985) The biology of mycorrhiza in the Ericaceae. IX. Peptides as nitrogen sources for the ericoid endophyte and for mycorrhizal and non-mycorrhizal plants. *New Phytologist*, **101**, 459–467.

Balan, J., Križková, L., Nemec, P. & Kolosváry, A. (1976) A qualitative method for detection of nematode attracting substances and proof of production of three different attractants by the fungus *Monacrosporium rutgeriensis. Nematologica*, **22**, 306–311.

Banihashemi, Z. & Mitchell, J.E. (1989) Effects of flavin inhibitors on photoactivation of oospores of *Phytophthora cactorum. Journal of Phyto-*

pathology, **126**, 167–174.

Barak, R., Elad, Y., Mirelman, D. & Chet, I. (1985) Lectins: a possible basis for specific recognition in the interaction of *Trichoderma* and *Sclerotium rolfsii*. *Phytopathology*, **75**, 458–462.

Bärlocher, F. (1985) The role of fungi in the nutrition of stream invertebrates. *Botanical Journal of the Linnean Society*, **91**, 83–94.

Barnett, H.J.A. & Sims, A.P. (1976) Some physiological observations on the uptake of D-glucose and 2-deoxy-D-glucose by starving exponentially-growing yeasts. *Archives of Microbiology*, **111**, 185–192.

Barnett, H.L. & Lilly, V.G. (1947) The effects of biotin upon the formation and development of perithecia, asci and ascospores by *Sordaria fimicola* Ces. and de Not. *American Journal of Botany*, **34**, 196–204.

Barr, D.J.S. (1978) Taxonomy and phylogeny of chytrids. *BioSystems*, **10**, 153–165.

Barran, L.R., Schneider, E.F. & Seamen, W.L. (1977) Requirements for the rapid conversion of macroconidia of *Fusarium sulphureum* to chlamydospores. *Canadian Journal of Microbiology*, **23**, 148–151.

Barron, G.L. (1976) *The Nematode Destroying Fungi*, Guelph, Ontario, Biological Publications.

Barron, G.L. (1985) Fungal parasites of bdelloid rotifers: Diheterospora. *Canadian Journal of Botany*, **63**, 211–222.

Barron, G.L. (1988) Microcolonies of bacteria as a nutrient source for lignicolous and other fungi. *Canadian Journal of Botany*, **66**, 2505–2510.

Barron, G.L. (1989) New species and new records of fungi that attack microscopic animals. *Canadian Journal of Botany*, **67**, 267–271.

Barron, G.L. (1990) A new predatory Hyphomycete capturing copepods. *Canadian Journal of Botany*, **68**, 691–696.

Barron, G.L. (1991) A new genus, Rotiferophthora, to accommodate the Diheterospora-like endoparasites of rotifers. *Canadian Journal of Botany*, **69**, 494–502.

Barron, G.L. & Thorn, R.G. (1987) Destruction of nematodes by species of *Pleurotus*. *Canadian Journal of Botany*, **65**, 774–778.

Bartnicki-Garcia, S. (1968) Cell wall chemistry, morphogenesis, and taxonomy of fungi. *Annual Review of Microbiology*, **22**, 87–108.

Bartnicki-Garcia, S. (1970) Cell wall composition and other biochemical markers in fungal phylogeny. In *Phytochemical Phylogeny*, ed. Harborne, J.B., pp. 81–103. London, Academic Press.

Bartnicki-Garcia, S. (1981) Cell wall construction during spore germination in Phycomycetes. In *The Fungal Spore: Morphogenetic Controls*, eds Turian, G. & Hohl, H.R., pp. 533–556. New York, Academic Press.

Bartnicki-Garcia, S. (1987) The cell wall: a crucial structure in fungal evolution. In *Evolutionary Biology of the Fungi*, eds Rayner, A.D.M., Brasier, C.M., & Moore, D., pp. 389–403. Cambridge, Cambridge University Press.

Bartnicki-Garcia, S. (1990) Role of vesicles in apical growth and a new mathematical model of hyphal morphogenesis. In *Tip Growth in Plant and Fungal Cells*, ed. Heath, I.B., pp. 211–232. London, Academic Press.

Bartnicki-Garcia, S., Hergert, F. & Gierz, G. (1989) Computer simulation of fungal morphogenesis and mathematical basis for hyphal (tip) growth. *Protoplasma*, **153**, 46–57.

Bartnicki-Garcia, S. & Lippman, E. (1969) Fungal morphogenesis: cell wall constuction in *Mucor rouxii*. *Science (New York)*, **165**, 302–304.

Bashi, E. & Aylor, D.E. (1983) Survival of detached sporangia of *Peronospora destructor* and *Peronospora tabacina*. *Phytopathology*, **73**, 1135–1139.

Bashi, E., Ben-Joseph, Y. & Rotem, J. (1982) Inoculation potential of *Phytophthora infestans* and the development of potato late blight epidemics. *Phytopathology*, **72**, 1043–1047.

Batra, L.R. & Batra, S.W.T. (1979) Termite–fungus mutualism. In *Insect–Fungus Symbioses*, ed. Batra, L.R., pp. 117–163. Montclair, New Jersey, Allanheld Osmun.

Battley, E.H. & Bartlett, E.J. (1966) A convenient pH-gradient method for the determination of the maximum and minimum pH for microbial growth. *Antonie van Leeuwenhoek*, **32**, 245–255.

Beakes, G.W. (1980) Electron microscopic study of oospore maturation and germination in an emasculate isolate of *Saprolegnia ferax*. 2. Wall differentiation. *Canadian Journal of Botany*, **58**, 195–208.

Beakes, G.W. & Bartnicki-Garcia, S. (1989) Ultrastructure of mature oogonium–oospore wall complexes in *Phytophthora megasperma*: a comparison of *in vivo* and *in vitro* dissolution of the oospore wall. *Mycological Research*, **93**, 321–334.

Bean, B. (1979) Chemotaxis in unicellular Eukaryotes. In *Physiology of Movements, Encyclopedia of Plant Physiology New Series*, Vol. 7, eds Haupt W. & Feinleib, M.E., pp. 335–354. Berlin, Springer.

Bean, B. (1984) Microbial geotaxis. In *Membranes and Sensory Transduction*, eds Colombetti, G. & Lenci, F., pp. 163–198. New York, Plenum.

Beattie, A.J. (1986) *The Evolutionary Ecology of Ant–Plant Mutualisms*, Cambridge, Cambridge University Press.

Beattie, A.J., Turnbull, C., Hough, T., Jobson, S. & Knox, R.B. (1985) The vulnerability of pollen and fungal spores to ant secretion: evidence and some evolutionary implications. *American Journal of Botany*, **72**, 606–614.

Beaver, R.A. (1989) Insect–fungus relationships in the bark and ambrosia beetles. In *Insect–Fungus Interactions*, eds Wilding, N., Collins, N.M., Hammond, P.M. & Webber, J.F., pp. 121–143. London, Academic Press.

Becker, J.O. & Cook, R.J. (1988) Role of siderophores in suppression of *Pythium* species and production of increased-growth response of wheat by fluorescent pseudomonads. *Phytopathology*, **78**, 778–782.

Beilby, J.P. & Kidby, D.K. (1982) The early synthesis of RNA, protein and some associated metabolic events in germinating vesicular–arbuscular mycorrhizal spores of *Glomus caledonius*. *Canadian Journal of Microbiology*, **28**, 623–628.

Bell, A.A. & Wheeler, M.W. (1986) Biosynthesis and function of fungal melanins. *Annual Review of Phytopathology*, **24**, 411–451.

Bengtsson, G., Erlandsson, A. & Rundgren, S. (1988) Fungal odour attracts soil Collembola. *Soil Biology & Biochemistry*, **20**, 25–30.

Benjamin, R.K. (1959) The merosporangiferous Mucorales. *Aliso*, **4**, 321–433.

Bennett, J.W. & Ciegler, A. (eds) (1983) *Secondary Metabolism and Differentiation in Fungi*, New York, Marcel Dekker.

Bennink, G.J.H. (1972) Photomorphogenesis in *Penicillium isariiforme*. II. The action spectrum for light-induced formation of coremia. *Acta Botanica Neerlandica*, **21**, 535–538.

Bergmann, K., Burke, P.V., Cerdá-Olmedo, E., David, C.N., Delbrük, M., Foster, K.W., Goodell, E.W., Heisenberg, M., Meissner, G., Zalokar, M., Dennison, D.S. & Shropshire Jr, W. (1969) *Phycomyces*. *Bacteriological Reviews*, **33**, 99–157.

Berkeley, R.C.W. (1979) Chitin, chitosan and their degradative enzymes. In *Microbial Polysaccharides and Polysaccharases*, eds Berkeley, R.C.W., Gooday, G.W. & Ellwood, D.C., pp. 205–236. London, Academic Press.

Berryman, A.A. (1989) Adaptive pathways in scolytus–fungus association. In *Insect–Fungus Interactions*, eds Wilding, N., Collins, N.M., Hammond, P.M. & Webber, J.F., pp. 145–159. London, Academic Press.

Beuchat, L.R. (1988) Thermal tolerance of *Talaromyces flavus* ascospores as affected by growth medium and temperature, age and sugar content in the inactivation medium. *Transactions of the British Mycological Society*, **90**, 359–364.

Bignell, D.E. (1984) The arthropod gut as an environment for microorganisms. In *Invertebrate–Microbial Interactions*, eds Anderson, J.M., Rayner, A.D.M. & Walton, D.W.H., pp. 205–227. Cambridge, Cambridge University Press.

Binder, R.G., Klisiewicz, J.M. & Waiss, A.C. (1977) Stimulation of germination of *Puccinia carthami* teliospores by polyacetylenes from safflower. *Phytopathology*, **67**, 472–474.

Binder, R.G., Lundin, R.E., Kint, S., Klisiewicz, J.M. & Waiss, A.C. (1978) Polyacetylenes from *Carthamus tinctorius*. *Phytopathology*, **17**, 315–317.

Bird, B.A., Remaley, A.T. & Campbell, I.M. (1981) Brevianamides A and B are formed only after conidiation has begun in solid cultures of *Penicillium brevicompactum*. *Applied and Environmental Microbiology*, **42**, 521–525.

Birraux, D. & Fries, N. (1981) Germination of *Thelephora terrestris* basidiospores. *Canadian Journal of Botany*, **59**, 2062–2064.

Bissett, J. & Borkent, A. (1988) Ambrosia galls: the significance of fungal nutrition in the evolution of the Cecidomyiidae (Diptera). In *Coevolution of Fungi with Plants and Animals*, eds Pirozynski, K.A. & Hawksworth, D.L., pp. 203–225. London, Academic Press.

Bistis, G. (1956) Sexuality in *Ascobolus stercorarius*. I. Morphology of the ascogonium; plasmogamy; evidence for a sexual hormone mechanism. *American Journal of Botany*, **43**, 389–394.

Bistis, G. (1957) Sexuality in *Ascobolus stercorarius*. II. Preliminary experiments on various aspects of the mating process. *American Journal of Botany*, **44**, 436–443.

Bistis, G.N. (1970) Dikaryotization in *Clitocybe truncicola*. *Mycologia*, **62**, 911–924.

Bistis, G.N. (1983) Evidence for diffusible, mating-type-specific trichogyne attractants in *Neurospora crassa*. *Experimental Mycology*, **7**, 292–295.

Black, R.L.B. & Dix, N.J. (1976) Spore germination and germ hyphal growth of microfungi from litter and soil in the presence of ferulic acid. *Transactions of the British Mycological Society*, **66**, 305–311.

Blackwell, M. & Malloch, D. (1989) Pyxidiophora: life histories and arthropod associations of two species. *Canadian Journal of Botany*, **67**, 2552–2562.

Blakeman, J.P. & Fokkema, N.J. (1982) Potential for biological control of plant diseases on the phylloplane. *Annual Review of Phytopathology*, **20**, 167–192.

Bloomfield, B.J. & Alexander, M. (1967) Melanin and resistance of fungi to lysis. *Journal of Bacteriology*, **93**, 1276–1280.

Blumental, H.J. (1976) Reserve carbohydrates in fungi. In *The Filamentous Fungi, Vol. II. Biosynthesis and Metabolism*, eds Smith, J.E. & Berry, D.R., pp. 292–307. London, Edward Arnold.

Boddy, L., Coates, D. & Rayner, A.D.M. (1983) Attraction of fungus gnats to zones of intraspecific antagonism on agar plates. *Transactions of the British Mycological Society*, **81**, 149–151.

Boddy, L. & Rayner, A.D.M. (1983) Mycelial interactions, morphogenesis and ecology of *Phlebia radiata* and *Phlebia rufa* from oak. *Transactions of the British Mycological Society*, **80**, 437–448.

Borgeson, C.E. & Bowman, B.J. (1985) Blue light-reducible cytochromes in membrane fractions from *Neurospora crassa. Plant Physiology*, **78**, 433–437.

Borneman, W.S., Akin, D.E. & Ljungdahl, L.G. (1989) Fermentation products and plant cell wall-degrading enzymes produced by monocentric and polycentric anaerobic ruminal fungi. *Applied and Environmental Microbiology*, **55**, 1066–1073.

Borrebaeck, C.A.K., Mattiasson, B. & Nordbring-Hertz, B. (1985) A fungal lectin and its apparent receptors on a nematode surface. *FEMS Microbiology Letters*, **27**, 35–39.

Botton, B. & Dexheimer, J. (1977) Ultrastructure des rhizomorphes du *Sphaerostilbe repens* B. et Br. (The ultrastructure of the rhizomorphs of *Sphaerostilbe repens* B. and Br.). *Zeitschrift für Planzenphysiologie*, **85**, 429–443.

Boucias, D.G. & Pendland, J.C. (1984) Nutritional requirements for conidial germination of several host range pathotypes of the entomopathogenic fungus *Nomuraea rileyi. Journal of Invertebrate Pathology*, **43**, 288–293.

Boucias, D.G. & Pendland. J.C. (1991) Attachment of mycopathogens to cuticle. In *The Fungal Spore and Disease Initiation in Plants and Animals*, eds Cole, G.T. & Hoch, H.C., pp. 101–127. New York, Plenum.

Bourett, T., Hoch, H.C. & Staples, R.C. (1987) Association of the microtubule cytoskeleton with the thigmotropic signal for appressorium formation in *Uromyces. Mycologia*, **79**, 540–545.

Bourret, J.A. (1986) Evidence that a glucose-mediated rise in cyclic AMP triggers germination of *Pilobolus longipes* spores. *Experimental Mycology*, **10**, 60–66.

Bourret, J.A., Flora, L.L. & Ferrer, L.M. (1989) Trehalose mobilization during early germination of *Pilobolus longipes* sporangiospores. *Experimental Mycology*, **13**, 140–148.

Bourret, J.A. & Smith, C.M. (1987) Cyclic AMP regulation of glucose transport in germinating *Pilobolus longipes* spores. *Archives of Microbiology*, **148**, 29–33.

Boyd, P.E. & Kohlmeyer, J. (1982) The influence of temperature on the seasonal and geographic distribution of three marine fungi. *Mycologia*, **74**, 894–902.

Bradley, R., Burt, A.J. & Read, D.J. (1981) Mycorrhizal infection and resistance to heavy metal toxicity in *Calluna vulgaris. Nature (London)*, **292**, 335–337.

Brambl, R.M. (1980) Mitochondrial biogenesis during fungal spore germination: biosynthesis and assembly of cytochrome *c* oxidase in *Botryodiplodia theobromae. Journal of Biological Chemistry*, **255**, 7673–7680.

Brambl, R.M. (1981) Respiration and mitochondrial biogenesis during fungal spore germination. In *The Fungal Spore: Morphogenetic Controls*, eds Turian, G. & Hohl, H.R., pp. 585–604. New York, Academic Press.

Brasier, C.M. (1978) Mites and reproduction in *Ceratocystis ulmi* and other fungi. *Transactions of the British Mycological Society*, **70**, 81–89.

Breton, A. (1978) Evidence for the existence of an intramycelial morphogenetic factor controlling the development of coremia of *Doratomyces purpureofuscus. Canadian Journal of Botany*, **56**, 1533–1536.

Brisbane, P.G., Harris, J.R. & Moen, R. (1989) Inhibition of fungi from wheat roots by *Pseudomonas fluorescens* 2-79 and fungicides. *Soil Biology & Biochemistry*, **21**, 1019–1025.

Brodie, L.D.S & Blakeman, J.P. (1977) Effect on nutrient leakage, respiration and germination of *Botrytis cinerea* conidia caused by leaching with water. *Transactions of the British Mycological Society*, **68**, 445–447.

Brownlee, C.D., Duddridge, J.A., Malibari, A. & Read, D.J. (1983) The structure and function of mycelial systems of ectomycorrhizal roots with special reference to their role in forming inter-plant connections and providing pathways for assimilate and water transport. In *Proceedings of the IUFRO Conference on Tree Roots and Their Mycorrhizas*, eds Atkinson, D., Bhat, K.K.S., Coutts, M.P., Mason, P.A. & Read, D.J. *Plant & Soil*, **71**, 433–443.

Bruns, T.D. (1984) Insect mycophagy in the Boletales: fungivore diversity and mushroom habitat. In *Fungus–Insect Relationships; Perspectives in Ecology and Evolution*, eds Wheeler, Q.D. & Blackwell, M., pp. 91–129. New York, Columbia University Press.

Brunt, S.A. & Silver, J.C. (1987) Steroid hormone-regulated proteins in *Achlya ambisexualis. Experimental Mycology*, **11**, 65–69.

Bull, A.T. (1970) Inhibition of polysaccharidases by melanin. Enzyme inhibition in relation to mycolysis. *Archives of Biochemistry and*

Biophysics, **137**, 343–356.

Bull, A.T. & Bushell, M.E. (1976) Environmental control of fungal growth. In *The Filamentous Fungi. Vol II. Biosynthesis and Metabolism*, eds Smith, J.E. & Berry, D.B., pp. 1–31. London, Edward Arnold.

Bull, A.T. & Trinci, A.P.J. (1977) The physiology and metabolic control of fungal growth. *Advances in Microbial Physiology*, **50**, 1–84.

Buller, A.H.R (1909) *Researches on Fungi*, Vol. I, London, Longman, Green.

Buller, A.H.R. (1934) *Researches on Fungi*, Vol. VI, London, Longman, Green.

Bu'Lock, J.D., Jones, B.E. & Winskill, N. (1976) The apocarotenoid system of sex hormones and prohormones in Mucorales. *Proceedings of Applied Chemistry*, **47**, 191–202.

Burge, M.N. (ed.) (1988) *Fungi in Biological Control Systems*, Manchester, Manchester University Press.

Burton, R.J. & Coley-Smith, J.R. (1985) Antibiotics in sclerotia and mycelium of *Rhizoctonia* species. *Transactions of the British Mycological Society*, **85**, 447–453.

Buston, H.W. & Khan, A.H. (1956) The influence of certain micro-organisms on the formation of perithecia by *Chaetomium globosum*. *Journal of General Microbiology*, **14**, 655–660.

Butler, G.M. (1958) The development and behaviour of mycelial strands in *Merulius lacrymans* (Wolf.) Fr. II. Hyphal behaviour during strand formation. *Annals of Botany*, **22**, 219–236.

Buxton, E.W., Last, F.T. & Nour, M.A. (1957) Some effects of ultraviolet radiation on the pathogenicity of *Botrytis fabae, Uromyces fabae* and *Erysiphe graminis*. *Journal of General Microbiology*, **16**, 764–773.

Byrt, P.N., Irving, H.R. & Grant, B.R. (1982) The effect of cations on zoospores of the fungus *Phytophthora cinnamomi*. *Journal of General Microbiology*, **128**, 1189–1198.

Caesar, A.J. & Pearson, R.C. (1983) Environmental factors affecting survival of ascospores of *Sclerotinia sclerotiorum*. *Phytopathology*, **73**, 1024–1030.

Caesar, J.C. & Clerk, G.C. (1985) Germinability of *Leveillula taurica* (powdery mildew) conidia obtained from water-stressed pepper plants. *Canadian Journal of Botany*, **63**, 1681–1684.

Cairney, J.W.G. (1990) Internal structure of mycelial cords of *Agaricus carminescens* from Heron Island, Great Barrier Reef. *Mycological Research*, **94**,117–135.

Cairney, J.W.G., Jennings, D.H. & Veltkamp, C.J. (1989) A scanning electron microscope study of the internal structure of mature linear mycelial organs of four basidiomycete species. *Canadian Journal of Botany*, **67**, 2266–2271.

Callow, J.A. (1984) Cellular and molecular recognition between higher plants and fungal pathogens. In *Cellular Interactions. Encyclopedia of Plant Physiology, New Series*, Vol. 17, eds Linskens, H.F. & Heslop-Harrison, J., pp. 212–237. Berlin, Springer.

Cano, C. & Ruiz-Herrera, J. (1988) Developmental stages during the germination of *Mucor* sporangiospores. *Experimental Mycology*, **12**, 47–59.

Capaccio, L.C.M. & Callow, J.A. (1982) The enzymes of polyphosphate metabolism in vesicular–arbuscular mycorrhizas. *New Phytologist*, **91**, 81–91.

Carlile, M.J. (1970) The photoreceptors of fungi. In *Photobiology of Microorganisms*, ed. Halldal, P., pp. 309–344. London, Wiley Interscience.

Carlile, M.J. (1975) Taxes and tropisms: diversity, biological significance and evolution. In *Primitive Sensory and Communication Systems: The Taxes and Tropisms of Micro-organisms and Cells*, ed. Carlile, M.J., pp. 1–28. London, Academic Press.

Carlile, M.J. (1979) Bacteria, fungal and slime mould colonies. In *Biology and Systematics of Colonial Organisms*, eds Larwood, G. & Rosen, B.R., pp. 3–27. London, Academic Press.

Carlile, M.J. (1980) From prokaryote to eukaryote: gains and losses. In *The Eukaryote Microbial Cell*, eds Gooday, G.W., Lloyd, D. & Trinci, A.P.J. *Symposium of the Society for General Microbiology*, **30**, 1–40.

Carlile, M.J. (1983) Motility, taxis and tropisms in *Phytophthora*. In *Phytophthora: Its Biology, Taxonomy, Ecology and Pathology*, eds Erwin, D.C., Bartnicki-Garcia, S. & Tsao, P.H., pp. 95–107. St Paul, Minnesota, The American Phytopathological Society.

Carlile, M.J. (1986) The zoospore and its problems. In *Water, Fungi and Plants*, eds Ayres, P.G. & Boddy, L., pp. 105–118. Cambridge, Cambridge University Press.

Carlile, M.J. & Gooday, G.W. (1978) Cell fusion in myxomycetes and fungi. In *Membrane Fusions*, eds Poste, G. & Nicholson, G.L., pp. 219–265. New York, Elsevier/North Holland Biomedical Press.

Carlile, M.J. & Matthews, S.L. (1988) Chemotropism of germ-tubes of *Chaetomium globosum* to volatile factors from wood. *Transactions of the British Mycological Society*, **90**, 643–644.

Carlile, M.J. & Tew, P.M. (1988) Chemotropism of germ-tubes of *Phytophthora citricola*. *Transactions of the British Mycological Society*, **90**, 644–646.

Carroll, G.C. (1986) The biology of endophytism

in plants with particular reference to woody perennials. In *Microbiology of the Phyllosphere*, eds Fokkema, N.J. & van den Heuvel, J., pp. 205–222. Cambridge, Cambridge University Press.

Carter, M.V. & Banyer, R.J. (1964) Periodicity of basidiospore release in *Puccinia malvacearum*. *Australian Journal of Biological Sciences*, **17**, 801–802.

Cavalier-Smith, T. (1987) The origin of fungi and pseudofungi. In *Evolutionary Biology of the Fungi*, eds Rayner, A.D.M., Brasier, C.M. & Moore, D., pp. 339–353. Cambridge, Cambridge University Press.

Champe, S.P., Rao, P. & Chang, A. (1987) An endogenous inducer of sexual development in *Aspergillus nidulans*. *Journal of General Microbiology*, **133**, 1383–1387.

Chang, H.S. (1980) Inheritance of light dependence for sporulation in *Cochliobolus miyabeanus*. *Transactions of the British Mycological Society*, **74**, 642–643.

Chanter, D.O. (1979) Harvesting the mushroom crop: a mathematical model. *Journal of General Microbiology*, **115**, 79–87.

Charnley, A.K. (1984) Physiological aspects of destructive pathogenesis in insects by fungi: a speculative review. In *Invertebrate–Microbial Interactions*, eds Anderson, J.M., Rayner, A.D.M. & Walton, D.W.H., pp. 229–270. Cambridge, Cambridge University Press.

Charnley, A.K. (1989) Mechanisms of fungal pathogenesis in insects. In *Biotechnology of Fungi For Improving Plant Growth*, eds Whipps, J.M. & Lumsden, R.D., pp. 84–125. Cambridge, Cambridge University Press.

Charnley, A.K. & St Leger, R.G. (1991) The role of cuticle-degrading enzymes in fungal pathogenesis in insects. In *The Fungal Spore and Disease Initiation in Plants and Animals*, eds Cole, G.T. & Hoch, H.C., pp. 267–286. New York, Plenum.

Chee, K.H. (1976) Factors affecting discharge, germination and variability of spores of *Microcyclus ulei*. *Transactions of the British Mycological Society*, **66**, 499–504.

Cherrett, J.M., Powell, R.J. & Stradling, D.J. (1989) The mutualism between leaf-cutting ants and their fungus. In *Insect–Fungus Interactions*, eds Wilding, N., Collins, N.M., Hammond, P.M. & Webber, J.F., pp. 93–120. London, Academic Press.

Chi, C.C. & Sabo, F.E. (1978) Chemotaxis of zoospores of *Phytophthora megasperma* to roots of alfalfa seedlings. *Canadian Journal of Botany*, **56**, 795–800.

Chilvers, G.A., Lapeyrie, F.F. & Douglass, P.A. (1985) A contrast between Oomycetes and other taxa of mycelial fungi in regard to metachromatic granule formation. *New Phytologist*, **99**, 203–210.

Cho, C.W. & Fuller, M.S. (1989) Ultrastructural studies of encystment and germination in *Phytophthora palmivora*. *Mycologia*, **81**, 539–548.

Ciegler, A. (1983) Evolution, ecology, and mycotoxins: some musings. In *Secondary Metabolism and Differentiation in Fungi*, eds Bennett, J.W. & Ciegler, A., pp. 429–439. New York, Marcel Dekker.

Clarke, R.W., Jennings, D.H. & Coggins, C.R. (1980) Growth of *Serpula lacrimans* in relation to water potential of substrate. *Transactions of the British Mycological Society*, **75**, 271–280.

Clay, K. (1986) Grass endophytes. In *Microbiology of the Phyllosphere*, eds Fokkema, N.J. & van den Heuvel, J., pp. 188–204. Cambridge, Cambridge University Press.

Clay, K. (1988) Clavicipitaceous fungal endophytes of grasses: coevolution and the change from parasitism to mutualism. In *Coevolution of Fungi with Plants and Animals*, eds Pirozynski, K.A. & Hawksworth, D.L., pp. 79–105. London, Academic Press.

Clay, K., Hardy, T.N. & Hammond, A.M. (1985a) Fungal endophytes of *Cyperus* and their effect on an insect herbivore. *American Journal of Botany*, **72**, 1284–1289.

Clay, K., Hardy, T.N. & Hammond, A.M. (1985b) Fungal endophytes of grasses and their effects on an insect herbivore. *Oecologia*, **66**, 1–6.

Clerk, G.C. & Madelin, M.F. (1965) The longevity of conidia of three insect-parasitizing Hyphomycetes. *Transactions of the British Mycological Society*, **48**, 193–209.

Coates, D. & Rayner, A.D.M. (1985a) Genetic control and variation in expression of the 'bow-tie' reaction between homokaryons of *Stereum hirsutum*. *Transactions of the British Mycological Society*, **84**, 191–205.

Coates, D. & Rayner, A.D.M. (1985b) Evidence for a cytoplasmically transmissible factor affecting recognition and somato-sexual differentiation in the basidiomycete, *Stereum hirsutum*. *Journal of General Microbiology*, **131**, 207–219.

Coates, D. & Rayner, A.D.M. (1985c) Induction of rhizomorphic organs in a non-rhizomorphic fungus, *Stereum hirsutum*. *Transactions of the British Mycological Society*, **84**, 527–530.

Cochrane, V.W. (1974) Dormancy in spores of fungi. *Transactions of the American Microscopical Society*, **93**, 599–609.

Coffell, R., Hudspeth, M.E.S. & Meganathan, R. (1990) Biochemical evidence for the exclusion of *Zoophagus insidians* from the Oomycetes. *Mycologia*, **82**, 326–331.

Cohen, B.L. (1981) Regulation of protease production in *Aspergillus*. *Transactions of the British Mycological Society*, **76**, 447–450.
Cohen, R.J. (1974) Cyclic AMP levels in *Phycomyces* during a response to light. *Nature (London)*, **251**, 144–146.
Cohen, R.J. (1976) The avoidance behaviour of *Phycomyces blakesleeanus*: is mediation by natural convection significant? *Journal of Theoretical Biology*, **58**, 207–217.
Cole, G.T. & Hoch, H.C. (eds) (1991) *The Fungal Spore and Disease Initiation in Plants and Animals*, New York, Plenum.
Coley-Smith, J.R. & Cooke, R.C. (1971) Survival and germination of fungal sclerotia. *Annual Review of Phytopathology*, **9**, 65–92.
Coley-Smith, J.R., Ghaffar, A. & Javed, Z.U.R. (1974) The effect of dry conditions on subsequent leakage and rotting of fungal sclerotia. *Soil Biology & Biochemistry*, **6**, 307–312.
Collinge, A.J. & Trinci, A.P.J. (1974) Hyphal tips of wild-type and spreading colonial mutants of *Neurospora crassa*. *Archiv für Mikrobiologie*, **99**, 353–368.
Collins, M.A. (1976) Periodicity of spore liberation in *Chrysomyxa abietis*. *Transactions of the British Mycological Society*, **67**, 336–339.
Comerford, J.G., Spencer-Phillips, P.T.H. & Jennings, D.H. (1985) Membrane-bound ATPase activity, properties of which are altered by growth in saline conditions, isolated from the marine yeast *Debaryomyces hansenii*. *Transactions of the British Mycological Society*, **85**, 431–438.
Conner, D.E., Beuchat, L.R. & Chang, C.J. (1987) Age-related changes in ultrastructure and chemical composition associated with changes in heat resistance of *Neosartorya fischeri*. *Transactions of the British Mycological Society*, **89**, 539–550.
Cook, R.J. & Duniway, J.M. (1981) Water relations in life-cycles of soilborne plant pathogens. In *Water Potential Relations in Soil Microbiology. Soil Science Society of America Special Publication*, Vol. 9, eds Parr, J.F., Gardner, W.R. & Elliott, L.F., pp. 119–140. Madison, Wisconsin, Soil Science Society of America.
Cooke, R.C. (1977) *The Biology of Symbiotic Fungi*, Chichester, Wiley.
Cooke, R.C. (1983) Morphogenesis of sclerotia. In *Fungal Differentiation. A Contemporary Synthesis*, ed. Smith, J.E., pp. 397–418. New York, Marcel Dekker.
Cooke, R.C. & Al-Hamdani, A.M. (1986) Water relations of sclerotia and other infective structures. In *Water, Fungi and Plants*, eds Ayres, P.G. & Boddy, L., pp. 49–63. Cambridge, Cambridge University Press.
Cooke, R.C. & Rayner, A.D.M. (1984) *Ecology of Saprotrophic Fungi*, London, Longman.
Cooke, R.C. & Whipps, J.M. (1980) The evolution of modes of nutrition in fungi parasitic on terrestrial plants. *Biological Reviews*, **55**, 341–362.
Cooke, R.C. & Whipps, J.M. (1987) Saprotrophy, stress and symbiosis. In *Evolutionary Biology of the Fungi*, eds Rayner, A.D.M., Brasier, C.M. & Moore, D., pp. 137–148. Cambridge, Cambridge University Press.
Cooper, M.R. & Morita, R.Y. (1972) Interaction of salinity and temperature on net protein synthesis and viability of *Vibrio marinus*. *Limnology and Oceanography*, **17**, 556–565.
Correa, L.C. & Lochi, W.R. (1986) Induction of sporulation in *Blastocladiella emersonii*: influence of nutritional variables. *Experimental Mycology*, **10**, 270–280.
Cotter, D.A. (1981) Spore activation. In *The Fungal Spore–Morphological Controls*, eds Turian, G. & Hohl, H.R., pp. 385–411. London, Academic Press.
Cotty, P.J. (1987) Modulation of sporulation of *Alternaria tagetica* by carbon dioxide. *Mycologia*, **79**, 508–513.
Coughlan, M.P. (1985) The properties of fungal and bacterial cellulases with comment on their production and application. *Biotechnology and Genetic Engineering Reviews*, **3**, 39–109.
Craig, E., Kang, P.J. & Boorstein, W. (1990) A review of the role of 70 kDa heat shock proteins in protein translocation across membranes. *Antonie van Leeuwenhoek*, **58**, 137–146.
Crandall, M., Egel, R. & MacKay, V.L. (1977) Physiology of mating in three yeasts. *Advances in Microbial Physiology*, **15**, 307–398.
Crawford, I.P. (1975) Gene rearrangements in the evolution of the tryptophan synthetic pathway. *Bacteriological Reviews*, **39**, 87–120.
Curtis, C.R. (1972) Action spectrum of photoinduced sexual stage in the fungus *Nectria haematococca* Berk. and Br. var. *cucurbitae* (Snyder and Hansen) Dingley. *Plant Physiology*, **49**, 235–239.
Curtis, F.C., Evans, G.H., Lillis, V., Lewis, D.H. & Cooke, R.C. (1978) Studies on mucoralean mycoparasites. I. Some effects of *Piptocephalis* species on host growth. *New Phytologist*, **80**, 157–165.
Daft, G.C. & Tsao, P.H. (1984) Parasitism of *Phytophthora cinnamomi* and *P. parasitica* spores by *Catenaria anguillulae* in a soil environment. *Transactions of the British Mycological Society*, **82**, 485–490.
Dahlberg, K.R. & van Etten, J.L. (1982) Physiology and biochemistry of fungal sporulation. *Annual*

Review of Phytopathology, **20**, 281–301.

Day, A.W. (1976) Communication through fimbriae during conjugation in a fungus. *Nature (London)*, **262**, 583–584.

Dekker, R.F.H. (1985) Biodegradation of the hemicelluloses. In *Biosynthesis and Biodegradation of Wood Components*, ed. Higuchi, T., pp. 505–533. London, Academic Press.

Delbrück, M., Katzir, A. & Presti, D. (1976) Response of *Phycomyces* indicating optical excitation of the lowest triplet state of riboflavin. *Proceedings of the National Academy of Sciences of the USA*, **73**, 1969–1973.

Delvalle, J.A. & Asenio, C. (1978) Distribution of adenosine 5′-triphosphate (ATP)-dependent hexose kinases in microorganisms. *BioSystems*, **10**, 265–282.

Dennis, C. & Blijtham, J.M. (1980) Effect of temperature on viability of sporangiospores of *Rhizopus* and *Mucor* species. *Transactions of the British Mycological Society*, **74**, 89–94.

Dennis, C. & Höcker, J. (1981) Effect of relative humidity on chilling sensitivity of sporangiospores of *Rhizopus* species. *Transactions of the British Mycological Society*, **77**, 179–222.

Dennison, D.S. (1979) Phototropism. In *Physiology of Movement. Encyclopedia of Plant Physiology, New Series*, Vol. 7, eds Haupt, W. & Fenleib, M.E., pp. 506–566. Berlin, Springer.

Denny, H.J. & Wilkins, D.A. (1987) Zinc tolerance in *Betula* spp. IV. The mechanism of ectomycorrhizal amelioration of zinc toxicity. *New Phytologist*, **106**, 545–553.

Derksen, J. & Emons, A.M. (1990) Microtubules in tip growth systems. In *Tip Growth in Plant and Fungal Cells*, ed. Heath, I.B., pp. 147–181. London, Academic Press.

Dernoeden, P.H. & Jackson, N. (1981) Enhanced germination of *Sclerophthora macrospora* oospores in response to various chemical or physical treatments. *Transactions of the British Mycological Society*, **76**, 337–341.

De Weille, G.A. (1960) Blister blight (*Exobasidium vexans*) of tea and its relationship with environmental conditions. *Netherlands Journal of Agricultural Science*, **8**, 183–210.

Dexter, Y. & Cooke, R.C. (1984a) Fatty acids, sterols and carotenoids of the psychrophile *Mucor strictus* and some mesophilic *Mucor* species. *Transactions of the British Mycological Society*, **83**, 455–461.

Dexter, Y. & Cooke, R.C. (1984b) Temperature-determined growth and sporulation in the psychrophile *Mucor strictus*. *Transactions of the British Mycological Society*, **83**, 561–568.

Dexter, Y. & Cooke, R.C. (1985) Effect of temperature on respiration, nutrient uptake and potassium leakage in the psychrophile *Mucor strictus*. *Transactions of the British Mycological Society*, **84**, 131–136.

Dickinson, C.H. & Bottomley, D. (1980) Germination and growth of *Alternaria* and *Cladosporium* in relation to their activity in the phylloplane. *Transactions of the British Mycological Society*, **74**, 309–319.

Dijksterhuis, J., Veenhuis, M. & Harder, W. (1990) Ultrastructural study of adhesion and initial stages of infection of nematodes by conidia of *Drechmeria coniospora*. *Mycological Research*, **94**, 1–8.

Dillard, H.R. (1988) Influence of temperature, pH, osmotic potential, and fungicide sensitivity on germination of conidia and growth from sclerotia of *Colletotrichum coccodes in vitro*. *Phytopathology*, **78**, 1357–1361.

Dillon, R.J. & Charnley, A.K. (1991) The fate of fungal spores in the insect gut. In *The Fungal Spore and Disease Initiation in Plants and Animals*, eds Cole, G.T. & Hoch, H.C., pp. 129–156. New York, Plenum.

Dix, N.J. (1985) Changes in the relationship between water content and water potential after decay and its significance for fungal successions. *Transactions of the British Mycological Society*, **85**, 649–653.

Dix, N.J. & Frankland, J.C. (1988) Tolerance of litter-decomposing agarics to water stress in relation to habitat. *Transactions of the British Mycological Society*, **88**, 127–129.

Dixon, R.A. (1988) Response of ectomycorrhizal *Quercus rubra* to soil cadmium, nickel and lead. *Soil Biology & Biochemistry*, **20**, 555–559.

Douglas, A.E. (1989) Mycetocyte symbiosis in insects. *Biological Reviews*, **64**, 409–434.

Dowding, P. (1984) The evolution of insect–fungus relationships in the primary invasion of forest timber. In *Invertebrate–Microbial Interactions*, eds Anderson, J.M., Rayner, A.D.M. & Walton, D.W.H., pp. 133–153. Cambridge, Cambridge University Press.

Dowsett, J.A., Reid, J. & Hopkin, A.A. (1984) Microscopic observations on the trapping of nematodes by the predaceous fungus *Dactylella cionopaga*. *Canadian Journal of Botany*, **62**, 674–679.

Dowson, C.D., Rayner, A.D.M. & Boddy, L. (1986) Outgrowth patterns of mycelial cord-forming Basidiomycetes from and between woody resource units in soil. *Journal of General Microbiology*, **132**, 203–211.

Drechsler, C. (1936) A *Fusarium*-like species of *Dactylella* capturing and consuming testaceous rhizopods. *Journal of the Washington Academy of Science*, **26**, 397–404.

Drechsler, C. (1937) New Zoopagaceae destructive to soil rhizopods. *Mycologia*, **29**, 229–249.

Drinkard, L.C., Nelson, G.E. & Sutter, R.P. (1982) Growth arrest: a prerequisite for sexual development in *Phycomyces blakesleeanus*. *Experimental Mycology*, **6**, 52–59.

Duncan, J.M. (1985) Effect of temperature and other factors on the *in vitro* germination of oospores of *Phytophthora fragariae*. *Transactions of the British Mycological Society*, **85**, 455–462.

Dupler, M., Smilanick, J.L. & Hoffmann, J.A. (1987) Effect of matric and osmotic potential on teliospore germination of *Tilletia indica*. *Phytopathology*, **77**, 594–598.

Durbin, R.D. (1959) Factors affecting the vertical distribution of *Rhizoctonia solani* with special reference to CO_2 concentration. *American Journal of Botany*, **46**, 22–25.

Durrell, L.W. & Shields, L.M. (1960) Fungi isolated in culture from soils of the Nevada test site. *Mycologia*, **52**, 636–641.

Dute, R.R., Weete, J.D. & Rushing, A.E. (1989) Ultrastructure of dormant and germinating conidia of *Aspergillus ochraceus*. *Mycologia*, **81**, 772–782.

Eamus, D. & Jennings, D.H. (1986a) Turgor and fungal growth: studies on water relations of mycelia of *Serpula lacrimans* and *Phallus impudicus*. *Transactions of the British Mycological Society*, **86**, 527–535.

Eamus, D. & Jennings, D.H. (1986b) Water, turgor and osmotic potentials of fungi. In *Water, Fungi and Plants*, eds Ayres, P.G. & Boddy, L., pp. 27–48. Cambridge, Cambridge University Press.

Eddy, A.A. (1982) Mechanisms of solute transport in selected eukaryotic micro-organisms. *Advances in Microbial Physiology*, **23**, 1–78.

Edwards, D. & Fanning, U. (1985) Evolution and environment in the late Silurian–early Devonian: the rise of the pteridophytes. *Philosophical Transactions of the Royal Society of London, Series B*, **309**, 147–165.

El-Abyad, M.S.H. & Webster, J. (1968a) Studies on pyrophilous Discomycetes. I. Comparative physiological studies. *Transactions of the British Mycological Society*, **51**, 353–367.

El-Abyad, M.S.H. & Webster, J. (1968b) Studies on pyrophilous Discomycetes. II. Competition. *Transactions of the British Mycological Society*, **51**, 369–375.

Elad, Y. & Baker, R. (1985a) Influence of trace amounts of cations and siderophore-producing Pseudomonads on chlamydospore germination of *Fusarium oxysporum*. *Phytopathology*, **75**, 1047–1052.

Elad, Y. & Baker, R. (1985b) The role of competition for iron and carbon in suppression of chlamydospore germination of *Fusarium* spp. by *Pseudomonas* spp. *Phytopathology*, **75**, 1053–1059.

Elad, Y., Barak, K. & Chet, I. (1983a) Possible role of lectins in mycoparasitism. *Journal of Bacteriology*, **154**, 1431–1435.

Elad, Y. & Chet, I. (1987) Possible role of competition for nutrients in biocontrol of *Pythium* damping-off by bacteria. *Phytopathology*, **77**, 190–195.

Elad, Y., Chet, I., Boyle, P. & Henis, Y. (1983b) Parasitism of *Trichoderma* spp. on *Rhizoctonia solani* and *Sclerotium rolfsii*–scanning electron microscopy and fluorescence microscopy. *Phytopathology*, **73**, 85–88.

Elad, Y., Lifshitz, R. & Baker, R. (1985) Enzymatic activity of the mycoparasite *Pythium nunn* during interaction with host and non-host fungus. *Physiological Plant Pathology*, **27**, 131–148.

Ellenbogen, B.B., Aaronson, S., Goldstein, S. & Belsky, M. (1969) Polyunsaturated fatty acids of aquatic fungi: possible phylogenetic significance. *Comparative Biochemistry and Physiology*, **29**, 805–811.

Elliott, C.G. (1977) Sterols in fungi: their functions in growth and reproduction. *Advances in Microbial Physiology*, **15**, 121–173.

Elliott, C.G. (1986) Inhibition of reproduction of *Phytophthora* by the calmodulin interacting compounds triluoperazine and ophiobolin A. *Journal of General Microbiology*, **132**, 2781–2785.

Elliott, C.G. (1988) Stages in oosporogenesis of *Phytophthora* sensitive to inhibitors of calmodulin and phosphodiesterase. *Transactions of the British Mycological Society*, **90**, 187–192.

Elliott, C.G. (1989) Some aspects of nitrogen nutrition and reproduction in *Phytophthora*. *Mycological Research*, **92**, 34–44.

Ellison, P.J., Harrower, K.M., Chilvers, G.A. & Owens, J.D. (1981) Patterns of sporulation in *Trichoderma viride*. *Transactions of the British Mycological Society*, **76**, 441–445.

Eppstein, D.A. & Tainter, F.H. (1976) Germination self-inhibitor from *Cronartium comandrae* aeciospores. *Phytopathology*, **66**, 1395–1397.

Epstein, L. & Lockwood, J.L. (1983) The role of exudation in the germination of *Cochliobolus victoriae* conidia. *Journal of General Microbiology*, **129**, 3629–3635.

Epstein, L. & Lockwood, J.L. (1984) Effect of soil microbiota on germination of *Bipolaris victoriae* conidia. *Transactions of the British Mycological Society*, **82**, 63–69.

Eriksson, K.E. & Wood, T.M. (1985) Biodegradation

of cellulose. In *Biosynthesis and Biodegradation of Wood Components*, ed. Higuchi, T., pp. 469–503. London, Academic Press.

Esler, G. & Coley-Smith, J.R. (1983) Flavour and odour characteristics of species of *Allium* in relation to their capacity to stimulate germination of sclerotia of *Sclerotium cepivorum*. *Plant Pathology*, **32**, 13–22.

Esser, K., Kück, U., Stahl, U. & Tudzynski, P. (1984) Senescence in *Podospora anserina* and its implication for genetic engineering. In *The Ecology and Physiology of the Fungal Mycelium*, eds Jennings, D.H. & Rayner, A.D.M., pp. 343–352. Cambridge, Cambridge University Press.

Evans, G.H. & Cooke, R.C. (1982) Studies on mucoralean mycoparasites. III. Diffusible factors from *Mortierella vinacea* Dixon-Stewart that direct germ tube growth in *Piptocephalis fimbriata* Richardson and Leadbeater. *New Phytologist*, **91**, 245–253.

Evans, G.H., Lewis, D.H. & Cooke, R.C. (1978) Studies on mucoralean mycoparasites. II. Persistent yeast-phase growth on *Mycotypha microspora* Fenner when infected by *Piptocephalis fimbriata* Richardson & Leadbeater. *New Phytologist*, **81**, 629–635.

Evans, H.C. (1988) Coevolution of entomogenous fungi and their insect hosts. In *Coevolution of Fungi with Plants and Animals*, eds Pirozynski, K.A. & Hawksworth, D.L., pp. 149–171. London, Academic Press.

Faison, B.D., Kirk, T.K. & Farrell, R.L. (1986) Role of veratryl alcohol in regulating ligninase activity in *Phanaerochaete chrysosporium*. *Applied and Environmental Microbiology*, **52**, 251–254.

Faraj Salman, A.G. (1971) Das Wirkungsspektrum der lichtabhangigen Zonierung der Koremian von zwei Mutanten von *Penicillium claviforme* Bainier (An action spectrum of photo-induced coremia-zonation of two mutants of *Penicillium claviforme* Bainier). *Planta (Berlin)*, **101**, 117–121.

Faull, J.L. (1988) Competitive antagonism of soil-borne plant pathogens. In *Fungi in Biological Control Systems*, ed. Burge, M.N., pp. 125–140. Manchester, Manchester University Press.

Febvay, G., Bourgeois, P. & Kermarrec, A. (1985) Antiappétants pour la fourmi attine, *Acromyrmex octospinosus* (Reich) (*Hymenoptera-Formicidae*), chez certaines espèces d'igname (*Discoreaceae*) cultivées aux Antilles (Antifeedants for an attine ant, *Acromyrmex octospinosus* (Reich) (*Hymenoptera-Formicidae*), in yam (*Discoreaceae*) leaves grown in the Antilles). *Agronomie (Paris)*, **5**, 439–444.

Fermor, T.R., Wood, D.A., Lincoln, S.P. & Fenlon, J.S. (1991) Bacteriolysis by *Agaricus bisporus*. *Journal of General Microbiology*, **137**, 15–22.

Field, J.I. & Webster, J. (1977) Traps of predacious fungi attract nematodes. *Transactions of the British Mycological Society*, **68**, 467–469.

Filonow, A.B., Akueshi, C.O. & Lockwood, J.L. (1983) Decreased virulence of *Cochliobolus victoriae* conidia after incubation on soils or on leached sand. *Phytopathology*, **73**, 1632–1636.

Filonow, A.B. & Arora, D.K. (1987) Influence of soil matric potential on ^{14}C exudation from fungal propagules. *Canadian Journal of Botany*, **65**, 2084–2089.

Filonow, A.B. & Lockwood, J.L. (1983a) Mycostasis in relation to the microbial nutrient sinks of five soils. *Soil Biology & Biochemistry*, **15**, 557–565.

Filonow, A.B. & Lockwood, J.L. (1983b) Loss of nutrient-independence for germination by fungal propagules incubated on soils or on a model system imposing diffusive stress. *Soil Biology & Biochemistry*, **15**, 567–573.

Finlay, R.D. (1985) Interaction between soil micro-arthropods and endomycorrhizal associations of higher plants. In *Ecological Interactions in Soil: Plants, Microbes and Animals*, eds Fitter, A.H., Atkinson, D., Read, D.J. & Usher, M.B., pp. 319–331. Oxford, Blackwell Scientific.

Fischer, F.G. & Werner, G. (1955) Eine Analyse des Chemotropismus einiger Pilze, insbesondere der Saprolegniaceen (An analysis of the chemotrophism of several fungi, especially of the Saprolegniaceae). *Hoppe-Seyler's Zeitschrift für Physiologische Chemie*, **300**, 211–236.

Fisher, R.F. (1977) Nitrogen and phosphorus mobilization by the fairy ring fungus, *Marasmius oreades* (Bolt.) Fr. *Soil Biology & Biochemistry*, **9**, 239–241.

Fitt, B.D.L., McCartney, H.A. & Walklate, P.J. (1989) The role of rain in dispersal of pathogen inoculum. *Annual Review of Phytopathology*, **27**, 241–270.

Flegg, P.B., Spencer, D.M. & Wood, D.A. (eds) (1985) *The Biology and Technology of the Cultivated Mushroom*, Chichester, Wiley.

Fogarty, W.M. (1983) Microbial amylases. In *Microbial Enzymes and Biotechnology*, ed. Fogarty, W.M., pp. 1–92. London, Applied Science.

Fogarty, W.M. & Kelly, C.T. (1983) Pectic enzymes. In *Microbial Enzymes and Biotechnology*, ed. Fogarty, W.M., pp. 131–182. London, Applied Science.

Förster, H., Coffey, M.D., Elwood, H. & Sogin, M.L. (1990) Sequence analysis of the small subunit ribosomal RNAs of three zoosporic fungi and implications for fungal evolution. *Mycologia*, **82**, 306–312.

Förster, H., Ribeiro, O.K. & Erwin, D.C. (1983) Factors affecting oospore germination of *Phyto-*

phthora megasperma f. sp. *medicaginis*. *Phytopathology*, **73**, 442–448.

Fox, F.M. (1987) Ultrastructure of mycelial strands of *Leccinum scabrum*, ectomycorrhizal on birch (*Betula* spp.). *Transactions of the British Mycological Society*, **89**, 551–560.

Fradkin, A. & Patrick, Z.A. (1982) Fluorescence microscopy to study colonization of conidia and hyphae of *Cochliobolus sativus* by soil microorganisms. *Soil Biology & Biochemistry*, **14**, 543–548.

French, R.C. (1985) The bioregulatory action of flavor compounds on fungal spores and other propagules. *Annual Review of Phytopathology*, **23**, 173–200.

Fries, N. (1978) Basidiospore germination in some mycorrhiza-forming Hymenomycetes. *Transactions of the British Mycological Society*, **70**, 319–324.

Fries, N. (1979) The taxon-specific spore germination reaction in *Leccinum*. *Transactions of the British Mycological Society*, **73**, 337–341.

Fries, N. (1981) Recognition reactions between basidiospores and hyphae in *Leccinum*. *Transactions of the British Mycological Society*, **77**, 9–14.

Fries, N. (1983) Spore germination, homing reaction, and intersterility groups in *Laccaria laccata* (Agaricales). *Transactions of the British Mycological Society*, **75**, 221–227.

Fries, N. & Swedjemark, G. (1985) Sporophagy in Hymenomycetes. *Experimental Mycology*, **9**, 74–79.

Frisvad, J.C. (1986) Taxonomic approaches to mycotoxin identification (Taxonomic indication of mycotoxin content in foods). In *Modern Methods in the Analysis and Structural Elucidation of Mycotoxins*, ed. Cole, R.J., pp. 415–457. New York, Academic Press.

Frisvad, J.C., Filtenborg, O. & Wicklow, D.T. (1987) Terverticillate penicillia isolated from underground seed caches and cheek pouches of banner-tailed kangaroo rats (*Dipodomyces spectabilis*). *Canadian Journal of Botany*, **68**, 765–773.

Funk, C.R., Halisky, P.M., Johnson, M.C., Siegel, M.R., Stewart, A.V., Ahmad, S., Hurley, R.H. & Harvey, I.C. (1983) An endophytic fungus and resistance to sod webworms: association in *Lolium perenne* L. *BioSystems*, **1**, 189–191.

Furch, B. (1981) Spore germination: heat mediated events. In *The Fungal Spore: Morphogenetic Controls*, eds Turian, G. & Hohl, H.R., pp. 413–433. New York, Academic Press.

Gadd, G.M. (1986) Fungal responses towards heavy metals. In *Microbes in Extreme Environments*, eds Herbert, R.A. & Codd, G.A., pp. 83–110. London, Academic Press.

Gadd, G.M., Chudek, J.A., Foster, R. & Reed, R.H. (1984) The osmotic responses of *Penicillium ochro-chloron*: changes in internal solute levels in response to copper and salt stress. *Journal of General Microbiology*, **130**, 1969–1975.

Gadd, G.M. & Griffiths, A.J. (1980) Influence of pH on toxicity and uptake of copper in *Aureobasidium pullulans*. *Transactions of the British Mycological Society*, **75**, 91–96.

Gadgil, M. & Solbrig, O.M. (1972) The concept of *r*- and *K*-selection: evidence from wild flowers and some theoretical considerations. *American Naturalist*, **106**, 14–31.

Gain, E.R. & Barnett, H.L. (1970) Parasitism and axenic growth of the mycoparasite *Gonatorhodiella highlei*. *Mycologia*, **62**, 1122–1129.

Galland, P. & Lipson, E.D. (1987) Light physiology of *Phycomyces* sporangiophores. In *Phycomyces*, eds Cerdá-Olmedo, E. & Lipson, E.D., pp. 49–92. Cold Spring Harbor, New York State, Cold Spring Harbor Laboratory.

Galvagno, M.A., Forchiassin, F., Cantore, M.L. & Passeron, S. (1984) The effect of light and cyclic AMP metabolism on fruiting body formation in *Saccobolus platensis*. *Experimental Mycology*, **8**, 334–341.

Gardiner, R.B., Jarvis, W.R. & Shipp, J.L. (1990) Ingestion of *Pythium* spp. by larvae of the fungus gnat *Bradysia impatiens* (Diptera: Sciaridae). *Annals of Applied Biology*, **116**, 205–212.

Garnier-Sillam, E., Toutain, F., Villemin, G. & Renoux, J. (1988) Transformation de la matière organique végétale sous l'action du termite *Macrotermes mülleri* (Sjöstedt) et de son champignon symbiotique (Transformation of plant organic matter under the influence of the termite *Macrotermes mülleri* (Sjöstedt) and its symbiotic fungus). *Canadian Journal of Microbiology*, **34**, 1247–1255.

Garrett, M.K. & Robinson, P.M. (1969) A stable inhibitor of spore germination produced by fungi. *Archiv für Mikrobiologie*, **67**, 370–377.

Garty, J. & Delarea, J. (1988) Evidence of liberation of lichen ascospores in clusters and reports on contact between free-living algal cells and germinating lichen ascospores under natural conditions. *Canadian Journal of Botany*, **66**, 2171–2177.

Gaspard, J.T. & Mankau, R. (1987) Density-dependence and host-specificity of the nematode-trapping fungus *Monacrosporium ellipsosporum*. *Revue de Nématologie*, **10**, 241–246.

Geller, A. (1983) Growth of bacteria in inorganic medium at different levels of airborne substances. *Applied and Environmental Microbiology*, **46**, 1258–1262.

Gemma, J.N. & Koske, R.E. (1988) Pre-infection

interactions between roots and the mycorrhizal fungus *Gigaspora gigantea*: chemotropism of germ-tubes and root growth response. *Transactions of the British Mycological Society*, **91**, 123–132.

Ghabrial, S.A. (1986) A transmissible disease of *Helminthosporium victoriae*–evidence for a viral etiology. In *Fungal Virology*, ed. Buck, K.W., pp. 163–176. Boca Raton, Florida, CRC.

Gianinazzi-Pearson, V. & Gianinazzi, S. (1986) The physiology of improved phosphate nutrition in mycorrhizal plants. In *Mycorrhizae: Physiology and Genetics*, eds Gianinazzi-Pearson, V. & Gianinazzi, S., pp. 101–109. Paris, INRA.

Gibb, E. & Walsh, J.H. (1980) Effect of nutritional factors and carbon dioxide on growth of *Fusarium moniliforme* and other fungi in reduced oxygen concentrations. *Transactions of the British Mycological Society*, **74**, 111–118.

Gildon, A. & Tinker, P.B. (1981) A heavy metal-tolerant strain of mycorrhizal fungus. *Transactions of the British Mycological Society*, **77**, 648–649.

Gillespie, A.T. & Moorhouse, E.R. (1989) The use of fungi to control pests of agricultural and horticultural importance. In *Biotechnology of Fungi for Improving Plant Growth*, eds Whipps, J.M. & Lumsden, R.D., pp. 55–84. Cambridge, Cambridge University Press.

Giuma, A.Y. & Cooke, R.C. (1971) Nematotoxin production by *Nematoctonus haptocladus* and *N. concurrens*. *Transactions of the British Mycological Society*, **56**, 89–94.

Gladders, P. & Coley-Smith, J.R. (1980) Interactions between *Rhizoctonia tuliparum* sclerotia and some soil microorganisms. *Transactions of the British Mycological Society*, **74**, 579–586.

Gochenaur, S.E. (1987) Evidence suggests that grazing regulates ascospore density in soil. *Mycologia*, **79**, 445–450.

Gold, M.H. & Cheng, T.M. (1979) Conditions for fruit body formation in white rot Basidiomycetes. *Archives of Microbiology*, **121**, 37–42.

Goldberg, N.P., Hawes, M.C. & Stanghellini, M.E. (1989) Specific attraction to and infection of cotton root cap cells by zoospores of *Pythium dissotocum*. *Canadian Journal of Botany*, **67**, 1760–1767.

Goldberg, N.P. & Stanghellini, M.E. (1990) Ingestion–egestion and aerial transmission of *Pythium aphanidermatum* by shore flies (Ephydrinae: *Scatella stagnalis*). *Phytopathology*, **80**, 1244–1246.

Gooday, G.W. (1975) Chemotaxis and chemotropism in fungi and algae. In *Primitive Sensory and Communication Systems: The Taxes and Tropisms of Micro-organisms and Cells*, ed. Carlile, M.J., pp. 155–204. London, Academic Press.

Gooday, G.W. (1981) Biogenesis of sporopollenin in fungal spore walls. In *The Fungal Spore: Morphogenetic Controls*, eds Turian, G. & Hohl, H.R., pp. 487–505. New York, Academic Press.

Gopalan, R. & Manners, J.G. (1984) Environmental and other factors affecting germination of urediniospores of *Puccinia striiformis*. *Transactions of the British Mycological Society*, **82**, 239–243.

Gough, F.J. & Lee, T.S. (1985) Moisture effects on the discharge and survival of conidia of *Septoria tritici*. *Phytopathology*, **75**, 180–182.

Gow, N.A.R. (1989) Circulating ionic currents in micro-organisms. *Advances in Microbial Physiology*, **30**, 89–123.

Gow, N.A.R. & Gooday, G.W. (1987) Effects of antheridiol on growth, branching and electrical currents of hyphae of *Achlya ambisexualis*. *Journal of General Microbiology*, **133**, 3531–3535.

Grant, B.R., Irving, H.R. & Radda, M. (1985) The effect of pectin and related compounds on encystment and germination of *Phytophthora palmivora* zoospores. *Journal of General Microbiology*, **131**, 669–676.

Gray, J. (1985) The microfossil record of early land plants: advances in understanding of early terrestrialization, 1970–1984. *Philosophical Transactions of the Royal Society of London, Series B*, **309**, 167–195.

Greenslade, P.J.M. (1983) Adversity selection and the habitat templet. *American Naturalist*, **122**, 352–365.

Gregory, P.H. (1966) The fungus spore: what it is and what it does. In *The Fungus Spore*, ed. Madelin, M.F., pp. 1–13. London, Butterworths.

Gregory, P.H. (1984) The fungal mycelium–an historical perspective. In *The Ecology and Physiology of the Fungal Mycelium*, eds Jennings, D.H. & Rayner, A.D.M., pp. 1–22. Cambridge, Cambridge University Press.

Gresik, M., Kolarova, N. & Farkas, V. (1988) Membrane potential, ATP, and cyclic AMP changes induced by light in *Trichoderma viride*. *Experimental Mycology*, **12**, 295–301.

Griffin, D.M. (1972) *Ecology of Soil Fungi*, London, Chapman and Hall.

Griffin, D.M. (1981) Water and microbial stress. In *Advances in Microbial Ecology*, Vol. 9, ed. Alexander, M., pp. 91–136. New York, Plenum.

Griffin, H., Dintzis, F.R., Krull, L. & Baker, F.L. (1984) A microfibril generating factor from the enzyme complex of *Trichoderma reesii*. *Biotechnology & Bioengineering*, **26**, 296–300.

Griffiths, D.A. (1982) Structure of *Verticillium*

nigrescens and *V. nubilum* chlamydospores. *Transactions of the British Mycological Society*, **78**, 141–145.

Griffiths, E. & Peverett, H. (1980) Effects of humidity and cirrhus extract on survival of *Septoria nodorum* spores. *Transactions of the British Mycological Society*, **75**, 147–150.

Grime, J.P. (1977) Evidence for the existence of three primary strategies in plants and its relevance to ecological and evolutionary theory. *American Naturalist*, **111**, 1169–1194.

Grime, J.P. (1979) *Plant Strategies and Vegetation Processes*, Chichester, Wiley.

Grogan, R.G. & Abawi, G.S. (1975) Influence of water potential on growth and survival of *Whetzelinia sclerotiorum*. *Phytopathology*, **65**, 122–128.

Grønvold, J., Wolstrup, J., Henriksen, A. & Nansen, P. (1987) Field experiments on the ability of *Arthrobotrys oligospora* (Hyphomycetales) to reduce the number of larvae of *Cooperia ancophora* (Trichostrongylidae) in cow pats and surrounding grass. *Journal of Helminthology*, **61**, 65–71.

Haber, J.E., Wejksnora, P.J., Wygall, D.D. & Lai, E.Y. (1977) Controls of sporulation in *Saccharomyces cerevisiae*. In *Eukaryotic Microbes as Model Development Systems*, eds O'Day, D.H. & Horgen, P.A., pp. 129–154. New York, Marcel Dekker.

Hale, M.D. & Eaton, R.A. (1985) Oscillatory growth of fungal hyphae in wood cells. *Transactions of the British Mycological Society*, **84**, 277–288.

Hallbauer, D.K. & van Warmelo, K.T. (1974) Fossilized plants in thucolite from Precambrian rocks in the Witwatersrand, South Africa. *Precambrian Research*, **1**, 199–212.

Halsall, D.M. (1976) Zoospore chemotaxis in Australian isolates of *Phytophthora* species. *Canadian Journal of Microbiology*, **22**, 409–472.

Hammond, J.B.W. & Nichols, R. (1979) Carbohydrate metabolism in *Agaricus bisporus*: changes in non-structural carbohydrates during periodic fruiting (flushing). *New Phytologist*, **83**, 723–730.

Hammond, P.M. & Lawrence, J.F. (1989) Appendix: mycophagy in insects: a summary. In *Insect–Fungus Interactions*, eds Wilding, N., Collins, N.M., Hammond, P.M. & Webber, J.F., pp. 275–324. London, Academic Press.

Hammonds, P. & Smith, S.N. (1986) Lipid composition of a psychrophilic, a mesophilic and a thermophilic *Mucor* species. *Transactions of the British Mycological Society*, **86**, 551–560.

Handelsman, J., Raffel, S., Mester, E.H., Wunderlich, L. & Grau, C.R. (1990) Biological control of damping-off of alfalfa seedlings with *Bacillus cereus* UW85. *Applied and Environmental Microbiology*, **56**, 713–718.

Hannau, R.M., Franklin, D., Yang, Z. & Nicholson, R.L. (1989) Conidiogenous development in *Helminthosporium carbonum*. *Experimental Mycology*, **13**, 337–347.

Hanski, I. (1989) Fungivory: fungi, insects and ecology. In *Insect–Fungus Interactions*, eds Wilding, N., Collins, N.M., Hammond, P.M. & Webber, J.F., pp. 25–68. London, Academic Press.

Hanson, M.A. & Marzluf, G.A. (1975) Control of synthesis of a single enzyme by multiple regulatory circuits in *Neurospora crassa*. *Proceedings of the National Academy of Sciences of the USA*, **72**, 1240–1244.

Hardie, K. (1979) Germination of *Chaetomium globosum* ascospores on hardwoods. *Transactions of the British Mycological Society*, **73**, 81–84.

Harding, R.W. (1973) Inhibition of conidiation and photoinduced carotenoid biosynthesis by cyclic-AMP, *Neurospora Newsletter*, **20**, 20–21.

Hardy, T.N., Clay, K. & Hammond, A.M. (1986) The effect of leaf age and related factors on endophyte-mediated resistance to fall armyworm (Lepidoptera: Noctuidae) in tall fescue. *Journal of Experimental Entomology*, **15**, 1083–1089.

Harman, G.E., Mattick, L.R., Nash, G.A. & Nedrow, B.L. (1980) Stimulation of fungal spore germination and inhibition of sporulation in fungal vegetative thalli by fatty acids and their volatile peroxidation products. *Canadian Journal of Botany*, **58**, 1541–1547.

Harold, F.M. & Caldwell, J.H. (1990) Tips and currents: electrobiology of apical growth. In *Tip Growth in Plant and Fungal Cells*, ed. Heath, I.B., pp. 59–90. London, Academic Press.

Harold, R.L. & Harold, F.M. (1980) Oriented growth of *Blastocladiella emersonii* in gradients of ionophores and inhibitors. *Journal of Bacteriology*, **144**, 1159–1167.

Harper, D.S., Swinburne, T.R., Moore, S., Brown, A.E. & Graham, H. (1980) A role for iron in germination of conidia of *Collectotrichum musae*. *Journal of General Microbiology*, **121**, 169–174.

Harper, J.E. & Webster, J. (1964) An experimental analysis of the coprophilous fungus succession. *Transactions of the British Mycological Society*, **47**, 511–530.

Harper, J.L. & Ogden, J. (1970) The reproductive strategy of higher plants. I. The concept of strategy with special reference to *Senecio vulgaris* L. *Journal of Ecology*, **58**, 681–689.

Harris, D.C. & Cole, D.M. (1982) Germination of *Phytophthora syringae* oospores. *Transactions of the British Mycological Society*, **79**, 527–530.

Harris, S.S. & Dennison, D.S. (1979) *Phycomyces*: interference between the light growth response and the avoidance response. *Science (New York)*, **206**, 357–358.

Harrison, J.G. (1983) Survival of *Botrytis fabae* conidia in air. *Transactions of the British Mycological Society*, **80**, 263–269.

Hartman, R.E., Keen, N.T. & Long, M. (1972) Carbon dioxide fixation by *Verticillium albo-atrum*, *Journal of General Microbiology*, **73**, 29–34.

Harvey, P.J., Schoemaker, H.E. & Palmer, J.M. (1985) Enzymic degradation of lignin and its potential to supply chemicals. *Annual Proceedings of the Phytochemical Society of Europe*, **26**, 249–266.

Haselwandter, K., Bobleter, O. & Read, D.J. (1990) Degradation of ^{14}C-labelled lignin and dehydropolymer of coniferyl alcohol by ericoid and ectomycorrhizal fungi. *Archives of Microbiology*, **153**, 352–354.

Hashiba, T. (1987) An improved system for biological control of damping-off by using plasmids in fungi. In *Innovative Approaches to Plant Disease Control*, ed. Chet, I., pp. 337–351. New York, Wiley Interscience.

Hawes, C.R. (1980) Conidium germination in the *Chalara* state of *Ceratocystis adiposa*: a light and electron microscope study. *Transactions of the British Mycological Society*, **74**, 321–328.

Hawker, L.E. & Madelin, M.F. (1976) The dormant spore. In *The Fungus Spore*, eds Weber, D.J. & Hess, W.M., pp. 1–72. New York, Wiley.

Hawksworth, D.L. (1987) The evolution and adaptation of sexual reproductive structures in the Ascomycotina. In *Evolutionary Biology of the Fungi*, eds Rayner, A.D.M., Brasier, C.M. & Moore, D., pp. 179–189. Cambridge, Cambridge University Press.

Hawksworth, D.L. (1991) The fungal dimension of biodiversity: magnitude, significance and conservation. *Mycological Research*, **95**, 641–655.

Hawksworth, D.L., Sutton, B.C. & Ainsworth, G.C. (eds) (1983) *Dictionary of the Fungi*, 7th edn, Kew, Commonwealth Mycological Institute.

Hearth, J.H. & Padgett, D.E. (1990) Salinity tolerance of an *Aphanomyces* isolate (Oomycetes) and its possible relationship to ulcerative mucosis (UM) of Atlantic menhaden. *Mycologia*, **82**, 364–369.

Hecker, L.I. & Sussman, A.S. (1973) Localization of trehalase in the ascospores of *Neurospora*: relation to ascospore dormancy and germination. *Journal of Bacteriology*, **115**, 592–599.

Hedger, J. (1990) Fungi in the tropical forest canopy. *Mycologist*, **4**, 200–202.

Heggo, A., Angle, J.S. & Chaney, R.L. (1990) Effects of vesicular–arbuscular mycorrhizal fungi on heavy metal uptake by soybeans. *Soil Biology & Biochemistry*, **22**, 865–869.

Held, A.A. (1970) Nutrition and fermentative energy metabolism of the water mould *Aqualinderella fermentans. Mycologia*, **62**, 339–358.

Held, A.A. (1973) Encystment and germination of the parasitic chytrid *Rozella allomycis* on host hyphae. *Canadian Journal of Botany*, **51**, 1825–1835.

Held, A.A., Emerson, R., Fuller, M.S. & Gleason F.H. (1969) *Blastocladiella* and *Aqualinderella*: fermentative water molds with high carbon dioxide optima. *Science (New York)*, **165**, 706–709.

Hemmes, D.E. & Hohl, H.R. (1969) Ultrastructural changes in directly germinating sporangia of *Phytophthora parasitica. American Journal of Botany*, **56**, 300–313.

Hemmes, D.E. & Lerma Jr, A.P. (1985) The ultrastructure of developing and germinating chlamydospores of *Phytophthora palmivora. Mycologia*, **77**, 743–755.

Hemmes, D.E. & Stasz, T.E. (1984) Ultrastructure of dormant, converted, and germinating oospores of *Pythium ultimum. Mycologia*, **76**, 924–935.

Herman, R.P. & Luchini, M.M. (1989) Lipoxygenase activity in the Oomycete *Saprolegnia* in developmental cues and reproductive competence. *Experimental Mycology*, **13**, 372–379.

Hervey, A. & Nair, M.S.R. (1979) Antibiotic metabolite of a fungus cultivated by gardening ants. *Mycologia*, **71**, 1064–1066.

Hess, S.L. Allen, P.J., Nelson, D. & Lester, H.H. (1975) Mode of action of methyl *cis*-ferulate, the self-inhibitor of stem rust uredospore germination. *Physiological Plant Pathology*, **5**, 107–112.

Hickman, C.J. (1970) Biology of *Phytophthora* zoospores. *Phytopathology*, **60**, 1128–1135.

Hinch, J.M. & Clarke, A.E. (1980) Adhesion of fungal zoospores to root surfaces is mediated by carbohydrate determinants of the root slime. *Physiological Plant Pathology*, **16**, 303–307.

Hintikka, V. & Korhonen, K. (1970) Effects of carbon dioxide on the growth of lignicolous and soil-inhabiting Hymenomycetes. *Comunicaciones Instituti Forestalis Fenniae*, **69**, 1–29.

Ho, C.S. & Smith, M.D. (1986) Morphological alterations in *Penicillium chrysogenum* caused by carbon dioxide. *Journal of General Microbiology*, **132**, 3479–3484.

Ho, W.C. & Ko, W.H. (1982) Characteristics of soil microbiostasis. *Soil Biology & Biochemistry*, **14**, 589–593.

Ho, W.C. & Ko, W.H. (1985) Soil microbiostasis: effect of environmental and edaphic factors. *Soil*

Biology & Biochemistry, **17**, 167–170.

Ho, W.C. & Ko, W.H. (1986) Microbiostasis by nutrient deficiency shown in natural and synthetic soils. *Journal of General Microbiology*, **132**, 2807–2815.

Ho, Y.W. & Bauchop, T. (1991) Morphology of three polycentric rumen fungi and description of a procedure for the induction of zoosporogenesis and release of zoospores in cultures. *Journal of General Microbiology*, **137**, 213–217.

Hobot, J.A. & Gull, K. (1980) The identification of a self-inhibitor from *Syncephalastrum racemosum* and its effects upon sporangiospore germination. *Antonie van Leeuwenhoek*, **46**, 435–441.

Hobot, J.A. & Jennings, D.H. (1981) Growth of *Debaryomyces hansenii* and *Saccharomyces cerevisiae* in relation to pH and salinity. *Experimental Mycology*, **5**, 217–228.

Hoch, H.C. & Staples, R.C. (1987) Structural and chemical changes among the rust fungi during appressorium formation. *Annual Review of Phytopathology*, **25**, 231–247.

Hock, B., Bahn, M., Walk, R.-A. & Nitsche, U. (1978) The control of fruiting body formation in the Ascomycete *Sordaria macrospora* Auersw. by regulation of hyphal development. *Planta*, **141**, 93–103.

Hocking, A.D. (1986) Effects of water activity and culture age on the glycerol accumulation patterns of five fungi. *Journal of General Microbiology*, **132**, 269–275.

Hocking, A.D. & Pitt, J.I. (1979) Water relations of some *Penicillium* species at 25°C. *Transactions of the British Mycological Society*, **73**, 141–145.

Hocking, D. & Cook, F.D. (1972) Myxobacteria exert partial control of damping-off and root diseases in container-grown tree seedlings. *Canadian Journal of Microbiology*, **18**, 1557–1560.

Holland, D.M. (1988) *Studies on the biology of* Mycogone perniciosa *(Magnus) Delacroix*. PhD Thesis, University of Sheffield, UK.

Holland, D.M., Parker, H.L., Cooke, R.C. & Evans, G.H. (1985) Germination of bicellular conidia of *Mycogone perniciosa* the wet bubble pathogen of the cultivated mushroom. *Transactions of the British Mycological Society*, **85**, 730–735.

Homma, Y. (1984) Perforation and lysis of hyphae of *Rhizoctonia solani* and conidia of *Cochliobolus miyabeanus* by soil Myxobacteria. *Phytopathology*, **74**, 1234–1239.

Homma, Y. & Cook, R.J. (1985) Influence of matric and osmotic water potentials and soil pH on the activity of giant vampyrellid amoebae. *Phytopathology*, **75**, 243–246.

Homma, Y & Ishii, M. (1984) Perforation of hyphae and sclerotia of *Rhizoctonia solani* Kühn by mycophagous soil amoebae from vegetable field soil in Japan. *Annals of the Phytopathological Society of Japan*, **50**, 229–240.

Homma, Y., Sato, Z., Hirama, F., Konno, K., Shirahama, H. & Suzui, T. (1989) Production of antibiotics by *Pseudomonas cepacia* as an agent for biological control of soil-borne plant pathogens. *Soil Biology & Biochemistry*, **21**, 723–728.

Homma, Y., Sitton, J.W., Cook, R.J. & Old, K.M. (1979) Perforation and destruction of pigmented hyphae of *Gaeumannomyces graminis* by vampyrellid amoebae from Pacific Northwest wheat field soils. *Phytopathology*, **69**, 1118–1122.

Honda, Y. & Aragaki, M. (1983) Light-dependence for fruiting body formation and its inheritance in *Phoma caricae-papayae*. *Mycologia*, **75**, 22–29

Horn, B.W. (1989) Requirement for potassium and pH shift in host-mediated sporangiospore extrusion from trichospores of *Smittium culisetae* and other *Smittium* species. *Mycological Research*, **93**, 303–313.

Horton, J.S. & Horgen, P.A. (1989) Molecular cloning of cDNAs regulated during steroid-induced sexual differentiation in the aquatic fungus *Achlya*. *Experimental Mycology*, **13**, 263–273.

Horwitz, B.A., Trad, C.H. & Lipson, E.D. (1986) Modified light-induced changes in *dim Y* photoresponse mutants of *Trichoderma*. *Plant Physiology*, **81**, 726–730.

Howard, J.J., Cazin Jr, J. & Wiemer, D.F. (1988) Toxicity of terpenoid deterrents to the leafcutting ant *Atta cephalotes* and its mutualistic fungus. *Journal of Chemical Ecology*, **14**, 59–69.

Howard, R.J. & Aist, J.R. (1980) Cytoplasmic microtubules and fungal morphogenesis: ultrastructural effects of methyl benzimidazole-2-yl carbamate determined by freeze-substitution of hyphal tips. *Journal of Cell Biology*, **87**, 55–64.

Howell, C.R., Beier, R.C. & Stipanovic, R.D. (1988) Production of ammonia by *Enterobacter cloacae* and its possible role in the biological control of *Pythium* pre-emergence damping-off by the bacterium. *Phytopathology*, **78**, 1075–1078.

Howell, C.R. & Stipanovic, R.D. (1980) Suppression of *Pythium ultimum*-induced damping-off of cotton seedlings by *Pseudomonas fluorescens* and its antibiotic, pyoluteorin. *Phytopathology*, **70**, 712–715.

Hubbell, S.P., Wiemer, D.F. & Adeboye, A. (1983) An antifungal terpenoid defends a neotropical tree (Hymenaea) against attack by fungus-growing ants (Atta). *Oecologia*, **60**, 321–327.

Humpherson-Jones, F.M. & Cooke, R.C. (1977a)

Morphogenesis in sclerotium-forming fungi. I. Effects of light on *Sclerotinia sclerotiorum*, *Sclerotium delphinii* and *S. rolfsii*. *New Phytologist*, **78**, 171–180.
Humpherson-Jones, F.M. & Cooke, R.C. (1977b) Morphogenesis in sclerotium-forming fungi. II. Rhythmic production of sclerotia by *Sclerotinia sclerotiorum* (Lib.) de Bary. *New Phytologist*, **78**, 181–187.
Humpherson-Jones, F.M. & Cooke, R.C. (1977c) Induction of sclerotium formation by acid staling compounds in *Sclerotinia sclerotiorum* and *Sclerotium rolfsii*. *Transactions of the British Mycological Society*, **78**, 413–420.
Hunsley, D. & Gooday, G.W. (1974) The structure and development of septa in *Neurospora crassa*. *Protoplasma*, **82**, 125–146.
Hunsley, D. & Kay, D. (1976) Wall structure of the *Neurospora* hyphal apex: immunoflurescent localization of wall surface antigens. *Journal of General Microbiology*, **95**, 233–248.
Hutchinson, S.A., Sharma, P., Clarke, K.R. & MacDonald, I. (1980) Control of hyphal orientation in colonies of *Mucor hiemalis*. *Transactions of the British Mycological Society*, **75**, 177–191.
Hütter, R. & DeMoss, J.A. (1967) Organization of the tryptophan pathway: a phylogenetic study of the fungi. *Journal of Bacteriology*, **94**, 1896–1907.
Hwang, K., Stelzig, D.A., Barnett, H.L., Roller, P.P. & Kelsey, M.I. (1985) Partial purification of the growth factor mycotrophein. *Mycologia*, **77**, 109–113.
Hyakumachi, M. & Lockwood, J.L. (1989) Relation of carbon loss from sclerotia of *Sclerotium rolfsii* during incubation in soil to decreased germinability and pathogenic aggressiveness. *Phytopathology*, **79**, 1059–1063.
Hyde, H.A. & Williams, D.A. (1953) The incidence of *Cladosporium herbarum* in the outdoor air at Cardiff, 1949–50. *Transactions of the British Mycological Society*, **36**, 260–266.
Ikediugwu, F.E.O., Dennis, C. & Webster, J. (1970) Hyphal interference by *Peniophora gigantea* against *Heterobasidion annosum*. *Transactions of the British Mycological Society*, **54**, 307–309.
Ikediugwu, F.E.O. & Webster, J. (1970a) Antagonism between *Coprinus heptemerus* and other coprophilous fungi. *Transactions of the British Mycological Society*, **54**, 181–204.
Ikediugwu, F.E.O. & Webster, J. (1970b) Hyphal interference in a range of coprophilous fungi. *Transactions of the British Mycological Society*, **54**, 205–210.
Ilott, T.W., Ingram, D.S. & Rawlinson, C.J. (1986) Evidence of a chemical factor in the control of sexual development in the light leaf spot fungus, *Pseudopeziza brassicae* (Ascomycotina). *Transactions of the British Mycological Society*, **87**, 303–308.
Inch, J.M.M. & Trinci, A.P.J. (1987) Effects of water activity on growth and sporulation of *Paecilomyces farinosus* in liquid and solid media. *Journal of General Microbiology*, **133**, 247–252.
Ingold, C.T. (1933) Spore discharge in Ascomycetes. I. Pyrenomycetes. *New Phytologist*, **32**, 175–196.
Ingold, C.T. (1959) Spore discharge in Pyrenomycetes. *Friesia*, **6**, 148–163.
Ingold, C.T. (1965) *Spore Liberation*, Oxford, Clarendon Press/London, Oxford University Press.
Ingold, C.T. (1978) Role of mucilage in dispersal of certain fungi. *Transactions of the British Mycological Society*, **70**, 137–173.
Ingold, C.T. (1985) Observations on spores and their germination in certain Heterobasidiomycetes. *Transactions of the British Mycological Society*, **85**, 417–423.
Ingold, C.T. (1988) Chlamydospore formation during basidiospore germination in *Marasmius oreades*. *Transactions of the British Mycological Society*, **90**, 495–496.
Ingold, C.T. & Dring, V.J. (1957) An analysis of spore discharge in *Sordaria*. *Annals of Botany*, **21**, 465–477.
Ingold, C.T. & Marshall, B. (1963) Further observations on light and spore discharge in certain Pyrenomycetes. *Annals of Botany*, **27**, 481–491.
Ingold, C.T. & Peach, J. (1970) Further observations on fruiting in *Sphaerobolus* in relation to light. *Transactions of the British Mycological Society*, **54**, 211–220.
Irvine, J.A., Dix, N.J. & Warren, R.C. (1978) Inhibitory substances in *Acer platanoides* leaves: seasonal activity and effects on growth of phylloplane fungi. *Transactions of the British Mycological Society*, **70**, 363–371.
Irving, H.R. & Grant, B.R. (1984) The effects of pectin and plant root surface carbohydrates on encystment and development of *Phytophthora cinnamomi* zoospores. *Journal of General Microbiology*, **130**, 1015–1018.
Iwanami, Y. (1978) Myrmicacin, a new inhibitor for mitotic progression after metaphase. *Protoplasma*, **95**, 267–271.
Iwata, Y. (1957) On the chemotaxis of zoospores of *Pseudoperonospora cubensis* (B. & C.) Rostow. *Annals of the Phytopathological Society of Japan*, **22**, 108–110.
Jaenike, J. (1985) Parasite pressure and the evolution of amanita tolerance in *Drosophila*. *Evolution*, **39**, 1295–1301.
Jaffee, B.A. & Zehr, E.I. (1982) Parasitism of the

nematode *Criconemella xenoplax* by the fungus *Hirsutella rhossiliensis. Phytopathology*, **72**, 1378–1381.

Jagger, J. (1958) Photoreactivation. *Bacteriology Reviews*, **22**, 99–142.

Jan, Y.N. (1974) Properties and cellular localization of chitin synthetase in *Phycomyces blakesleeanus. Journal of Biological Chemistry*, **249**, 1973–1979.

Jansson, H.-B., Jeyaprakash, A. & Zuckerman, B.M. (1985) Differential adhesion and infection of nematodes by the endoparasitic fungus *Meria coniospora* (Deuteromycetes). *Applied and Environmental Microbiology*, **49**, 552–555.

Jansson, H.-B., Johansson, T., Nordbring-Hertz, B., Tunlid, A. & Odham, G. (1988) Chemotropic growth of germ-tubes of *Cochliobolus sativus* to barley roots or root exudates. *Transactions of the British Mycological Society*, **90**, 647–650.

Jansson, H.-B. & Nordbring-Hertz, B. (1979) Attraction of nematodes to living mycelium of nematophagous fungi. *Journal of General Microbiology*, **112**, 89–91.

Jansson, H.-B. & Nordbring-Hertz, B. (1980) Interactions between nematophagous fungi and plant-parasitic nematodes: attraction, induction of trap formation and capture. *Nematologica*, **26**, 383–389.

Jansson, H.-B. & Nordbring-Hertz, B. (1983) The endoparasitic nematophagous fungus *Meria coniospora* infects nematodes specifically at the chemosensory organs. *Journal of General Microbiology*, **129**, 1121–1126.

Jansson, H.-B. & Nordbring-Hertz, B. (1984) Involvement of sialic acid in nematode chemotaxis and infection by an endoparasitic nematophagous fungus. *Journal of General Microbiology*, **130**, 39–43.

Janzen, D.H. (1977) Why fruits rot, seeds mold and meat spoils. *American Naturalist*, **111**, 691–713.

Jasalavich, C.A., Hyakumachi, M. & Lockwood, J.L. (1990) Loss of endogenous carbon by conidia of *Cochliobolus sativus* exposed to soil and its effect on conidial germination and pathogenic aggressiveness. *Soil Biology and Biochemistry*, **22**, 761–767.

Jaworski, A.J. & Harrison, J.A. (1986) RNA synthesised during late sporulation is required for germtube formation in *Blastocladiella emersonii. Experimental Mycology*, **10**, 42–51.

Jeffries, P. (1985) Mycoparasitism within the Zygomycetes. *Botanical Journal of the Linnean Society*, **91**, 135–150.

Jennings, D.H. (1979) Membrane transport and hyphal growth. In *Fungal Walls and Hyphal Growth*, eds Burnett, J.H. & Trinci, A.P.J., pp. 279–294. Cambridge, Cambridge University Press.

Jennings, D.H. (1984a) Water flow through mycelia. In *The Ecology and Physiology of the Fungal Mycelium*, eds Jennings, D.H. & Rayner, A.D.M., pp. 143–164. Cambridge, Cambridge University Press.

Jennings, D.H. (1984b) Polyol metabolism in fungi. *Advances in Microbial Physiology*, **25**, 149–193.

Jennings, D.H. (1987) Translocation of solutes in fungi. *Biological Reviews*, **62**, 215–243.

Jensen, J.D. (1983) The development of *Diaporthe phaseolorum* variety *sojae* in culture. *Mycologia*, **75**, 1074–1091.

Jesaitis, A.J. (1974) Linear dichroism and orientation of the *Phycomyces* photopigment. *Journal of General Physiology*, **63**, 1–21.

Jeves, T.M. & Coley-Smith, J.R. (1980) Germination of sclerotia of *Stromatinia gladioli. Transactions of the British Mycological Society*, **74**, 13–18.

Joblin, K.N., Naylor, G.E. & Williams, A.G. (1990) Effect of *Methanobrevibacter smithii* on xylanolytic activity of anaerobic ruminal fungi. *Applied and Environmental Microbiology*, **56**, 2287–2295.

Johnson, C.L. & Preece, T.F. (1979) Natural history of *Pilobolus kleinii*: experiments with sporangia in the cow digestive tract and the early stages of growth. *Transactions of the British Mycological Society*, **72**, 453–457.

Johnson, F.S., Mo, T. & Green, A.E. (1976) Average latitudinal variation in ultraviolet radiation at the Earth's surface. *Photochemistry and Photobiology*, **23**, 179–188.

Johnson, G.I. (1988) Inhibition of germination of sporangia of *Peronospora hyoscyami* by cation deprivation: the effects of substrate and chelating agents. *Plant Pathology*, **87**, 125–130.

Johnson, G.L. & Gamow, R.I. (1971) The avoidance response in *Phycomyces. Journal of General Microbiology*, **57**, 41–49.

Johnson, L.F. & Osborne, T.S. (1964) Survival of fungi in soil exposed to gamma radiation. *Canadian Journal of Botany*, **42**, 105–113.

Johnson, M.C., Dahlman, D.L., Siegel, M.R., Bush, L.P., Latch, G.C.M., Potter, D.A. & Varney, D.R. (1985) Insect feeding deterrents in endophyte-infected tall fescue. *Applied and Environmental Microbiology*, **49**, 568–571.

Johnson, S.A. & Lovett, J.S. (1984) Gene expression during development of *Blastocladiella emersonii. Experimental Mycology*, **8**, 132–145.

Jones, E.B.G. & Byrne, P.J. (1976) Physiology of higher marine fungi. In *Recent Advances in Aquatic Mycology*, ed. Jones, E.B.G., pp. 135–175. London, Elek Science.

Jones, R.P. & Gadd, G.M. (1990) Ionic nutrition of yeast–physiological mechanisms involved and implications for biotechnology. *Enzyme and Microbial Technology*, **12**, 402–418.

Jong, S.-C. & Donovick, R. (1989) Antitumor and antiviral substances from fungi. *Advances in Applied Microbiology*, **34**, 183–262.

Jouvenaz, D.P. & Kimbrough, J.W. (1991) *Myrmecomyces annellisae* gen. nov., sp. nov. (Deuteromycotina:Hyphomycetes), an endoparasitic fungus of fire ants, *Solenopsis* spp. (Hymenoptera:Formicidae). *Mycological Research*, **95**, 1395–1401.

Jurzitza, G. (1979) The fungi symbiotic with anobiid beetles. In *Insect–Fungus Symbioses*, ed. Batra, L.R., pp. 65–79. Montclair, New Jersey, Allanheld, Osmun.

Kalisz, H.M., Wood, D.A. & Moore, D. (1987) Production, regulation and release of extracellular proteinase activity in Basidiomycete fungi. *Transactions of the British Mycological Society*, **88**, 221–227.

Kaplan, J.D. & Goos, R.D. (1982) The effect of water potential on zygospore formation in *Syzygites megalocarpus*. *Mycologia*, **74**, 684–686.

Kasatkina, T.B., Zharikova, G.G. & Rubin, A.B. (1980) Morphogenesis of *Polyporus ciliatus* Fr. ex Fr. fruit bodies during space travel. *Mikologia i Fitopatologia*, **14**, 193–197.

Kemp, R.F.C. (1970) Inter-specific sterility in *Coprinus bisporus*, *C. congregatus* and other Basidiomycetes. *Transactions of the British Mycological Society*, **54**, 488–489.

Kennedy, M.J. (1991) *Candida* blastospore adhesion, association, and invasion of the gastrointestinal tract of vertebrates. In *The Fungus Spore and Disease Initiation in Plants and Animals*, eds Cole, G.T. & Hoch, H.C., pp. 157–180. New York, Plenum.

Kerekes, R. & Nagy, G. (1980) Membrane lipid composition of a mesophilic and psychrophilic yeast. *Acta Alimentaria*, **9**, 93–98.

Kerwin, J.L. (1984) Fatty acid regulation of the germination of *Erynia variabilis* conidia on adults and puparia of the lesser housefly, *Fannia canicularis*. *Canadian Journal of Microbiology*, **30**, 158–161.

Kerwin, J.L. & Washino, R.K. (1984) Cyclic nucleotide regulation of oosporogenesis by *Lagenidium giganteum* and related fungi. *Experimental Mycology*, **8**, 215–224.

Kerwin, J.L. & Washino, R.K. (1986) Oosporogenesis by *Lagenidium giganteum*: induction and maturation are regulated by calcium and calmodulin. *Canadian Journal of Microbiology*, **32**, 663–672.

Khairi, S.M. & Preece, T.F. (1978) Hawthorn powdery mildew: diurnal and seasonal distribution of conidia in air near infected plants. *Transactions of the British Mycological Society*, **71**, 395–397.

Khew, K.L. & Zentmeyer, G.A. (1973) Chemotactic response of zoospores of five species of *Phytophthora*. *Phytopathology*, **63**, 1511–1517.

King, J.E. & Coley-Smith, J.R. (1969) Production of volatile alkyl sulphides by microbial degradation of alliin and alliin-like compounds in relation to germination of sclerotia of *Sclerotium cepivorum* Berk. *Annals of Applied Biology*, **64**, 303–314.

Kirfman, G.W., Brandenburg, R.L. & Garner, G.B. (1986) Relationship between insect abundance and endophyte infestation level in tall fescue in Missouri. *Journal of the Kansas Entomological Society*, **59**, 552–554.

Kirk, P.W. & Gordon, A.S. (1988) Hydrocarbon degradation by filamentous marine higher fungi. *Mycologia*, **80**, 776–782.

Kirk, T.K. & Fenn, P. (1982) Formation and action of the ligninolytic system in basidiomycetes. In *Decomposer Basidiomycetes: their Biology and Ecology*, eds Frankland, J.C., Hedger, J.N. & Swift, M.J., pp. 67–90. Cambridge, Cambridge University Press.

Kirk, T.K., Schultz, E., Connors, W.J., Lorenz, L.F. & Zeikus, J.G. (1978) Influence of culture parameters on lignin metabolism by *Phanerochaete chrysosporium*. *Archives of Microbiology*, **117**, 277–285.

Klionsky, D.J., Herman, P.K. & Emr, S.D. (1990) The fungal vacuole: composition, function and biogenesis. *Microbiological Reviews*, **54**, 266–292.

Knights, I.K. & Lucas, J.A. (1980) Photosensitivity of *Puccinia graminis* f. sp. *tritici* urediniospores *in vitro* and on the leaf surface. *Transactions of the British Mycological Society*, **74**, 543–549.

Knights, I.K. & Lucas, J.A. (1981) Photocontrol of *Puccinia graminis* f. sp. *tritici* urediniospore germination in the field. *Transactions of the British Mycological Society*, **77**, 519–527.

Ko, W.H. (1985) Stimulation of sexual reproduction of *Phytophthora cactorum* by phospholipids. *Journal of General Microbiology*, **131**, 2591–2594.

Koch, E. & Hoppe, H.H. (1987) Effect of light on uredospore germination and germ tube growth of soybean rust (*Phakopsora pachyrhizi* Syd.). *Journal of Phytopathology*, **119**, 64–74.

Kolattukudy, P.E. (1985) Enzymic penetration of the plant cuticle by fungal pathogens. *Annual Review of Phytopathology*, **23**, 223–250.

Koltin, Y., Finkler, A. & Ben-Zvi. B.-S. (1987) Double-stranded RNA viruses of pathogenic fungi: virulence and plant protection. In *Fungal*

Infection of Plants, eds Pegg, G.F. & Ayres, P.G., pp. 334–348. Cambridge, Cambridge University Press.

Korn, E.D., Greenblatt, C.L. & Lees, A.M. (1965) Synthesis of unsaturated fatty acids in the slime mould *Physarum polycephalum* and the zooflagellates *Leishmania tarentolae*, *Trypanosoma lewisi* and *Crithidia* sp.: a comparative study. *Journal of Lipid Research*, **6**, 43–50.

Koske, R.E. (1981) Multiple germination by spores of *Gigaspora gigantea*. *Transactions of the British Mycological Society*, **76**, 328–330.

Koske, R.E. (1982) Evidence for a volatile attractant from plant roots affecting germ tubes of a VA mycorrhizal fungus. *Transactions of the British Mycological Society*, **79**, 305–310.

Kramer, C.L., Pady, S.M., Clary, R. & Haard, R. (1968) Diurnal periodicity in aeciospore release of certain rusts. *Transactions of the British Mycological Society*, **51**, 679–687.

Krátký, Z. & Biely, P. (1980) Inducible β-xylosidase permease as a constituent of the xylan-degrading enzyme system of the yeast *Cryptococcus albidus*. *European Journal of Biochemistry*, **112**, 367–373.

Kueh, T.K. & Khew, K.L. (1982) Survival of *Phytophthora palmivora* in soil and after passing through alimentary canals of snails. *Plant Disease*, **66**, 897–899.

Kuhlman, E.G. (1987) Temperature, inoculum type, and leaf maturity affect urediospore production by *Cronartium quercuum* f. sp. *fusiforme*. *Mycologia*, **79**, 405–409.

Kukor, J.J. & Martin, M.M. (1987) Nutritional ecology of fungus-feeding arthropods. In *Nutritional Ecology of Insects, Mites, Spiders and Related Invertebrates*, eds Slansky Jr, F. & Rodriguez, J.G., pp. 791–814. New York, Wiley.

Kumagai, T. (1988) Mycochrome system in the reversible photoinduction of conidiation of *Helminthosporium oryzae*, with special reference to intermittent photo stimuli. *Experimental Mycology*, **12**, 28–34.

Kumagai, T. & Oda, Y. (1969) An action spectrum for photoinduced sporulation in the fungus *Trichoderma viride*. *Plant and Cell Physiology (Tokyo)*, **10**, 387–392.

Kuthubutheen, A.J. & Webster, J. (1986a) Water availability and the coprophilous fungus succession. *Transactions of the British Mycological Society*, **86**, 63–76.

Kuthubutheen, A.J. & Webster, J. (1986b) Effects of water availability on germination, growth and sporulation of coprophilous fungi. *Transactions of the British Mycological Society*, **86**, 77–91.

Lange, L. & Olson, L.W. (1983) The fungal zoospore – its structure and biological significance. In *Zoosporic Plant Pathogens – A Modern Perspective*, ed. Buczacki, S.T., pp. 1–42. London Academic Press.

Lapp, M.S. & Skoropad, W.P. (1978) Location of appressoria of *Colletotrichum graminicola* on natural and artifical barley leaf surfaces. *Transactions of the British Mycological Society*, **70**, 225–228.

Latham, M.J. (1979) The animal as an environment. In *Microbial Ecology: A Conceptual Approach*, eds Lynch, J.M. & Poole, N.J., pp. 115–137. Oxford, Blackwell Scientific.

Latorre, B.A., Yanez, P. & Rauld, E. (1985) Factors affecting the release of ascospores by the pear scab fungus (*Venturia pirina*). *Plant Disease*, **69**, 213–216.

Lawrey, J.D. (1980) Correlations between lichen secondary chemistry and grazing activity by *Pallifera varia*. *Bryologist*, **83**, 328–334.

Lax, A.R., Templeton, G.C. & Meyer, W.L. (1985) Isolation, purification and biological activity of a self-inhibitor from conidia of *Colletotrichum gloeosporioides*. *Phytopathology*, **75**, 386–390.

Lazarus, C.M., Earl, A.J., Turner, G. & Küntzel, H. (1980) Amplification of a mitochondrial DNA sequence in the cytoplasmically inherited 'ragged' mutant of *Aspergillus amstelodami*. *European Journal of Biochemistry*, **106**, 633–641.

Leach, C.M. & Trione, E.J. (1965) An action spectrum for light-induced sporulation in the fungus *Ascochyta pisi*. *Plant Physiology*, **40**, 808–812.

Leach, C.M. & Trione, E.J. (1966) Action spectra for light-induced sporulation of the fungi *Pleospora herbarum* and *Alternaria dauci*. *Photochemistry and Photobiology*, **5**, 621–630.

Leake, J.R. & Read, D.J. (1990) Proteinase activity in mycorrhizal fungi. I. The effect of extracellular pH on the production and activity of proteinase by ericoid endophytes from soils of contrasted pH. *New Phytologist*, **115**, 243–250.

Leatham, G.F. & Stahmann, M.A. (1987) Effect of light and aeration on fruiting of *Lentinula edodes*. *Transactions of the British Mycological Society*, **88**, 9–20.

Leathers, C.R. (1961) Comparative survival of rehydrated and nonrehydrated wheat stem rust uredospores on dry leaf surfaces. *Phytopathology*, **51**, 410–411.

Leggett, M.E. & Rahe, J.E. (1985) Factors affecting the survival of sclerotia of *Sclerotium cepivorum* in the Fraser Valley of British Columbia. *Annals of Applied Biology*, **106**, 255–263.

Leggett, M.E., Rahe, J.E. & Utkhede, R.S. (1983) Survival of sclerotia of *Sclerotium cepivorum* in muck soil as affected by drying and location of sclerotia in the soil. *Soil Biology & Biochemistry*, **15**, 325–327.

Léjohn, H.B. (1974) Biochemical parameters of fungal phylogenetics. *Evolutionary Biology*, **7**, 79–125.

Léjohn, H.B. & Braithwaite, C.E. (1984) Heat and nutritional shock-induced proteins of the fungus *Achlya* are different and under independent transcriptional control. *Canadian Journal of Biochemistry and Cell Biology*, **62**, 837–846.

Leong, J. (1986) Siderophores: their biochemistry and possible role in the biocontrol of plant pathogens. *Annual Review of Phytopathology*, **24**, 189–209.

Lesuisse, E. & Labbe, P. (1989) Reductive and non-reductive mechanisms of iron assimilation by the yeast *Saccharomyces cerevisiae. Journal of General Microbiology*, **135**, 257–263.

Lewis, B.G. & Day, J.R. (1972) Behaviour of uredospore germ-tubes of *Puccinia graminis tritici* in relation to the fine structure of wheat leaf surfaces. *Transactions of the British Mycological Society*, **58**, 139–145.

Lewis, D.H. (1973) Concepts in fungal nutrition and the origin of biotrophy. *Biological Reviews*, **48**, 261–268.

Lewis, D.H. (1974) Micro-organisms and plants: the evolution of parasitism and mutualism. *Symposia of the Society for General Microbiology*, **24**, 367–392.

Lewis, D.H. (1987) Evolutionary aspects of mutualistic associations between fungi and photosynthetic organisms. In *Evolutionary Biology of the Fungi*, eds Rayner, A.D.M., Brasier, C.M. & Moore, D., pp. 161–178. Cambridge, Cambridge University Press.

Lewis, K., Whipps, J.M. & Cooke, R.C. (1989) Mechanisms of biological disease control with special reference to the case study of *Pythium oligandrum* as an antagonist. In *Biotechnology of Fungi for Improving Plant Growth*, eds Whipps, J.M. & Lumsden, R.D., pp. 191–217. Cambridge, Cambridge University Press.

Li, J., Heath, I.B. & Cheng, K.-J. (1991) The development and zoospore ultrastructure of a polycentric chytridiomycete gut fungus, *Orpinomyces joyonii* comb. nov. *Canadian Journal of Botany*, **69**, 580–589.

Liddell, C.M. & Burgess, L.W. (1985) Survival of *Fusarium moniliforme* at controlled temperature and relative humidity. *Transactions of the British Mycological Society*, **84**, 121–130.

Lilly, V.G. & Barnett, H.L. (1949) The influence of concentrations of nutrients, thiamin and biotin upon growth and formation of perithecia and ascospores by *Chaetomium convolutum. Mycologia*, **41**, 186–196.

Lim, W.C. & Lockwood, J.L. (1988) Chemotaxis of some phytopathogenic bacteria to fungal propagules *in vitro* and in soil. *Canadian Journal of Microbiology*, **34**, 196–199.

Linz, J.E. & Orlowski, M. (1987) Regulation of gene expression during aerobic germination of *Mucor racemosus* sporangiospores. *Journal of General Microbiology*, **133**, 141–148.

Lockwood, J.L. (1988) Evolution of concepts associated with soilborne plant pathogens. *Annual Review of Phytopathology*, **26**, 93–121.

Lockwood, J.L. & Filonow, A.B. (1981) Responses of fungi to nutrient-limiting conditions and to inhibitory substances in natural habitats. In *Advances in Microbial Ecology*, Vol. 5, ed. Alexander, M., pp. 1–61. New York, Plenum.

Loffler, H.J.M. & Schippers, B. (1984) Ammonia-induced mycostasis is not mediated by enhanced release of carbon compounds. *Canadian Journal of Botany*, **30**, 1038–1041.

Long, P.E. & Jacobs, L. (1974) Aseptic fruiting of the cultivated mushroom, *Agaricus bisporus. Transactions of the British Mycological Society*, **63**, 99–107.

Louis, I. & Cooke, R.C. (1983) Influence of the conidial matrix of *Sphaerellopsis filum* (*Darluca filum*) on spore germination. *Transactions of the British Mycological Society*, **81**, 667–670.

Louis, I. & Cooke, R.C. (1985) Conidial matrix and spore germination in some plant pathogens. *Transactions of the British Mycological Society*, **84**, 661–667.

Luard, E.J. (1982a) Accumulation of intracellular solutes by two filamentous fungi in response to growth at low steady state osmotic potential. *Journal of General Microbiology*, **128**, 2563–2574.

Luard, E.J. (1982b) Growth and accumulation of solutes by *Phytophthora cinnamomi* and other lower fungi in response to changes in external solute potential. *Journal of General Microbiology*, **128**, 2583–2590.

Lucas, J.A., Kendrick, R.E. & Givan, C.V. (1975) Photocontrol of fungal spore germination. *Plant Physiology*, **56**, 847–849.

Lutchmeah, R.S. & Cooke, R.C. (1984) Aspects of antagonism by the mycoparasite *Pythium oligandrum. Transactions of the British Mycological Society*, **83**, 696–700.

Luttrell, E.S. (1974) Parasitism of fungi on vascular plants. *Mycologia*, **66**, 1–15.

Lutz, A. & Menge, J. (1986) Breaking winter dormancy of *Phytophthora parasitica* propagules using heat shock. *Mycologia*, **78**, 148–150.

Lyons, P.C., Plattner, R.D. & Bacon, C.W. (1986) Occurrence of peptide and clavine ergot alkaloids in tall fescue grass. *Science (New York)*, **232**, 487–489.

Lysek, G. (1974) Zonierungen und Hexenringe,

Morphologische Differenzierungen bei Pilzen (Zonation and fairy rings, morphological differentiation in fungi). *Naturwissenschaftliche Runschau*, **27**, 449–455.

Lysek, G. (1984) Physiology and ecology of rhythmic growth and sporulation in fungi. In *The Ecology and Physiology of the Fungal Mycelium*, eds Jennings, D.H. & Rayner, A.D.M., pp. 323–342. Cambridge, Cambridge University Press.

Lysek G. & Nordbring-Hertz, B. (1981) An endogenous rhythm of trap formation in the nematophagous fungus *Arthrobotrys oligospora. Planta*, **152**, 50–53.

McBride, R.P. (1972) Larch leaf waxes utilized by *Sporobolomyces roseus in situ. Transactions of the British Mycological Society*, **58**, 329–352.

McClaren, D.L., Huang, H.C., Rimmer, S.R. & Kokko, E.G. (1989) Ultrastructural studies on infection of sclerotia of *Sclerotinia sclerotiorum* by *Talaromyces flavus. Canadian Journal of Botany*, **67**, 2199–2205.

McCracken, A.R. & Swinburne, T.R. (1979) Siderophores produced by saprophytic bacteria as stimulants of germination of conidia of *Colletotrichum musae. Physiological and Molecular Plant Pathology*, **15**, 331–340.

McCracken, A.R. & Swinburne, T.R. (1980) Effect of bacteria isolated from surface of banana fruits on germination of *Colletotrichum musae* conidia. *Transactions of the British Mycological Society*, **74**, 212–214.

Macfarlane, I. (1970) Germination of resting spores of *Plasmodiophora brassicae. Transactions of the British Mycological Society*, **55**, 97–112.

McIlveen, W.D. & Cole Jr, H. (1976) Spore dispersal of Endogonaceae by worms, ants, wasps and birds. *Canadian Journal of Botany*, **54**, 1486–1489.

McKee, N.D. & Robinson, P.M. (1988) Production of volatile inhibitors of germination and hyphal extension by *Geotrichum candidum. Transactions of the British Mycological Society*, **91**, 157–160.

McKeen, W.E. (1970) Lipid in *Erysiphe graminis hordei* and its possible role during germination. *Canadian Journal of Microbiology*, **16**, 1041–1044.

McKeen, W.E., Mitchell, N. & Smith, R. (1967) The *Erysiphe cichoracearum* conidium. *Canadian Journal of Botany*, **48**, 1489–1496.

Macko, V. (1981) Inhibitors and stimulants of spore germination and infection structure formation in fungi. In *The Fungal Spore: Morphogenetic Controls*, eds Turian, G. & Hohl, H.R., pp. 565–584. London, Academic Press.

McLean, K.M. & Prosser, J.I. (1987) Development of vegetative mycelium during colony growth of *Neurospora crassa. Transactions of the British Mycological Society*, **88**, 489–495.

McMillan, J.H. (1976) Laboratory observations on the food preference of *Onychiurus armatus* (Tullb.) Gisin (Collembola, Family Onychiuridae). *Révue d'Écologie et de Biologie du Sol*, **13**, 353–364.

Macrae, A.R. (1983) Extracellular microbial lipases. In *Microbial Enzymes and Biotechnology*, ed. Fogarty, W.M., pp. 225–250. London, Elsevier Applied Science.

Madelin, M.F. (ed.) (1966) *The Fungus Spore*, London, Butterworths.

Madelin, M.F. (1984) Myxomycetes, microorganisms and animals: a model of diversity in animal–microbial interactions. In *Invertebrate–Microbial Interactions*, eds Anderson, J.M., Rayner, A.D.M. & Walton, D.H., pp. 1–33. Cambridge, Cambridge University Press.

Madelin, M.F., Toomer, D.K. & Ryan, J. (1978) Spiral growth of fungus colonies. *Journal of General Microbiology*, **106**, 73–80.

Magan, N. (1988) Effects of water potential and temperature on spore germination and germ-tube growth *in vitro* and on straw leaf sheaths. *Transactions of the British Mycological Society*, **90**, 97–107.

Magan, N. & Lacey, J. (1984a) Effect of temperature and pH on water relations of field and storage fungi. *Transactions of the British Mycological Society*, **82**, 71–81.

Magan, N. & Lacey, J. (1984b) Effects of gas composition and water activity on growth of field and storage fungi and their interactions. *Transactions of the British Mycological Society*, **82**, 305–314.

Maheshwari, R. & Hildebrandt, A.C. (1967) Directional growth of the urediospore germ tubes and stomatal penetration. *Nature (London)*, **214**, 1145–1146.

Malloch, D.H., Pirozynski, K.A. & Raven, P.H. (1980) Ecological and evolutionary significance of mycorrhizal symbioses in vascular plants (a review). *Proceedings of the National Academy of Sciences of the USA*, **77**, 2113–2118.

Manachère, G. (1980) Conditions essential for controlled fruiting of macromycetes–a review. *Transactions of the British Mycological Society*, **75**, 255–270.

Manachère, G., Robert, J.-C., Durand, R., Bret, J.-P. & Fevre, M. (1983) Differentiation in the Basidiomycetes. In *Fungal Differentiation. A Contemporary Synthesis*, ed. Smith, J.E., pp. 481–514. New York, Marcel Dekker.

Manavathu, E.K. & Thomas, D. des S. (1985) Chemotropism of *Achlya ambisexualis* to methionine and methionyl compounds. *Journal of General Microbiology*, **131**, 751–756.

Mani, K. & Swamy, R.N. (1981) Light-induced changes in permeability in two groups of fungi. *Experimental Mycology*, **5**, 292–294.

Manocha, M.S. (1985) Specificity of mycoparasite attachment to the host cell surface. *Canadian Journal of Botany*, **63**, 772–778.

Manocha, M.S. & McCullough, C.M. (1985) Suppression of host cell wall synthesis at penetration sites in a compatible interaction with a mycoparasite. *Canadian Journal of Botany*, **63**, 967–973.

Marek, L.E. (1984) Light affects *in vitro* development of gametangia and sporangia of *Monoblepharis macandra* (Chytridiomycetes, Monoblepharidales). *Mycologia*, **76**, 420–425.

Maresca, B. & Kobayashi, G.S. (1989) Dimorphism in *Histoplasma capsulatum*: a model for the study of cell differentiation in pathogenic fungi. *Microbiological Reviews*, **53**, 186–209.

Martin, M.M. (1984) The role of ingested enzymes in the digestive processes of insects. In *Invertebrate–Microbial Interactions*, eds Anderson, J.M., Rayner, A.D.M. & Walton, D.W.H., pp. 155–172. Cambridge, Cambridge University Press.

Mayo, K., Davis, R.E. & Motta, J. (1986) Stimulation of germination of spores of *Glomus versiforme* by spore-associated bacteria. *Mycologia*, **78**, 426–431.

Medina, J.R. & Cerdá-Olmedo, E. (1977) A quantitative model of *Phycomyces* phototropism. *Journal of Theoretical Biology*, **69**, 709–719.

Mehrotra, B.S. & Mehrotra, B.M. (1980) Induction of zygospores at high temperatures in the thermophilic species *Mucor miehei* with aspartic acid and phenylalanine. *Experientia*, **36**, 53–54.

Merek, E.L. & Fergus, C.L. (1954) The effect of temperature and relative humidity on, the longevity of spores of the oak wilt fungus. *Phytopathology*, **44**, 61–64.

Merrill, W. & Cowling, E.B. (1966) The role of nitrogen in wood deterioration: amount and distribution of nitrogen in fungi. *Phytopathology*, **56**, 1083–1090.

Meyer, J.R. & Linderman, R.G. (1986a) Response of subterranean clover to dual inoculation with vesicular–arbuscular mycorrhizal fungi and a plant growth-promoting bacterium *Pseudomonas putida. Soil Biology & Biochemistry*, **18**, 185–190.

Meyer, J.R. & Linderman, R.G. (1986b) Selective influence on populations of rhizoplane bacteria and Actinomycetes by mycorrhizas formed by *Glomus fasciculatum. Soil Biology & Biochemistry*, **18**, 191–196.

Milne, A., Theodorou, M.K., Jordan, M.G.C., King-Spooner, C. & Trinci, A.P.J. (1989) Survival of anaerobic fungi in feces, in saliva, and in pure culture. *Experimental Mycology*, **13**, 27–37.

Mirocha, C.J. & Devay, J.E. (1971) Growth of fungi on an inorganic medium. *Canadian Journal of Microbiology*, **17**, 1373–1378.

Mitchell, D.T. & Cooke, R.C. (1968) Some effects of temperature on germination and longevity in sclerotia of *Claviceps purpurea. Transactions of the British Mycological Society*, **51**, 721–729.

Mitchell, J.E. (1976) The effect of roots on the activity of soil-borne plant pathogens. In *Encyclopedia of Plant Physiology, New Series, Vol. 4, Physiological Plant Pathology*, eds Heitefuss, R. & Williams, P.H., pp. 94–128. Berlin, Springer.

Miyakawa, T., Tachikawa, T., Akada, R., Tsuchiya, E. & Fukui, S. (1986) Involvement of Ca^{2+}/calmodulin in sexual differentiation induced by mating pheromone rhodotorucine A in *Rhodosporidium toruloides. Journal of General Microbiology*, **132**, 1453–1457.

Mok, Y.W. & Barreto da Silva, M.S. (1984) Mycoflora of human dermal surfaces. *Canadian Journal of Microbiology*, **30**, 1205–1209.

Mol, P.C., Vermeulen, C.A. & Wessels, J.G.H. (1990) Diffuse extension in stipes of *Agaricus bisporus* may be based on a unique wall structure. *Mycological Research*, **94**, 480–488.

Mol, P.C. & Wessels, J.G.H. (1990) Differences in wall structure between substrate hyphae and hyphae of fruit-body stipes in *Agaricus bisporus. Mycological Research*, **94**, 472–479.

Monoson, H.L., Galsky, A.G., Griffin, J.A. & McGrath, E.J. (1973) Evidence for and partial characterization of a nematode attraction substance. *Mycologia*, **65**, 78–86.

Monte, E. & Garcia-Acha, I. (1988) Germination of conidia in *Phoma betae. Transactions of the British Mycological Society*, **91**, 133–139.

Montencourt, B.S. & Eveleigh, D.E. (1979) Production and characterization of high yielding cellulase mutants of *Trichoderma reesei*. In *Annual Meetings Proceedings of TAPPI*, pp. 101–108. Atlanta, Georgia, TAPPI.

Moore, D. (1981) Developmental genetics of *Coprinus cinereus*: genetic evidence that sclerotia and carpophores share a common pathway of initiation. *Current Genetics*, **3**, 145–150.

Moore-Landecker, E. (1987) Effects of medium composition and light on formation of apothecia and sclerotia by *Pyronema domesticum. Canadian Journal of Botany*, **65**, 2276–2279.

Morgan, W.M. (1983) Viability of *Bremia lactucae* oospores and stimulation of their germination by lettuce seedlings. *Transactions of the British Mycological Society*, **80**, 403–408.

Morrison, M., Mackie, R.I. & Kistner, A. (1990) Evidence that cellulolysis by an anaerobic ruminal fungus is catabolite regulated by glucose, cellobiose and soluble starch. *Applied and Environmental Microbiology*, **56**, 3227–3229.

Mortimer, P.H. & di Menna, M.E. (1983) Ryegrass staggers: further substantiation of a *Lolium* endophyte aetiology and the discovery of weevil resistance of ryegrass pastures infected with *Lolium* endophyte. *Proceedings of the New Zealand Grassland Association*, **44**, 240–243.

Moser, J.C., Perry, T.J. & Solheim, H. (1989) Ascospores hyperphoretic on mites associated with *Ips typographus*. *Mycological Research*, **93**, 513–517.

Moss, S.T. (1970) Trichomycetes inhabiting the digestive tract of *Simulium equinum* larvae. *Transactions of the British Mycological Society*, **54**, 1–13.

Moss, S.T. (1979) Commensalism of the Trichomycetes. In *Insect–Fungus Symbioses*, ed. Batra, L.R., pp. 175–227. Montclair, New Jersey, Allenheld, Osmun.

Motta, J.J. (1969) Cytology and morphogenesis in the rhizomorph of *Armillaria mellea*. *American Journal of Botany*, **56**, 610–619.

Motta, J.J. & Peabody, D.C. (1982) Rhizomorph cytology and morphogenesis in *Armillaria tabescens*. *Mycologia*, **74**, 671–674.

Mountfort, D.O. & Asher, R.A. (1983) Role of catabolite regulating mechanisms in control of carbohydrate utilization by the rumen anaerobic fungus *Neocallimastix frontalis*. *Applied and Environmental Microbiology*, **46**, 1331–1338.

Mowe, G., King, B. & Senn, S.J. (1983) Tropic responses of fungi to wood volatiles. *Journal of General Microbiology*, **129**, 779–784.

Muchovej, J.J., Della Lucia, T.M. & Muchovej, R.M.C. (1991) *Leucoagaricus weberi* sp. nov. from a live nest of leaf-cutting ants. *Mycological Research*, **95**, 1308–1311.

Muehlstein, L.K., Amon, J.P. & Leftler, D.L. (1987) Phototaxis in the marine fungus *Rhizophydium littoreum*. *Applied and Environmental Microbiology*, **53**, 1819–1821.

Muehlstein, L.K., Amon, J.P. & Leftler, D.L. (1988) Chemotaxis in the marine fungus *Rhizophydium littoreum*. *Applied and Environmental Microbiology*, **54**, 1668–1672.

Mugnier, J. & Mosse, B. (1987) Spore germination and viability of a vesicular arbuscular mycorrhizal fungus, *Glomus mosseae*. *Transactions of the British Mycological Society*, **88**, 411–413.

Mulder, J.L., Ghannoum, M.A., Khamis, L. & Abu Elteen, K. (1989) Growth and lipid composition of some dematiaceous Hyphomycete fungi grown at different salinities. *Journal of General Microbiology*, **135**, 3393–3404.

Murrin, F. & Nolan, R.A. (1987) Ultrastructure of the infection of spruce budworm larvae by the fungus *Entomophaga aulicae*. *Canadian Journal of Botany*, **65**, 1694–1706.

Musgrave, A., Ero, L., Scheffer, R. & Oehlers, E. (1977) Chemotropism of *Achlya ambisexualis* germ hyphae to casein hydrolysate and amino acids. *Journal of General Microbiology*, **101**, 65–70.

Musgrave, A. & Nieuwenhuis, D. (1975) Metabolism of radioactive antheridiol by *Achlya* species. *Archives of Microbiology*, **105**, 313–317.

Muthkumar, G. & Nickerson, K.W. (1984) Ca(II)-calmodulin regulation of fungal dimorphism in *Ceratocystis ulmi*. *Journal of Bacteriology*, **159**, 390–392.

Naiki, T. & Ui, T. (1975) Ultrastructure of sclerotia of *Rhizoctonia solani* Kühn invaded and decayed by soil microorganisms. *Soil Biology & Biochemistry*, **7**, 301–304.

Nair, M.S.R. & Hervey, A. (1979) Structure of lepiochlorin, an antibiotic metabolite of a fungus cultivated by ants. *Phytochemistry*, **18**, 326–327.

Nelson, E.B. (1987) Rapid germination of sporangia of *Pythium* species in response to volatiles from germinating seeds. *Phytopathology*, **77**, 1108–1112.

Nelson, E.B. (1991) Exudate molecules initiating fungal responses to seeds and roots. In *The Rhizosphere and Plant Growth*, eds Keister, D.L. & Cregan, P.B., pp. 197–209. Dordrecht, Kluwer Academic Publishers.

Nelson, E.B., Chao, W.-L., Norton, J.M., Nash, G.T. & Harman, G.E. (1986) Attachment of *Enterobacter cloacae* to hyphae of *Pythium ultimum*: possible role in the biological control of *Pythium* preemergence damping-off. *Phytopathology*, **76**, 327–335.

Nesbitt, H.J., Malajczuk, N. & Glenn, A.R. (1981) Bacterial colonization and lysis of *Phytophthora cinnamomi*. *Transactions of the British Mycological Society*, **77**, 47–54.

Newell, K. (1984a) Interaction between two decomposer Basidiomycetes and a collembolan under Sitka spruce: distribution, abundance and selective grazing. *Soil Biology & Biochemistry*, **16**, 227–233.

Newell, K. (1984b) Interaction between two decomposer Basidiomycetes and a collembolan under Sitka spruce: grazing and its potential effects on fungal distribution and litter decomposition. *Soil Biology & Biochemistry*, **16**, 235–239.

Newhook, F.J., Young, B.R., Allen, S.D. & Allen, R.N. (1981) Zoospore motility of *Phytophthora cinnamomi* in particulate substrates. *Phyto-*

pathologische Zeitschrift, **101**, 202–209.

Newsted, W.J. & Huner, N.P.A. (1988) Major sclerotial polypeptides of psychrophilic fungi: temperature regulation of *in vivo* synthesis in vegetative hyphae. *Canadian Journal of Botany*, **66**, 1755–1761.

Nicholson, R.L. & Moraes, W.B.C. (1980) Survival of *Colletotrichum graminicola*: importance of the spore matrix. *Phytopathology*, **70**, 225–261.

North, M.J. (1982) Comparative biochemistry of the proteinases of eukaryotic microorganisms. *Microbiological Reviews*, **46**, 308–340.

Norton, J.M. & Harman, G.E. (1985) Responses of soil microorganisms to volatile exudates from germinating pea seeds. *Canadian Journal of Botany*, **63**, 1040–1045.

Nuss, D.L. & Koltin, Y. (1990) Significance of dsRNA genetic elements in plant pathogenic fungi. *Annual Review of Phytopathology*, **28**, 37–58.

Odds, F.C. (1991) Potential for penetration of passive barriers to fungal invasion in humans. In *The Fungal Spore and Disease Initiation in Plants and Animals*, eds Cole, G.T. & Hoch, H.C., pp. 287–295. New York, Plenum.

Ogunsanya, O.C. & Madelin, M.F. (1977) Sensitivity of *Botryodiplodia ricinicola* conidia to mild chilling. *Transactions of the British Mycological Society*, **69**, 191–195.

Old, K.M. (1977) Giant soil amoebae cause perforation of conidia of *Cochliobolus sativus*. *Transactions of the British Mycological Society*, **68**, 277–281.

Old, K.M. & Darbyshire, J.F. (1978) Soil fungi as food for giant amoebae. *Soil Biology & Biochemistry*, **10**, 93–100.

O'Leary, D.J. & Lockwood, J.L. (1988) Debilitation of conidia of *Cochliobolus sativus* at high soil matric potentials. *Soil Biology & Biochemistry*, **20**, 239–243.

Olthof, T.H.A. & Estey, R.H. (1963) A nematotoxin produced by the nematophagous fungus *Arthrobotrys oligospora* Fresenius. *Nature (London)*, **197**, 514–515.

Omar, M. & Heather, W.A. (1979) Effect of saprophytic phylloplane fungi on germination and development of *Melampsora larici-populina*. *Transactions of the British Mycological Society*, **72**, 225–231.

Oritsejafor, J.J. (1986) Carbon and nitrogen nutrition in relation to growth and sporulation of *Fusaruium oxysporum* f. sp. *elaeidis*. *Transactions of the British Mycological Society*, **87**, 519–524.

Orlowski, M. (1980) Cyclic adenosine 3′,5′ monophosphate and germination of sporangiospores from the fungus *Mucor*. *Archives of Microbiology*, **126**, 133–140.

Orlowski, M. (1991) *Mucor* dimorphism. *Microbiological Reviews*, **55**, 234–258.

Orlowski, M. & Sypher, P.S. (1978) Regulation of macromolecular synthesis during hyphal germ tube emergence from *Mucor racemosus* sporangiospores. *Journal of Bacteriology*, **134**, 76–83.

Orpin, C.G. (1978) Induction of zoosporogenesis in the rumen phycomycete *Neocallimastix frontalis* in rumen fluid after the addition of haem-containing compounds. *Proceedings of the Society for General Microbiology*, **5**, 46.

Orpin, C.G. & Bountiff, L. (1978) Zoospore chemotaxis in the rumen Phycomycete *Neocallimastix frontalis*. *Journal of General Microbiology*, **104**, 113–122.

Orpin, C.G. & Greenwood, Y. (1986a) The role of haems and related compounds in the nutrition and zoosporogenesis of the rumen chytridiomycete *Neocallimastix frontalis* H8. *Journal of General Microbiology*, **132**, 2179–2185.

Orpin, C.G. & Greenwood, Y. (1986b) Nutritional and germination requirements of the rumen Chytridiomycete *Neocallimastix patriciarum*. *Transactions of the British Mycological Society*, **86**, 103–109.

Osburn, R.M., Schroth, M.N., Hancock, J.G. & Hendson, M. (1989) Dynamics of sugar beet seed colonization by *Pythium ultimum* and *Pseudomonas* species: effects on seed rot and damping-off. *Phytopathology*, **79**, 709–716.

Osman, M. & Valadon, L.R.G. (1981) Effect of light (especially near-UV) on spore germination and ultrastructure of *Verticillium agaricinum*. *Transactions of the British Mycological Society*, **77**, 187–189.

Owens, O.V.H. & Krizek, D.T. (1980) Multiple effects of UV radiation (265–340 nm) on fungal spore emergence. *Photochemistry and Photobiology*, **32**, 41–49.

Pacioni, G. (1990) Scanning electron microscopy of *Tuber* sporocarps and associated bacteria. *Mycological Research*, **94**, 1086–1089.

Page, R.M., Sherf, A.F. & Morgan, T.L. (1947) The effect of temperature and relative humidity on the longevity of the conidia of *Helminthosporium oryzae*. *Mycologia*, **39**, 158–164.

Pall, M.L. & Robertson, C.K. (1986) Cyclic AMP control of hierarchical growth pattern of hyphae in *Neurospora crassa*. *Experimental Mycology*, **10**, 161–165.

Palmer, F.E., Staley, J.T. & Ryan, B. (1990) Ecophysiology of microconidial fungi and lichens on rocks in northeastern Oregon. *New Phytologist*, **116**, 613–620.

Papendick, R.I. & Mulla, D.J. (1986) Basic principles of cell and tissue water relations. In *Water, Fungi and Plants*, eds Ayres, P.G. & Boddy L., pp. 1–25, Cambridge, Cambridge University Press.

Parbery, D.G. & Emmett, R.W. (1977) Hypotheses regarding appressoria, spores, survival and phylogeny in parasitic fungi. *Revue de Mycologie*, **41**, 429–447.

Parke, J.L. (1990) Population dynamics of *Pseudomonas cepacia* in the pea spermosphere in relation to biocontrol of *Pythium*. *Phytopathology*, **80**, 1307–1311.

Parkinson, D., Visser, S. & Whittaker, J.B. (1979) Effects of collembolan grazing on fungal colonization of leaf litter. *Soil Biology & Biochemistry*, **11**, 529–535.

Paterson, R.R.M., Simmonds, M.S.J. & Blaney, W.M. (1987) Mycopesticidal effects of characterized extracts of *Penicillium* isolates and purified secondary metabolites (including mycotoxins) on *Drosophila melanogaster* and *Spodoptera littoralis*. *Journal of Invertebrate Pathology*, **50**, 124–133.

Pearce, M.H. & Malajczuk, N. (1990) Factors affecting growth of *Armillaria luteobubalina* rhizomorphs in soil. *Mycological Research*, **94**, 38–48.

Pemberton, C.M., Davey, R.A., Webster, J., Dick, M.W. & Clark, G. (1990) Infection of *Pythium* and *Phytophthora* species by *Olpidium gracilis* (Oomycetes). *Mycological Research*, **94**, 1081–1085.

Pennington, C.J., Iser, J.R., Grant, B.R. & Gayler, K.R. (1989) Role of RNA and protein synthesis in stimulated germination of zoospores of the phytopathogenic fungus *Phytophthora palmivora*. *Experimental Mycology*, **13**, 158–168.

Perry, D.F. & Fleming, R.A. (1989) *Erynia crustosa* zygospore germination. *Mycologia*, **81**, 154–158.

Perry, D.F. & Latge, J.-P. (1982) Dormancy and germination of *Cochliobolus obscurus* azygospores. *Transactions of the British Mycological Society*, **78**, 221–225.

Petersen, G.R., Russo, G.M. & van Etten, J.L. (1982) Identification of major proteins in sclerotia of *Sclerotinia minor* and *Sclerotinia trifoliorum*. *Experimental Mycology*, **6**, 268–273.

Peterson, R.C., Cummins, K.W. & Ward, G.M. (1989) Microbial and animal processing of detritus in a woodland stream. *Ecological Monographs*, **59**, 21–39.

Pezet, R. & Pont, V. (1975) Elemental sulfur responsible for self-inhibition of spores of *Phomopsis viticola*. *Experientia*, **31**, 439–440.

Pfyffer, G.E., Boraschi-Gaia, C., Weber, B., Hoesch, L., Orpin, C.G. & Rast, D.M. (1990) A further report on the occurrence of acyclic sugar alcohols in fungi. *Mycological Research*, **94**, 219–222.

Pfyffer, G.E., Pfyffer, B.U. & Rast, D.M. (1986) The polyol pattern, chemotaxonomy, and phylogeny of the fungi. *Sydowia*, **39**, 160–201.

Pianka, E.R. (1970) On *r*- and *K*-selection. *American Naturalist*, **104**, 592–597.

Pirozynski, K.A. (1976) Fossil fungi. *Annual Review of Phytopathology*, **14**, 237–246.

Pirozynski, K.A. (1981) Interactions between fungi and plants through the ages. *Canadian Journal of Botany*, **39**, 1824–1827.

Pirozynski, K.A. & Hawksworth, D.L. (eds) (1988) *Coevolution of Fungi with Plants and Animals* London, Academic Press.

Pirozynski, K.A. & Malloch, D.W. (1975) The origin of land plants: a matter of mycotrophism. *BioSystems*, **6**, 153–164.

Pirozynski, K.A. & Malloch, D.W. (1988) Seeds, spores and stomachs: coevolution in seed dispersal mutualisms. In *Coevolution of Fungi with Plants and Animals*, eds Pirozynski, K.A. & Hawksworth, D.L., pp. 227–246. London, Academic Press.

Pitt, D. & Mosley, M.J. (1985) Pathways of glucose catabolism and the origin and metabolism of pyruvate during calcium-induced conidiation of *Penicillium notatum*. *Antonie van Leeuwenhoek*, **51**, 365–384.

Plassard, C., Martin, F., Mousain, D. & Salsac, L. (1986) Physiology of nitrogen assimilation by mycorrhiza. In *Mycorrhizae: Physiology and Genetics*, eds Gianinazzi-Pearson, V. & Gianinazzi, S., pp. 111–120. Paris, INRA.

Plesofsky-Vig, N. & Brambl, R. (1985) The heat shock response of fungi. *Experimental Mycology*, **9**, 187–194.

Plunkett, B.E. (1961) The change of tropism of *Polyporus brumalis* stipes and the effect of directional stimuli on pileus differentiation. *Annals of Botany*, **25**, 206–223.

Pluorde, D.F. & Green, R.J. (1982) Effect of monochromatic light on germination of oospores and formation of sporangia of *Phytophthora citricola*. *Phytopathology*, **72**, 58–61.

Pocock, K. & Duckett, J.G. (1985) The alternative mycorrhizas: fungi and hepatics. *Bulletin of the British Bryological Society*, **45**, 10–11.

Pommerville, J.C. & Olson, L.W. (1987) Negative chemotaxis of gametes and zoospores of *Allomyces*. *Journal of General Microbiology*, **133**, 2573–2579.

Ponge, J.F. & Charpentie, M.-J. (1981) Étude des relations microflore-microfaune: experiences sur *Pseudosinella alba* (Packard), Collembole mycophage (Study of microflora–microfauna relationships: experiments on *Pseudosinella alba* (Packard), a mycophagous collembolan). *Révue d'Écologie et de Biologie du Sol*, **18**, 291–303.

Poon, N.H., Martin, J. & Day, A.W. (1974) Conjugation in *Ustilago violacea*. I. Morphology. *Canadian Journal of Botany*, **20**, 187–191.

Potapova, T.V., Levina, N.N., Belozerskaya, T.A., Kritsky, M.S. & Chailkhian, L.M. (1984) Investigation of electrophysiological responses of *Neurospora crassa* to blue light. *Archives of Microbiology*, **137**, 262–265.

Powell, R.J. & Stradling, D.J. (1986) Factors influencing growth of *Attamyces brumatificus*, a symbiont of attine ants. *Transactions of the British Mycological Society*, **87**, 205–213.

Prosser, J.I. (1983) Hyphal growth patterns. In *Fungal Differentiation*, ed. Smith, J.E., pp. 357–396. New York, Marcel Dekker.

Pugh, G.J.F. (1980) Strategies in fungal ecology. *Transactions of the British Mycological Society*, **75**, 1–14.

Pugh, G.J.F. & Buckley, N.G. (1971) The leaf surface as a substrate for colonization by fungi. In *Ecology of Leaf Surface Micro-organisms*, eds Preece, T.F. & Dickinson, C.H., pp. 431–445. London, Academic Press.

Pugh, G.J.F. & Evans, M.D. (1970) Keratinophilic fungi associated with birds. I. Fungi isolated from feathers, nests and soils. *Transactions of the British Mycological Society*, **54**, 233–240.

Puhalla, J.E. (1973) Differences in sensitivity of *Verticillium* species to ultraviolet irradiation. *Phytopathology*, **63**, 1488–1492.

Punja, Z.K. (1986) Effect of carbon and nitrogen step-down on sclerotium biomass and cord development in *Sclerotium rolfsii* and *S. delphinii*. *Transactions of the British Mycological Society*, **86**, 537–544.

Punja, Z.K. & Grogan, R.G. (1981) Mycelial growth and infection without a food base by eruptively germinating sclerotia of *Sclerotium rolfsii*. *Phytopathology*, **71**, 1099–1103.

Punja, Z.K., Jenkins, S.F. & Grogan, R.G. (1984) Effect of volatile compounds, nutrients and source of sclerotia on eruptive sclerotial germination of *Sclerotium rolfsii*. *Phytopathology*, **74**, 1290–1295.

Pyssalo, H. (1976) Identification of volatile compounds in seven edible fresh mushrooms. *Acta Chemica Scandinavica*, **30**, 235–244.

Quinlan, R.J. & Cherrett, J.M. (1979) The role of fungus in the diet of the leaf-cutting ant *Atta cephalotes* (L.). *Ecological Entomology*, **4**, 151–160.

Quinn, M.A. (1987) The influence of saprophytic competition on nematode predation by nematode-trapping fungi. *Journal of Invertebrate Pathology*, **49**, 170–174.

Rabatin, S.C. & Stinner, B.C. (1985) Arthropods as consumers of vesicular–arbuscular mycorrhizal fungi. *Mycologia*, **77**, 320–322.

Ragan, M.A. & Chapman, D.J. (1978) *A Biochemical Phylogeny of the Protista*, New York, Academic Press.

Rainey, P.B., Cole, A.L.J., Fermor, T.R. & Wood, D.A. (1990) A model system for examining involvement of bacteria in basidiome initiation of *Agaricus bisporus*. *Mycological Research*, **94**, 191–195.

Ramadoss, C.L., Uhlig, J., Carlson, D.M., Butler, L.G. & Nicholson, R.L. (1985) Composition of the mucilaginous spore matrix of *Colletotrichum graminicola*, a pathogen of corn, sorghum and other grasses. *Journal of Agricultural and Food Chemistry*, **33**, 728–732.

Rao, P.S. & Niederpreum, D.J. (1969) Carbohydrate metabolism during morphogenesis of *Coprinus lagopus* (*sensu* Buller). *Journal of Bacteriology*, **100**, 1222–1228.

Raper, J.R. (1939) Sexual hormones in *Achlya*. I. Indicative evidence for a hormone coordinating mechanism. *American Journal of Botany*, **26**, 639–650.

Rast, D.M., Stussi, H., Hegenaur, H. & Nyhlen, L.E. (1981) Melanins. In *The Fungal Spore: Morphogenetic Controls*, eds Turian, G. & Hohl, H.R., pp. 507–531. New York, Academic Press.

Rathaiah, Y. (1977) Stomatal tropism of *Cercospora beticola* in sugar beet. *Phytopathology*, **67**, 358–362.

Rawlins, A. (1989) *Germination of the thick-walled conidia of* Mycogone perniciosa *the wet bubble pathogen of cultivated mushrooms*. PhD Thesis, University of Sheffield, UK.

Rayner, A.D.M. & Coates, D. (1987) Regulation of mycelial organisation and responses. In *Evolutionary Biology of the Fungi*, eds Rayner, A.D.M., Brasier, C.M. & Moore, D., pp. 115–136. Cambridge, Cambridge University Press.

Rayner, A.D.M., Coates, D., Ainsworth, A.M., Adams, T.J.H., Williams, E.N.D. & Todd, N.K. (1984) The biological consequences of the individualistic mycelium. In *The Ecology and Physiology of the Fungal Mycelium*, eds Jennings, D.H. & Rayner, A.D.M., pp. 509–540. Cambridge, Cambridge University Press.

Rayner, A.D.M., Powell, K.A., Thompson, W. & Jennings, D.H. (1985) Morphogenesis of vegetative organs. In *Developmental Biology of Higher Fungi*, eds Moore, D., Casselton, L.A., Wood, D.A. & Frankland, J.C., pp. 249–279. Cambridge, Cambridge University Press.

Rayner, A.D.M. & Webber, J.F. (1984) Interspecific mycelial interactions – an overview. In *The Ecology and Physiology of the Fungal Mycelium*, eds Jennings, D.H. & Rayner, A.D.M., pp. 383–417. Cambridge, Cambridge University Press.

Read, D.J. (1983) The biology of mycorrhiza in the Ericales. *Canadian Journal of Botany*, **61**, 985–1004.

Read, D.J. (1984) The structure and function of the vegetative mycelium of mycorrhizal roots. In *The Ecology and Physiology of the Fungal Mycelium*, eds Jennings, D.H. & Rayner, A.D.M., pp. 215–240. Cambridge, Cambridge University Press.

Read, N.D. (1983) A scanning electron microscope study of the external features of perithecium development in *Sordaria humana*. *Canadian Journal of Botany*, **61**, 3217–3229.

Read, N.D. & Beckett, A. (1985) The anatomy of the mature perithecium of *Sordaria humana* and its significance for fungal multicellular development. *Canadian Journal of Botany*, **63**, 281–296.

Recca, J. & Mrak, E.M. (1952) Yeast occurring in citrus products. *Food Technology*, **6**, 450–454.

Riffle, J.W. (1967) Effect of an *Aphelenchoides* species on the growth of a mycorrhizal and a pseudomycorrhizal fungus. *Phytopathology*, **57**, 541–544.

Ritchie, D. & Jacobsohn, M. (1963) The effects of osmotic and nutritional variation on growth of a salt-tolerant fungus, *Zalerion eistla*. In *Symposium on Marine Microbiology*, ed. Oppenheimer, C.H., pp. 286–299. Springfield, Illinois, C.C. Thomas.

Ritz, K. & Crawford, J. (1990) Quantification of the fractal nature of colonies of *Trichoderma viride*. *Mycological Research*, **94**, 1138–1141.

Robertson, M.D. (1991) Suppression of phagocytic cell responses by conidia and conidial products of *Aspergillus fumigatus*. In *The Fungal Spore and Disease Initiation in Plants and Animals*, eds Cole, G.T. & Hoch, H.C., pp. 461–480. New York, Plenum.

Robinson, P.M. (1973a) Chemotropism in fungi. *Transactions of the British Mycological Society*, **61**, 303–313.

Robinson, P.M. (1973b) Autotropism in fungal spores and hyphae. *Botanical Reviews*, **39**, 367–384.

Robinson, P.M. & Bolton, S.K. (1984) Autotropism in hyphae of *Saprolegnia ferax*. *Transactions of the British Mycological Society*, **83**, 237–263.

Robinson, P.M., McKee, N.D., Thompson, L.A.A., Harper, D.B. & Hamilton, J.T.G. (1989) Autoinhibition of germination and growth in *Geotrichum candidum*. *Mycological Research*, **93**, 214–222.

Robinson, P.M., Park, D. & Graham, T.A. (1968) Autotropism in fungal spores. *Journal of Experimental Botany*, **19**, 125–134.

Robinson, P.M., Thompson, L.A.A. (1982) Volatile promoter of germination and hyphal extension produced by *Geotrichum candidum*. *Transactions of the British Mycological Society*, **78**, 353–383.

Rogers, H.J., Buck, K.W. & Brasier, C.M. (1986) Transmission of double-stranded RNA and a disease-factor in *Ophiobolus ulmi*. *Plant Pathology*, **35**, 277–287.

Roncadori, R.W., Duffey, S.S. & Blum, M.S. (1985) Antifungal activity of defensive secretions of certain millipedes. *Mycologia*, **77**, 185–191.

Rosenweig, W.D. & Ackroyd, D. (1984) Influence of soil microorganisms on the trapping of nematodes by nematophagous fungi. *Canadian Journal of Microbiology*, **30**, 1437–1439.

Rosenweig, W.D., Premachandran, D. & Pramer, D. (1985) Role of trap lectins in the specificity of nematode capture by fungi. *Canadian Journal of Microbiology*, **31**, 693–695.

Rosin, I.V., Horner, J. & Moore, D. (1985) Differentiation and pattern formation in the fruit body cap of *Coprinus cinereus*. In *Developmental Biology of Higher Fungi*, eds Moore, D., Casselton, L.A., Wood, D.A. & Frankland, J.C., pp. 333–351. Cambridge, Cambridge University Press.

Ross, I.K. (1985) Determination of the initial steps in differentiation in *Coprinus congregatus*. In *Developmental Biology of Higher Fungi*, eds Moore, D., Casselton, L.A., Wood, D.A. & Frankland, J.C., pp. 353–373. Cambridge, Cambridge University Press.

Ross, I.S. (1975) Some effects of heavy metals on fungal cells. *Transactions of the British Mycological Society*, **64**, 175–193.

Ross, I.S. & Walsh, A.L. (1981) Resistance to copper in *Saccharomyces cerevisiae*. *Transactions of the British Mycological Society*, **77**, 27–32.

Rosset, J. & Bärlocher, F. (1985) Aquatic hyphomycetes: influence of pH, Ca^{2+} and HCO_3^- on growth *in vitro*. *Transactions of the British Mycological Society*, **84**, 137–145.

Rotem, J. (1968) Thermoxerophytic properties of *Alternaria porri* f. sp. *solani*. *Phytopathology*, **58**, 1284–1287.

Rotem, J., Wooding, B. & Aylor, D.E. (1985) The role of solar radiation, especially ultraviolet, in the mortality of fungal spores. *Phytopathology*, **75**, 510–514.

Roth, R.R. & James, W.D. (1988) Microbial ecology of the skin. *Annual Review of Microbiology*, **42**, 441–464.

Royle, D.J. & Hickman, C.J. (1964) Analysis of factors governing *in vitro* accumulation of zoospores of *Pythium aphanidermatum* on roots. II. Substances causing response. *Canadian Journal of Microbiology*, **10**, 201–219.

Royle, D.J. & Thomas, C.G. (1973) Factors affecting zoospore responses towards stomata in hop downy mildew (*Pseudoperonospora humuli*) including some comparisons with grapevine downy mildew (*Plasmopara viticola*). *Physiological Plant Pathology*, **3**, 405–417.

Rowan, D.D., Hunt, M.B. & Gaynor, D.L. (1986) Peramine, a novel insect feeding deterrent from ryegrass infected with the endophyte *Acremonium loliae. Journal of the Chemical Society, Chemical Communications*, **12**, 935–936.

Ruel, K. (1990) Ultrastructural alterations of wood cell walls during degradation by fungi. In *Advances in Biological Treatment of Lignocellulosic Materials*, eds Coughlan, M.P. & Amaral Collaço, M.T., pp. 117–128. London, Elsevier Applied Science.

Ruiters, M.H.J. & Wessels, J.G.H. (1989) *In situ* localization of specific RNAs in developing fruitbodies of the basidiomycete *Schizophyllum commune. Experimental Mycology*, **13**, 212–222.

Rush, C.M. & Lyda, S.D. (1982) Effects of anhydrous ammonia on mycelium and sclerotia of *Phymatotrichum omnivorum. Phytopathology*, **72**, 1085–1089.

Safar, H.M. & Cooke, R.C. (1988a) Exploitation of faecal resource units by coprophilous Ascomycotina. *Transactions of the British Mycological Society*, **90**, 593–599.

Safar, H.M. & Cooke, R.C. (1988b) Interactions between bacteria and coprophilous Ascomycotina and a *Coprinus* species on agar and in copromes. *Transactions of the British Mycological Society*, **91**, 73–80.

Sagara, N. (1989) European record of the presence of a mole's nest indicated by a particular fungus. *Mammalia*, **83**, 301–305.

Sagara, N., Kobayashi, S., Ota, H., Itsubo, T. & Okabe, H. (1989) Finding *Euroscaptor mizura* (Mammalia: Insectivora) and the nest from under *Hebeloma radicosum* (Fungi: Agaricales) in Ashiu, Kyoto, with data of possible contiguous occurrences of three alpine species in this region. *Contributions from the Biological Laboratory of Kyoto University*, **27**, 261–272.

Saikawa, M. & Morikawa, C. (1985) Electron microscopy on a nematode-trapping fungus, *Acaulopage pectospora. Canadian Journal of Botany*, **63**, 1386–1390.

Saikawa, M., Yamaguchi, K. & Morikawa, C. (1988) Capture of rotifers by *Acaulopage pectospora*, and further evidence of its similarity to *Zoophagus insidians. Mycologia*, **80**, 880–884.

Sakagami, Y., Isogai, A., Suzuki, A., Tamura, S., Kitada, C. & Fujino, M. (1979) Structure of tremellogen A-10, a peptidal hormone inducing conjugation tube formation in *Tremella mesenterica. Agricultural & Biological Chemistry*, **43**, 2643–2645.

Salvatore, M.A., Gray, F.A. & Hine, R.B. (1973) Enzymatically induced germination of oospores of *Phytophthora megasperma. Phytopathology*, **63**, 1083–1084.

Sargent, M.L. & Briggs, W.R. (1967) The effects of light on a circadian rhythm of conidiation in *Neurospora. Plant Physiology*, **42**, 1504–1510.

Savile, D.B.O. (1968) Possible interrelationships between fungal groups. In *The Fungi. An Advanced Treatise. Vol. III. The Fungal Population*, eds Ainsworth, G.C. & Sussman, A.S., pp. 649–675. New York, Academic Press.

Sayre, R.M. & Keeley, L.S. (1969) Factors influencing *Catenaria anguillulae* infections in a free-living and a plant-parasitic nematode. *Nematologia*, **15**, 492–502.

Scarborough, G.A. (1985) The mechanisms of energization of solute transport in fungi. In *Environmental Regulation of Microbial Metabolism*, eds Kulaev, I.S., Davies, E.A. & Tempest, D.W., pp. 39–51. London, Academic Press.

Schechter, S.E. & Gray, L.E. (1987) Oospore germination in *Phytophthora megasperma* f. sp. *glycinea. Canadian Journal of Botany*, **65**, 1465–1467.

Schein, R.D. & Rotem, J. (1965) Temperature and humidity effects on uredospore viability. *Mycologia*, **57**, 397–403.

Schenk, S., Chase, T., Rosenweig, W.D. & Pramer, D. (1980) Collagenase production by nematode trapping fungi. *Applied and Environmental Microbiology*, **40**, 567–570.

Schippers, B., Meijer, J.W. & Liem, J.I. (1982) Effect of ammonia and other soil volatiles on germination and growth of soil fungi. *Transactions of the British Mycological Society*, **79**, 253–259.

Schippers, B. & Palm, L.C. (1973) Ammonia, a fungistatic volatile in chitin-amended soil. *Netherlands Journal of Plant Pathology*, **79**, 279–281.

Schmidle, A. (1951) Die Tagesperiodizität der asexuellen Reproduktion von *Pilobolus sphaerosporus* (The photoperiod of asexual reproduction in *Pilobolus sphaerosporus*). *Archiv für Mikrobiologie*, **16**, 80–100.

Schmidt, D. (1986) La quenouille rend-elle le fourrage toxique? (Does choke disease make forage grasses toxic?) *Revue Suisse Agriculture*, **18**, 329–332.

Schroth, M.N. & Hildebrand, D.C. (1964) Influence of plant exudates on root infecting fungi. *Annual Review of Phytopathology*, **2**, 101–132.

Schrüfer, K. & Lysek, G. (1990) Rhythmic growth and sporulation in *Trichoderma* species: differences within a population of isolates. *Mycological Research*, **94**, 124–127.

Schwalb, M.N. (1978) Regulation of fruiting. In *Genetics and Morphogenesis in the Basidiomycetes*, eds Schwalb, M.N. & Miles, P.G., pp. 135–165. New York, Academic Press.

Shafer, S.R., Rhodes, L.H. & Riedel, R.M. (1981) *In vitro* parasitism of endomycorrhizal fungi of ericaceous plants by the mycophagous nematode *Aphelenchoides bicaudatus*. *Mycologia*, **73**, 141–149.

Sharland, P.R., Burton, J.L. & Rayner, A.D.M. (1986) Mycelial dimorphism, interactions and pseudosclerotial plate formation in *Hymenochaete corrugata*. *Transactions of the British Mycological Society*, **86**, 158–163.

Shavlovsky, G.M. & Sibirny, A.A. (1985) Riboflavin transport in yeasts and its regulation. In *Environmental Regulation of Microbial Metabolism*, eds Kulaev, I.S., Davies, E.A. & Tempest, D.W., pp. 385–392. London, Academic Press.

Shaw, G., Leake, J.R., Baker, A.J.M. & Read, D.J. (1990) The biology of mycorrhiza in the Ericacae XVII. The role of mycorrhizal infection in the regulation of iron uptake by ericaceous plants. *New Phytologist*, **115**, 251–258.

Shaw, P.J.A. (1985) Grazing preferences of *Onychiurus armatus* (Insecta: Collembola) for mycorrhizal and saprophytic fungi of pine plantations. In *Ecological Interactions in Soil: Plants, Microbes and Animals*, eds Fitter, A.H., Fitter, D., Atkinson, D., Read, D.J. & Usher, M.B., pp. 333–337. Oxford, Blackwell Scientific.

Shaw, R. (1965) The occurrence of γ-linolenic acid in fungi. *Biochimica et Biophysica Acta*, **98**, 230–237.

Shaw, R. (1966) The polyunsaturated fatty acids of micro-organisms. *Advances in Lipid Research*, **4**, 107–114.

Sheard, J. & Farrar, J.F. (1987) Transport of sugar in *Phytophthora palmivora* (Butl.) Butl. *New Phytologist*, **105**, 265–272.

Sheehan, P.L. & Gochenaur, S.E. (1984) Spore germination and microcycle conidiation of two penicillia in soil. *Mycologia*, **76**, 523–527.

Sherwood-Pike, M.A. & Gray, J. (1985) Silurian fungal remains: probable records of the class Ascomycetes. *Lethaia*, **18**, 1–20.

Shigo, A.L. (1960) Parasitism of *Gonatobotrys fuscum* on species of *Ceratocystis*. *Mycologia*, **52**, 584–598.

Shipton, W.A. (1983) Possible relationship of some growth and sporulation responses of *Pythium* to the occurrence of equine phycomycosis. *Transactions of the British Mycological Society*, **80**, 13–18.

Shishkoff, N. (1989) Zoospores encystment pattern and germination on onion roots, and the colonization of hypodermal cells by *Pythium coloratum*. *Canadian Journal of Botany*, **67**, 258–262.

Show, N.M. & Harding, R.W. (1987) Intracellular and extracellular cyclic nucleotides in wild-type and white collar mutant strains of *Neurospora crassa*. Temperature dependent efflux of cyclic AMP from mycelia. *Plant Physiology*, **83**, 377–383.

Siegel, M.R., Latch, G.C.M. & Johnson, M.C. (1987) Fungal endophytes of grasses. *Annual Review of Phytopathology*, **25**, 293–315.

Sietsma, J.H., Rast, D. & Wessels, J.G.H. (1977) The effect of carbon dioxide on fruiting and on the degradation of cell-wall glucan in *Schizophyllum commune*. *Journal of General Microbiology*, **102**, 385–389.

Sihtola, H. & Neimo, L. (1975) The structure and properties of cellulose. In *Symposium on Enzymatic Hydrolysis of Cellulose*, eds Bailey, M., Enari, T.-M. & Linko, M., pp. 9–21. Helsinki, SITRA.

Siqueira, J.O., Sylvia, D.M., Gibson, J. & Hubbell, D.H. (1985) Spores, germination, and germ tubes of vesicular–arbuscular mycorrhizal fungi. *Canadian Journal of Microbiology*, **31**, 965–972.

Sitton, J.W. & Cook, R.J. (1981) Comparative morphology and survival of chlamydospores of *Fusarium roseum* 'Culmorum' and 'Graminearum'. *Phytopathology*, **71**, 85–90.

Sivinski, J. (1981) Arthropods attracted to luminous fungi. *Psyche*, **88**, 383–390.

Slocum, R.D., Ahmadjian, V. & Hildreth, K.C. (1980) Zoosporogenesis in *Trebouxia gelatinosa*: ultrastructure, potential for zoospore release and implications for the lichen association. *Lichenologist*, **12**, 173–187.

Smith, A.M. (1972) Drying and wetting sclerotia promotes biological control of *Sclerotium rolfsii* Sacc. *Soil Biology & Biochemistry*, **4**, 119–123.

Smith, A.M. & Griffin, D.M. (1971) Oxygen and the ecology of *Armillariella elegans* Heim. *Australian Journal of Biological Sciences*, **24**, 231–262.

Smith, D.C. (1979) Is a lichen a good model of biological interactions in nutrient-limited environments? In *Strategies of Microbial Life in Extreme Environments*, ed. Shilo, M., pp. 291–303. Berlin, Dahlem Konferenzen/Weinheim, Verlag Chemie.

Smith, S.N., Armstrong, R.A. & Rimmer, J.J. (1984) Influence of environmental factors on zoospores of *Saprolegnia diclina*. *Transactions of the British Mycological Society*, **82**, 413–421.

Sneh, B., Dupler, M., Elad, Y. & Baker, R. (1984) Chlamydospore germination of *Fusarium oxysporum* f. sp. *cucumerinum* as affected by fluorescent and lytic bacteria from *Fusarium*-

suppressive soil. *Phytopathology*, **74**, 1115–1124.

Sneh, B., Humble, S.J. & Lockwood, J.L. (1977) Parasitism of oospores of *Phytophthora megasperma* var. *sojae*, *P. cactorum*, *Pythium* sp., and *Aphanomyces euteiches* in soil by Oomycetes, Chytridiomycetes, Hyphomycetes, Actinomycetes, and bacteria. *Phytopathology*, **67**, 622–628.

Southwood, T.R.E. (1977) Habitat, the templet for ecological strategies? *Journal of Animal Ecology*, **46**, 337–365.

Spencer, D.M. & Atkey, P.T. (1981) Parasitic effects of *Verticillium lecanii* on two rust fungi. *Transactions of the British Mycological Society*, **77**, 535–542.

Sreeramula, T. (1962) Aerial dissemination of barley loose smut (*Ustilago nuda*). *Transactions of the British Mycological Society*, **45**, 373–384.

Staples, R.C. & Hoch, H.C. (1987) Host infection structure–form and function. *Experimental Mycology*, **11**, 163–169.

Staub, T., Dahmen, H. & Schwinn, F.J. (1974) Light- and scanning electron microscopy of cucumber and barley powdery mildew on host and non-host plants. *Phytopathology*, **64**, 364–372.

Steer, M.W. (1990) Role of actin in tip growth. In *Tip Growth in Plant and Fungal Cells*, ed. Heath, I.B., pp. 119–145. London, Academic Press.

Stewart, E., Gow, N.A.R. & Bowen, D.V. (1988) Cytoplasmic alkalinization during germtube formation in *Candida albicans*. *Journal of General Microbiology*, **134**, 1079–1087.

Stewart, P.R. & Rogers, P.J. (1983) Fungal dimorphism. In *Fungal Differentiation. A Contemporary Synthesis*, ed. Smith, J.E., pp. 267–313. New York, Marcel Dekker.

Stirling, G.R. (1988) Prospects for the use of fungi in nematode control. In *Fungi in Biological Control Systems*, ed. Burge, M.N., pp. 188–210. Manchester, Manchester University Press.

Stirling, J.L., Cook, G.A. & Pope, A.M.S. (1979) Chitin and its degradation. In *Fungal Cell Walls and Hyphal Growth*, eds Burnett, J.H. & Trinci, A.P.J., pp. 169–188. Cambridge, Cambridge University Press.

Stubblefield, S.P. & Taylor, T.N. (1988) Recent advances in palaeomycology. *New Phytologist*, **108**, 3–25.

Stubblefield, S.P., Taylor, T.N. & Beck, C.B. (1985) Studies on Paleozoic fungi. IV. Wood-decaying fungi in *Callixylon newberryi* from the Upper Devonian. *American Journal of Botany*, **72**, 1765–1774.

Stump, R.F., Robinson, K.R., Harold, R.L. & Harold, F.M. (1980) Electrical currents in the water mold *Blastocladiella emersonii* during growth and sporulation. *Proceedings of the National Academy of Science of the USA*, **77**, 6673–6677.

Subrahmanyam, P., Reddy, P.M. & McDonald, D. (1988) Photosensitivity of urediniospore germination in *Puccinia arachidis*. *Transactions of the British Mycological Society*, **90**, 229–232.

Sussman, A.S. (1965a) Physiology of dormancy and germination in the propagules of cryptogamic plants. In *Encyclopedia of Plant Physiology. Vol. 15. Differentiation and Development*, ed. Lang, A., pp. 933–1025. Berlin, Springer.

Sussman, A.S. (1965b) Dormancy of soil-microorganisms in relation to survival. In *Ecology of Soil-borne Plant Pathogens*, eds Baker, K.F. & Snyder, W.C., pp. 99–110. London, John Murray.

Sussman, A.S. (1976) Activators of fungal spore germination. In *The Fungal Spore*, eds Weber, D.J. & Hess, W.M., pp. 101–139. New York, Wiley.

Sussman, A.S. & Halvorson, H.O. (1966) *Spores: Their Dormancy and Germination*, New York, Harper & Row.

Sutherland, E.D. & Lockwood, J. (1984) Hyperparasitism of oospores of some Peronosporales by *Actinoplanes missouriensis* and *Humicola fuscoatra* and other Actinomycetes and fungi. *Canadian Journal of Plant Pathology*, **6**, 139–145.

Sutter, R.P. & Whitaker J.P. (1981) Zygophore-stimulating precursors (pheromones) of trisporic acid active in (−)-*Phycomyces blakesleeanus*. *Journal of Biological Chemistry*, **256**, 2334–2341.

Swift, M.J. (1965) Loss of suberin from bark tissue rotted by *Armillaria mellea*. *Nature (London)*, **207**, 436–437.

Swift, M.J. & Boddy, L. (1984) Animal–microbial interaction in wood decomposition. In *Invertebrate–Microbial Interactions*, eds Anderson, J.M., Rayner, A.D.M. & Walton, D.W.H., pp. 89–131. Cambridge, Cambridge University Press.

Talou, T., Gaset, A., Delmas, M., Kulifaj, M. & Montant, C. (1990) Dimethyl sulphide: the secret of black truffle hunting by animals? *Transactions of the British Mycological Society*, **94**, 277–278.

Tan, K.K. (1978) Light-induced fungal development. In *The Filamentous Fungi. Vol. III. Developmental Mycology*, eds Smith, J.E. & Berry, D.R., pp. 334–357. London, Edward Arnold.

Tansey, M.R. & Brock, T.D. (1973) *Dactylaria gallopava*, a cause of avian encephalitis, in hot spring effluents, thermal soils and self-heated coal waste piles. *Nature (London)*, **242**, 202–203.

Tansey, M.R. & Brock, T.D. (1978) Microbial life at high temperatures: ecological aspects. In *Microbial Life in Extreme Environments*, ed. Kushner, D.J., pp. 159–216. London, Academic Press.

Tavares, I. (1985) *Laboulbeniales (Fungi, Asco-*

mycetes), Braunschweig, J. Cramer.

Teitell, L. (1958) Effects of relative humidity on viability of conidia of Aspergilli. *American Journal of Botany*, **45**, 748–753.

Teunissen, M.J., op den Camp, H.J.M., Orpin, C.G., Huis in 't Veld, J.H.J. & Vogels, G.D. (1991) Comparison of growth characteristics of anaerobic fungi isolated from ruminant and non-ruminant herbivores during cultivation in a defined medium. *Journal of General Microbiology*, **137**, 1407–1408.

Theodorou, C. & Bowen, G.D. (1987) Germination of basidiospores of mycorrhizal fungi in the rhizosphere of *Pinus radiata* D. Don. *New Phytologist*, **106**, 217–223.

Theodorou, M.K., Gill, M., King-Spooner, C. & Beever, D.E. (1990) Enumeration of anaerobic Chytridiomycetes as thallus-forming units: novel method for quantification of fibrolytic fungal populations from the digestive tract ecosystem. *Applied and Environmental Microbiology*, **56**, 1073–1078.

Thevelein, J.M., Beullens, M., Honshoven, F., Hoebeeck, G., Detremerie, K., Griewel, B., den Hollander, J.A. & Jans, A.W.H. (1987) Regulation of the cAMP level in the yeast *Saccharomyces cerevisiae*: the glucose-induced cAMP signal is not mediated by a transient drop in the intracellular pH. *Journal of General Microbiology*, **133**, 2197–2205.

Thomas, A.R. & Hill, E.C. (1976) *Aspergillus fumigatus* and supersonic aviation. I. Growth of *Aspergillus fumigatus*. *International Biodeterioration Bulletin*, **12**, 87–94.

Thomas, D.D. & Peterson, A.P. (1990) Chemotactic auto-aggregation in the water mould *Achlya*. *Journal of General Microbiology*, **136**, 847–853.

Thomas, R.J. (1987a) Distribution of *Termitomyces* Heim and other fungi in the nests and major workers of *Macrotermes bellicosus* (Smeathman) in Nigeria. *Soil Biology & Biochemistry*, **19**, 329–333.

Thomas, R.J. (1987b) Distribution of *Termitomyces* and other fungi in the nests and major workers of several Nigerian Macrotermitinae. *Soil Biology & Biochemistry*, **19**, 335–341.

Thomas, R. J. (1987c) Factors affecting the distribution and activity of fungi in the nests of Macrotermitinae (Isoptera). *Soil Biology & Biochemistry*, **19**, 343–349.

Thomashow, L.S., Weller, D.M., Bonsall, R.F. & Pierson, L.S. (1990) Production of the antibiotic phenazine-1-carboxylic acid by fluorescent *Pseudomonas* species in the rhizosphere of wheat. *Applied and Environmental Microbiology*, **56**, 908–912.

Thompson, W. (1984) Distribution, development and functioning of mycelial cord systems of decomposer basidiomycetes of the deciduous woodland floor. In *The Ecology and Physiology of the Fungal Mycelium*, eds Jennings, D.H. & Rayner, A.D.M., pp. 185–214. Cambridge, Cambridge University Press.

Thrower, L.B. (1966) Terminology for plant parasites. *Phytopathologische Zeitschrift*, **56**, 258–259.

Timberlake, W.E. (1980) Developmental gene regulation in *Aspergillus nidulans*. *Developmental Biology*, **78**, 497–510.

Tinline, R.D., Stauffer, J.F. & Dickson, J.G. (1960) *Cochliobolus sativus*. III. Effect of ultraviolet radiation. *Canadian Journal of Botany*, **38**, 275–282.

Tommerup, I.C. (1983) Spore dormancy in vesicular–arbuscular mycorrhizal fungi. *Transactions of the British Mycological Society*, **81**, 37–45.

Tommerup, I.C. (1984) Effect of soil water potential on spore germination by vesicular–arbuscular mycorrhizal fungi. *Transactions of the British Mycological Society*, **83**, 193–202.

Tommerup, I.C. (1985) Inhibition of spore germination of vesicular–arbuscular mycorrhizal fungi in soil. *Transactions of the British Mycological Society*, **85**, 267–278.

Tonon, F., Prior de Castro, C. & Odier, E. (1990) Nitrogen and carbon regulation of lignin peroxidase and enzymes of nitrogen metabolism in *Phanaerochaete chrysosporium*. *Experimental Mycology*, **14**, 243–254.

Torzilli, A.P., Vinroot, S. & West, C. (1985) Interactive effect of temperature and salinity on growth and activity of a salt marsh isolate of *Aureobasidium pullulans*. *Mycologia*, **77**, 278–284.

Trinci, A.P.J. (1969) A kinetic study of the growth of *Aspergillus nidulans* and other fungi. *Journal of General Microbiology*, **57**, 11–24.

Trinci, A.P.J. (1971) Influence of the width of the peripheral growth zone on the radial growth rate of fungal colonies on solid growth media. *Journal of General Microbiology*, **67**, 325–344.

Trinci, A.P.J. (1978) Wall and hyphal growth. *Science Progress (Oxford)*, **65**, 75–99.

Trinci, A.P.J. (1984) Regulation of hyphal branching and hyphal orientation. In *The Ecology and Physiology of the Fungal Mycelium*, eds Jennings, D.H. & Rayner, A.D.M., pp. 23–52. Cambridge, Cambridge University Press.

Trinci, A.P.J. & Collinge, A.J. (1975) Hyphal wall growth in *Neurospora crassa* and *Geotrichum candidum*. *Journal of General Microbiology*, **91**, 355–361.

Trione, E.J. (1977) Endogenous germination inhibitors in teliospores of the wheat bunt fungus. *Phytopathology*, **67**, 1245–1249.

Trione, E.J. & Ross, W.D. (1988) Lipids as bioregulators of teliospore germination and sporidial formation in the wheat bunt fungi, *Tilletia* species. *Mycologia*, **80**, 38–45.

Tsuru, T., Koga, K., Aoyama, H. & Ootaki, T. (1988) Optics in *Phycomyces blakesleeanus* sporangiophores relative to determination of phototropic orientation. *Experimental Mycology*, **12**, 302–312.

Tu, J.C. (1986) Hyperparasitism of *Streptomyces albus* on a destructive mycoparasite *Nectria inventa*. *Journal of Phytopathology*, **117**, 71–76.

Turhan, G. & Grossman, F. (1986) Investigation of a great number of Actinomycete isolates on their antagonistic effects against soil-borne fungal plant pathogens by an improved method. *Journal of Phytopathology*, **116**, 238–243.

Turian, G. (1955) Recherches sur l'action de l'acide borique sur la fructification de *Sordaria* (Research on the effect of boric acid on fruiting of *Sordaria*). *Phytopathologische Zeitschrift*, **25**, 181–189.

Turian, G. (1978) Sexual morphogenesis in the ascomycetes. In *The Filamentous Fungi. Vol. III. Developmental Mycology*, eds Smith, J.E. & Berry, D.R., pp. 315–333. London, Edward Arnold.

Turian, G. & Hohl, H.R. (eds) (1981) *The Fungal Spore: Morphogenetic Controls*, New York, Academic Press.

Turner, G.J. & Tribe, H.T. (1976) On *Coniothyrium minitans* and its parasitism of *Sclerotinia* species. *Transactions of the British Mycological Society*, **66**, 97–105.

Uebelmesser, E.R. (1954) Über den endonomen Tagesrhythmus der Sporangienträgerbildung von *Pilobolus* (Studies on the endogenous circadian rhythm of sporangiophore production by Pilobolus). *Archiv für Mikrobiologie*, **20**, 1–33.

Ulanowski, Z. & Ludlow, I.K. (1989) Water distribution, size and wall thickness in *Lycoperdon pyriforme* spores. *Mycological Research*, **93**, 28–32.

Uno, I. & Ishikawa, T. (1976) Effect of cyclic AMP on glycogen phosphorylase in *Coprinus macrorhizus*. *Biochimica et Biophysica Acta*, **452**, 112–120.

Upadhyay, R. & Pavgi, M.S. (1979) Biphasic diurnal cycle of ascus development of *Taphrina deformans* Butler. *Mycopathologia*, **69**, 33–41.

Vakalounakis, D.J. & Christias, C. (1985) Blue-light inhibition of conidiation in *Alternaria cichorii*. *Transactions of the British Mycological Society*, **85**, 285–289.

Vakalounakis, D.J. & Christias, C. (1986) Light quality, temperature and sporogenesis in *Alternaria cichorii*. *Transactions of the British Mycological Society*, **86**, 247–254.

Valder, P. (1958) The biology of *Helicobasidium purpureum*. *Transactions of the British Mycological Society*. **64**, 367–380.

van Alfen, N.K., Jaynes, R.A., Anagnostakis, S.L. & Day, P.R. (1975) Chestnut blight: Biological control by transmissible hypovirulence in *Endothia parasitica*. *Science (New York)*, **189**, 890–891.

van den Ende, H. (1983) Fungal pheromones. In *Fungal Differentiation: a Contemporary Synthesis*, ed. Smith, J.E., pp. 449–479. New York, Marcel Dekker.

van Etten, J.L., Dahlberg, K.R. & Russo, G.M. (1983) Fungal spore germination. In *Fungal Differentiation: A Contemporary Synthesis*, ed. Smith, J.E., pp. 235–266. New York, Marcel Dekker.

van Laere, A., Francois, A., Overloop, K., Verbeke, M. & van Geruen, L. (1987) Relation between germination, trehalose and the status of water in *Phycomyces blakesleeanus* spores as measured by proton-NMR. *Journal of General Microbiology*, **133**, 239–245.

van Laere, A.J. & Hulsmans, E. (1987) Water potential, glycerol synthesis, and water content of germinating *Phycomyces* spores. *Archives of Microbiology*, **147**, 257–262.

Vanniasingham, V.M. & Gilligan, C.A. (1988) Effects of biotic and abiotic factors on germination of pycnidiospores of *Leptosphaeria maculans in vitro*. *Transactions of the British Mycological Society*, **90**, 415–420.

Van Roermund, H.J.W., Perry, D.F. & Tyrrell, D. (1984) Influence of temperature, light, nutrients and pH in determination of the mode of conidial germination in *Zoophthora radicans*. *Transactions of the British Mycological Society*, **82**, 31–38.

van Zyl, P.J. & Prior, B.A. (1990) Water relations of polyol accumulation by *Zygosaccharomyces rouxii* in continuous culture. *Applied Microbial Biotechnology*, **33**, 12–17.

Venkateswerlu, G. & Stotzky, G. (1989) Binding of metals by cell walls of *Cunninghamella blakesleana* in the presence of copper or cobalt. *Applied Microbiology and Biotechnology*, **31**, 619–625.

Verbeke, M.N. & van Laere, A.J. (1986) The role of water in the activation of *Phycomyces blakesleeanus* sporangiospores. *Experimental Mycology*, **10**, 190–195.

Visser, S. (1985) Role of the soil invertebrates in determining the composition of soil microbial communities. In *Ecological Interactions in Soil:*

Plants, Microbes and Animals, eds Fitter, A.H., Atkinson, D., Read, D.J. & Usher, M.B., pp. 297–317. Oxford, Blackwell Scientific.

Visser, S. & Whittaker, J.B. (1977) Feeding preferences for certain litter fungi by *Onychiurus subtnuis* (Collembola). *Oikos*, **29**, 320–325.

Vogel, H.J. (1964) Distribution of lysine pathways among fungi: evolutionary implications. *American Naturalist*, **98**, 435–446.

Voorhees, D.A. & Peterson, J.L. (1986) Hypha–spore attractions in *Schizophyllum commune. Mycologia*, **78**, 762–765.

Waggoner, P.E. (1983) The aerial dispersal of the pathogens of plant disease. *Philosophical Transactions of Royal Society of London, Series B, Biological Sciences*, **302**, 451–462.

Waggoner, P.E. & Taylor, G.S. (1958) Dissemination by atmospheric turbulence: spores of *Peronospora tabacina. Phytopathology*, **48**, 46–51.

Wainwright, M. (1988) Metabolic diversity of fungi in relation to growth and mineral cycling in soil–a review. *Transactions of the British Mycological Society*, **90**, 159–170.

Waldorf, A.R. (1991) Cell-mediated host response to fungal aggression. In *The Fungal Spore and Disease Initiation in Plants and Animals*, eds Cole, G.T., & Hoch, H.C., pp. 445–460. New York, Plenum.

Wall, C.J. & Lewis, B.G. (1980) Survival of chlamydospores and subsequent development of *Mycocentrospora acerina* in soil. *Transactions of the British Mycological Society*, **75**, 207–211.

Wallace, D.R., Macleod, D.M., Sullivan, C.R., Tyrrell, D. & Delyzer, A.J. (1976) Induction of resting spore germination in *Entomophthora aphidis* by long day light conditions. *Canadian Journal of Botany*, **54**, 1410–1418.

Waller, D.A. (1982) Leaf-cutting ants and avoided plants: defences against *Atta texana* attack. *Oecologia*, **52**, 400–403.

Walsh, T.J. & Pizzo, P.A. (1988) Nosocomal fungal infections: a classification for hospital-acquired fungal infections and mycoses arising from endogenous flora or reactivation. *Annual Review of Microbiology*, **42**, 517–545.

Washington, W.S. (1988) Diurnal periodicity of ascospore discharge in *Venturia pirina. Transactions of the British Mycological Society*, **90**, 112–114.

Watkinson, S.C. (1977) Effect of amino acids on coremium development in *Penicillium claviforme. Journal of General Microbiology*, **101**, 269–275.

Watkinson, S.C. (1978) End-to-side fusions in hyphae of *Penicillium claviforme. Transactions of the British Mycological Society*, **70**, 451–453.

Watson, P. (1964) Spore germination in *Spinellus macrocarpus. Transactions of the British Mycological Society*, **47**, 239–245.

Weber, D.J. & Hess, W.M. (eds) (1976) *The Fungal Spore*, New York, Wiley.

Webster, J. (1970) Coprophilous fungi. *Transactions of the British Mycological Society*, **54**, 161–180.

Weiss, J. & Weisenseel, M.H. (1990) Blue light-induced changes in membrane potential and intracellular pH of *Phycomyces* hyphae. *Journal of Plant Physiology*, **136**, 78–85.

Weller, D.M. (1988) Biological control of soil-borne plant pathogens in the rhizosphere with bacteria. *Annual Review of Phytopathology*, **26**, 379–407.

Wells, J.M., Hughes, C. & Boddy, L. (1990) The fate of soil-derived phosphorus in mycelial cord systems of *Phanaerochaete velutina* and *Phallus impudicus. New Phytologist*, **114**, 595–606.

Wells, T.K., Hammond, J.B.W. & Dickerson, A.G. (1987) Variations in activities of glycogen phosphorylase and trehalase during periodic fruiting of the edible mushroom *Agaricus bisporus* (Lange) Imbach. *New Phytologist*, **105**, 273–280.

Wessels, J.G.H., de Vries, O.M.H., Ásgeirsdóttir, S.A. & Springer, J. (1991) The *thn* mutation of *Schizophyllum commune*, which suppresses formation of aerial hyphae, affects expression of the *Sc3* hydrophobin gene. *Journal of General Microbiology*, **137**, 2439–2445.

Wessels, J.G.H., Dons, J.J.M. & de Vries, O.M.H. (1985) Molecular biology of fruit body formation in *Schizophyllum commune*. In *Developmental Biology of Higher Fungi*, eds Moore, D., Casselton, L.A., Wood, D.A. & Frankland, J.C., pp. 485–497. Cambridge, Cambridge University Press.

Wethered, J.M. & Jennings, D.H. (1985) Major solutes contributing to solute potential of *Thraustochytrium aureum* and *T. roseum* after growth in media of different salinities. *Transactions of the British Mycological Society*, **85**, 439–446.

Whaley, J.W. & Barnett, H.L. (1963) Parasitism and nutrition of *Gonatobotrys simplex. Mycologia*, **55**, 199–210.

Wheeler, K.A., Hocking, A.D. & Pitt, J.I. (1988a) Effects of temperature and water activity on germination and growth of *Wallemia sebi. Transactions of the British Mycological Society*, **90**, 365–368.

Wheeler, K.A., Hocking, A.D. & Pitt, J.I. (1988b) Influence of temperature on the water relations of *Polypaecilum pisce* and *Basipetospora halophila*, two halophilic fungi. *Journal of General Microbiology*, **134**, 2255–2260.

Wheeler, K.A., Hocking, A.D. & Pitt, J.I. (1988c) Water relations of some *Aspergillus* species isolated from dried fish. *Transactions of the British Mycological Society*, **91**, 631–637.

Wheeler, Q. & Blackwell, M. (eds) (1984) *Fungus–Insect Relationships*. New York, Columbia University Press.

Whipps, J.M. & Cooke, R.C. (1978) Behaviour of zoosporangia and zoospores of *Albugo tragopogonis* in relation to infection of *Senecio squalidus*. *Transactions of the British Mycological Society*, **71**, 121–127.

Whipps, J.M., Lewis, K. & Cooke, R.C. (1988) Mycoparasitism and plant disease control. In *Fungi in Biological Control Systems*, ed. Burge, M.N., pp. 161–187. Manchester, Manchester University Press.

Whipps, J.M. & Lumsden, R.D. (eds) (1989) *Biotechnology of Fungi for Improving Plant Growth*, Cambridge, Cambridge University Press.

Whisler, H.C. & Travland, L.B. (1974) The rotifer trap of *Zoophagus*. *Archiv für Mikrobiologie*, **101**, 95–107.

White, J.F. & Cole, G.T. (1985) Endophyte–host associations in forage grasses. I. Distribution of fungal endophytes in some species of *Lolium* and *Festuca*. *Mycologia*, **77**, 323–327.

Whitney, H.S., Bandoni, R.J. & Oberwinkler, F. (1987) *Entomocorticium dendroctoni* gen. et sp. nov. (Basidiomycotina), a possible nutritional symbiote of the mountain pine beetle in lodgepole pine in British Columbia. *Canadian Journal of Botany*, **65**, 95–102.

Whitney, K.D. & Arnott, H.J. (1987) Calcium oxalate crystal morphology and development in *Agaricus bisporus*. *Mycologia*, **79**, 180–187.

Whitney, K.D. & Arnott, H.J. (1988) The effect of calcium on mycelial growth and calcium oxalate crystal formation in *Gilbertella persicaria* (Mucorales). *Mycologia*, **80**, 707–715.

Whittaker, R.H. (1975) The design and stability of plant communities. In *Unifying Concepts in Ecology*, eds van Dobben, W.H. & Lowe-McConnell, R.H., pp. 169–181. The Hague, Junk.

Whittaker, R.H. & Feeny, P. (1971) Allelochemics: chemical interactions between species. *Science (Washington)*, **171**, 757–770.

Wicklow, D.T. (1975) Fire as an environmental cue initiating ascomycete development in a tall grass prairie. *Mycologia*, **67**, 852–862.

Wicklow, D.T. (1979) Hair ornamentation and predator defence in *Chaetomium*. *Transactions of the British Mycological Society*, **72**, 107–110.

Wicklow, D.T. (1981) Interference competition. In *The Fungal Community*, eds Wicklow, D.T. & Carroll, G.C., pp. 351–378. New York, Marcel Dekker.

Wicklow, D.T. (1988a) Parallels in the development of post-fire fungal and herb communities. In *Fungi and Ecological Disturbance*, eds Boddy, L., Watling, R. & Lyon, A.J.E., *Proceedings of the Royal Society of Edinburgh, Section B*, **94**, 87–95.

Wicklow, D.T. (1988b) Metabolism in the co-evolution of fungal chemical defence systems. In *Coevolution of Fungi with Plants and Animals*, eds Pirozynski, K.A. & Hawksworth, D.L., pp. 173–207. London, Academic Press.

Wicklow, D.T., Dowd, P.F., Tepaske, M.R. & Gloer, J.B. (1988) Sclerotial metabolites of *Aspergillus flavus* toxic to a detritivorous maize insect (*Coprophilus hemipterus*, Nitidulidae). *Transactions of the British Mycological Society*, **91**, 433–438.

Wicklow, D.T. & Shotwell, O.L. (1983) Intrafungal distribution of aflatoxin among conidia and sclerotia of *Aspergillus flavus* and *Aspergillus parasiticus*. *Canadian Journal of Microbiology*, **29**, 1–5.

Widden, P. (1984) The effects of temperature on competition for spruce needles among sympatric species of *Trichoderma*. *Mycologia*, **76**, 873–883.

Widden, P. & Abitbol, J.J. (1980) Seasonality of *Trichoderma* species in a spruce-forest soil. *Mycologia*, **72**, 775–784.

Widden, P. & Hsu, D. (1984) Competition between *Trichoderma* species: effects of temperature and litter type. *Soil Biology & Biochemistry*, **19**, 89–93.

Wiebe, C. & Winkelmann, G. (1975) Kinetic studies on the specificity of chelate-iron uptake in *Aspergillus*. *Journal of Bacteriology*, **123**, 837–842.

Wiemken, A. (1990) Trehalose in yeast, stress protectant rather than reserve carbohydrate. *Antonie van Leeuwenhoek*, **58**, 209–217.

Wilding, N., Collins, M., Hammond, P.M. & Webber, J.F. (eds) (1988) *Insect–Fungus Interactions. 14th Symposium of the Royal Entomological Society of London in Collaboration with the British Mycological Society*. London, Academic Press.

Willetts, H.J. (1978) Sclerotium formation. In *The Filamentous Fungi. Vol. III. Developmental Mycology*, eds Smith, J.E. & Berry, D.R., pp. 197–213. London, Edward Arnold.

Williams, A.G. & Orpin, C.G. (1987a) Polysaccharide-degrading enzymes formed by three species of anaerobic rumen fungi grown on a range of carbohydrate substrates. *Canadian Journal of Microbiology*, **33**, 418–426.

Williams, A.G. & Orpin, C.G. (1987b) Glycoside hydrolase enzymes present in the zoospore and vegetative growth stages of the rumen fungi *Neocallimastix patriciarum*, *Piromonas com-*

munis, and an unidentified isolate, grown on a range of carbohydrates. *Canadian Journal of Microbiology*, **33**, 427–434.

Williams, M.A.J., Beckett, A. & Read, N.D. (1985) Ultrastructural aspects of fruit body differentiation in *Flammulina velutipes*. In *Developmental Biology of Higher Fungi*, eds Moore, D., Casselton, L.A., Wood, D.A. & Frankland, J.C., pp. 429–450. Cambridge, Cambridge University Press.

Willoughby, L.G. (1956) Studies on soil chytrids. I. *Rhizidium richmondense* sp. nov. and its parasites. *Transactions of the British Mycological Society*, **39**, 125–141.

Wimble, D.B. & Young, T.W.K. (1983) Structure of adhesive knobs in *Dactylella lysipaga*. *Transactions of the British Mycological Society*, **80**, 515–519.

Wingfield, M.J. & Gibbs, J.N. (1991) *Leptographium* and *Graphium* species associated with pine-infesting bark beetles in England. *Mycological Research*, **95**, 1257–1260.

Winkelmann, G. (1986) Iron complex products (siderophores). In *Biotechnology*, Vol. 4., eds Rehm, H.-J., & Reed, G. pp. 215–243. Weinheim, VCH Verlagsgesellschaft.

Wisniewski, M., Biles, C., Droby, S., McLaughlin, R., Wilson, C. & Chalutz, E. (1991) Mode of action of the postharvest biocontrol yeast, *Pichia guilliermondii*. I. Characterization of attachment to *Botrytis cinerea*. *Physiological and Molecular Plant Pathology*, **39**, 245–258.

Wisniewski, M., Wilson, C. & Hershberger, W. (1989) Characterization of inhibition of *Rhizopus stolonifer* germination and growth by *Enterobacter cloacae*. *Canadian Journal of Botany*, **67**, 2317–2323.

Wood, D.A., Claydon, N., Dudley, K.J., Stephens, S.K. & Allan, M. (1988) Cellulase production in the life cycle of the cultivated mushroom, *Agaricus bisporus*. In *Biochemistry and Genetics of Cellulose Degradation*, eds Aubert, J.-P., Beguin, P. & Millet, J., pp 53–70. London, Academic Press.

Wood, S.N. & Cooke, R.C. (1986) Effect of *Piptocephalis* species on growth and sporulation of *Pilaira anomala*. *Transactions of the British Mycological Society*, **86**, 672–674.

Wood, T.G. & Thomas, R.J. (1989) The mutualistic association between Macrotermitinae and *Termitomyces*. In *Insect–Fungus Interactions*, eds Wilding, N., Collins, N.M., Hammond, P.M. & Webber J.F., pp. 69–92. London, Academic Press.

Wood, T.M. (1985) Properties of cellulolytic enzyme systems. *Biochemical Society Transactions*, **13**, 407–440.

Wood, T.M. (1989) Mechanisms of cellulose degradation by enzymes from aerobic and anaerobic fungi. In *Enzyme Systems for Lignocellulose Degradation*, ed. Coughlan, M.P., pp. 17–35. London, Elsevier Applied Science.

Woodbridge, B. (1988) *The biology of* Typhula incarnata *causing snow rot disease in winter barley*. PhD Thesis, University of Hull, UK.

Woodin, T.S. & Wang, J.L. (1989) Sulfate permease of *Penicillium dupontii*. *Experimental Mycology*, **13**, 380–391.

Wright, R.M. & Cummings, D.J. (1983) Integration of mitochondrial gene sequences within the nuclear genome during senescence in a fungus. *Nature (London)*, **302**, 86–89.

Wright, V.F., Vesonder, R.F. & Ciegler, A. (1982) Mycotoxins and other fungal metabolites as insecticides. In *Microbial and Viral Pesticides*, ed. Kurstak, E., pp. 559–583. New York, Marcel Dekker.

Wright, V.P. (1985) The precursor environment for vascular plant colonization. *Philosophical Transactions of the Royal Society of London, Series B*, **309**, 143–145.

Wymore, L.A. & Lorbeer, J.W. (1987) Effect of cold treatment and drying on mycelial germination by sclerotia of *Sclerotinia minor*. *Phytopathology*, **77**, 851–856.

Wyness, L.E. & Ayres, P.G. (1985) Water or salt stress increases infectivity of *Erysiphe pisi* conidia taken from stressed plants. *Transactions of the British Mycological Society*, **85**, 471–476.

Wynn, A.R. & Epton, H.A.S. (1979) Parasitism of oospores of *Phytophthora erythroseptica* in soil. *Transactions of the British Mycological Society*, **73**, 235–248.

Wynn, W.K. (1976) Appressorium formation over stomates by the bean rust fungus: response to a surface contact stimulus. *Phytopathology*, **66**, 136–146.

Wynn, W.K. (1981) Tropic and taxic responses of pathogens to plants. *Annual Review of Phytopathology*, **19**, 237–296.

Yamanaka, K., Wakabayaski, K. & Saito, T. (1988) Capture of pine-wilt nematodes by *Arthrobotrys oligospora* Y4007. Mucin-specific hemagglutin and its role in the capture. *Agricultural Biological Chemistry*, **52**, 675–683.

Yanashigima, N. (1988) Sexual interactions in yeast. In *Eukaryote Cell Recognition: Concepts and Model Systems*, eds Chapman, G.P., Ainsworth, C.C. & Chatham, C.J., pp. 49–70. Cambridge, Cambridge University Press.

Yarwood, C.E. (1950) Water content of fungus spores. *American Journal of Botany*, **27**, 636–639.

Yarwood, C.E. (1956) Simultaneous self-stimulation and self-inhibition of urediospore germination. *Mycologia*, **48**, 20–24.

Young, B.R., Newhook, F.J. & Allen, R.N. (1979) Motility and chemotactic response of *Phytophthora cinnamomi* zoospores in 'ideal soils'. *Transactions of the British Mycological Society*, **72**, 395–401.

Zahari, P. & Shipton, W.A. (1988) Growth and sporulation responses of *Basidiobolus* to changes in environmental parameters. *Transactions of the British Mycological Society*, **91**, 141–148.

Zebold, S.L., Whisler, H.C., Schemanchuk, J.A. & Travland, L.B. (1979) Host specificity and penetration in the mosquito pathogen *Coelomyces psorophorae. Canadian Journal of Botany*, **57**, 2766–2770.

Zentmeyer, G.A. (1966) Role of amino acids in chemotaxis of zoospores of three species of *Phytophthora. Phytopathology*, **86**, 907.

Zoberi, M.H. (1961) Take-off of mould spores in relation to wind speed and humidity. *Annals of Botany*, **25**, 53–64.

Zoberi, M.H. (1964) Effect of temperature and humidity on ballistospore discharge. *Transactions of the British Mycological Society*, **47**, 109–114.

Zonneveld, B.J.M. (1975) Sexual differentiation in *Aspergillus nidulans*. The requirements for manganese and its effect on α-1,3 glucan synthesis and degradation. *Archives of Microbiology*, **105**, 101–104.

Zonneveld, B.J.M. (1988) Effect of carbon dioxide on fruiting in *Aspergillus nidulans. Transactions of the British Mycological Society*, **91**, 625–629.

Zuckerman, B.M. & Jansson, H.B. (1984) Nematode chemotaxis and possible mechanisms of host/prey recognition. *Annual Review of Phytopathology*, **22**, 95–113.

Index

Italic numerals denote illustrations or tables.